Lecture Notes of
the Unione Matematica Italiana

1

T0218403

Luc Tartar

An Introduction to Navier–Stokes Equation and Oceanography

Author

Luc Tartar
Department of Mathematical Sciences
Carnegie Mellon University
Pittsburgh, PA 15213-3890
USA

Library of Congress Control Number: 2006928046

Mathematics Subject Classification (2000): 35-xx, 76-xx

ISSN print edition: 1862-9113
ISSN electronic edition: 1862-9121
ISBN-10 3-540-35743-2 Springer Berlin Heidelberg New York
ISBN-13 978-3-540-35743-8 Springer Berlin Heidelberg New York

Springer is a part of Springer Science+Business Media
springer.com
© Springer-Verlag Berlin Heidelberg 2006
Printed in The Netherlands

Typesetting by the authors and SPi using a Springer LaTeX macro package
Cover design: *design & production* GmbH, Heidelberg

Printed on acid-free paper   SPIN: 11496601   VA 41/3100/SPi   5 4 3 2 1 0

In memory of Jean LERAY, Nov. 7, 1906 – Nov. 10, 1998
In memory of Olga LADYZHENSKAYA, Mar. 7, 1922 – Jan. 12, 2004

They pioneered the mathematical study of the Navier–Stokes equation, which is an important part of these lecture notes.

In memory of my father
Georges TARTAR, Oct. 9, 1915 – Aug. 5, 2003

He dedicated his life to what he believed God expected from him. As for me, once the doubt had entered my mind, what other choice did it leave me but to search for the truth, in all fields?

To my children
Laure, Michaël, André, Marta

Preface

In the spring of 1999, I taught (at CARNEGIE MELLON University) a graduate course entitled *Partial Differential Equations Models in Oceanography*, and I wrote lecture notes which I distributed to the students; these notes were then made available on the Internet, and they were distributed to the participants of a Summer School held in Lisbon, Portugal, in July 1999. After a few years, I feel it will be useful to make the text available to a larger audience by publishing a revised version.

To an uninformed observer, it may seem that there is more interest in the Navier–Stokes equation nowadays, but many who claim to be interested show such a lack of knowledge about continuum mechanics that one may wonder about such a superficial attraction. Could one of the Clay Millennium Prizes be the reason behind this renewed interest? Reading the text of the conjectures to be solved for winning that particular prize leaves the impression that the subject was not chosen by people interested in continuum mechanics, as the selected questions have almost no physical content. Invariance by translation or scaling is mentioned, but why is invariance by rotations not pointed out and why is Galilean invariance[1] omitted, as it is the essential fact which makes

---

[1] Velocities involved for ordinary fluids being much smaller than the velocity of light c, no relativistic corrections are necessary and Galilean invariance should then be used, but one should be aware that once the mathematical equation has been written it is not automatic that its solutions will only use velocities bounded by c. One should learn to distinguish between a mathematical property of an equation and a conjecture that some property holds which one guesses from the belief that the equation corresponds to a physical problem. One should learn about which defects are already known concerning how a mathematical model describes physical reality, but one should not forget that a mathematical model which is considered obsolete from the physical point of view may still be useful for mathematical reasons. I often wonder why so many forget to mention the defects of the models that they study.

the equation introduced by NAVIER[2] much better than that introduced later by STOKES? If one had used the word "turbulence" to make the donator believe that he would be giving one million dollars away for an important realistic problem in continuum mechanics, why has attention been restricted to unrealistic domains without boundary (the whole space R^3, or a torus for periodic solutions), as if one did not know that vorticity is created at the boundary of the domain? The problems seem to have been chosen in the hope that they will be solved by specialists of harmonic analysis, and it has given the occasion to some of these specialists to help others in showing the techniques that they use, as in a recent book by Pierre Gilles LEMARIÉ-RIEUSSET [17]; some of the techniques are actually very similar to those that I have learnt in the theory of interpolation spaces, on which I have already written some lecture notes which I plan to revise, and I hope that this particular set of lecture notes on the Navier–Stokes equation and another one not yet finished on kinetic theory may help the readers understand a little more about the physical content of the equation, and also its limitations, which many do not seem to be aware of.

Being a mathematician interested in science, and having learnt more than most mathematicians about various aspects of mechanics and physics,[3] one reason for teaching various courses and writing lecture notes is to help isolated researchers to learn about some aspects unknown to most mathematicians whom they could meet, or read. A consequence of this choice is then to make researchers aware that some who claim to work on problems of continuum

---

[2] The equation which one calls now after both NAVIER and STOKES was introduced by NAVIER, while STOKES only later introduced the equation without inertial effects, which is linear and does not present so much mathematical difficulty nowadays, but there are cases where the nonlinear term in the Navier equation disappears and the equation reduces to the Stokes equation, an example being irrotational flows. If one believes that the Stokes equation is a good model for small velocities (and bounded derivatives of the velocity) then using the Stokes equation in a frame moving at a local velocity and invoking Galilean invariance makes one discover the Navier equation (which I shall call the Navier–Stokes equation apart from this footnote); of course, I shall point out other defects of the model along the way.

[3] Classical mechanics is an 18th century point of view of mechanics, which requires ordinary differential equations as mathematical tools. Continuum mechanics is an 18th –19th century point of view of mechanics, which requires partial differential equations as mathematical tools; the same is true for many aspects of physics. However, 20th century aspects of mechanics (plasticity, turbulence) or physics (quantum effects) require mathematical tools which are beyond partial differential equations, similar to those that I have tried to develop in my research work, improving concepts such as *homogenization, compensated compactness* and *H-measures*.

mechanics or physics have forgotten to point out known defects of the models that they use.[4]

I once heard my advisor, Jacques-Louis LIONS, mention that once the detailed plan of a book is made, the book is almost written, and he was certainly speaking of experience as he had already written a few books at the time. He gave me the impression that he could write directly a very reasonable text, which he gave to a secretary for typing; maybe he then gave chapters to one of his students, as he did with me for one of his books [19], and very few technical details had to be fixed. His philosophy seemed to be that there is no need to spend too much time polishing the text or finding the best possible statement, as the goal is to take many readers to the front of research, or to be more precise to one front of research, because in the beginning he changed topics every two or three years. As for myself, I have not yet written a book, and the main reason is that I am quite unable to write in advance a precise plan of what I am going to talk about, and I have never been very good at writing even in my mother tongue (French). When I write, I need to read again and again what I have already written until I find the text acceptable (and that notion of acceptability evolves with time and I am horrified by my style of twenty years ago), so this way of writing is quite inefficient, and makes writing a book prohibitively long. One solution would be not to write books, and when I go to a library I am amazed by the number of books which have been written on so many subjects, and which I have not read, because I never read much. Why then should I add a new book? However, I am even more amazed by the number of books which are not in the library, and although I have access to a good inter-library loan service myself, I became concerned with how difficult it is for isolated students to have access to scientific knowledge (and I do consider mathematics as part of science, of course).

I also thought of a different question. It is clear that fewer and fewer students in industrialized countries are interested in studying mathematics, for various reasons, and as a consequence more and more mathematicians are likely to come from developing countries. It will therefore be of utmost importance that developing countries should not simply become a reservoir of good students that industrialized countries would draw upon, but that these countries develop a sufficiently strong scientific environment for the benefit of their own economy and people, so that only a small proportion of the new trained generations of scientists would become interested in going to work abroad. I have seen the process of decolonization at work in the early 1960s, and I have witnessed the consequences of too hasty a transition, which was not to the benefit of the former colonies, and certainly the creation of a scientific tradition is not something that can be done very fast. I see the development

---

[4] Of course, I also suffer from the same disease of not having learnt enough, but my hope is that by explaining what I have already understood and by showing how to analyze and criticize classical models, many will acquire my understanding and a few will go much further than I have on the path of discovery.

of mathematics as a good way to start building a scientific infrastructure, and inside mathematics the fields that I have studied should play an important role, where mathematics interacts with continuum mechanics and physics.

In the spring of 1999, I found the right solution for me, which is to give a course and to prepare lecture notes for the students, trying to write down after each lecture the two or three pages describing what I have just taught; for such short texts my problems about writing are not too acute. I could hardly have guessed at the beginning of the course how much an introduction to oceanography my course would become, and when after a short introduction and the description of some classical methods for solving the Navier–Stokes equation (in the over-simplified version which mathematicians usually consider), it was time to describe some of the models considered in oceanography, I realized that I did not believe too much in the derivation of these models, and I preferred to finish the course by describing some of the general mathematical tools for studying the nonlinear partial differential equations of continuum mechanics, some of which I have developed myself. The resulting set of lecture notes is not as good as I would have liked, but an important point was to make this introductory course available on the Internet. In the spring of 2000, I wrote similar lecture notes for a course divided into two parts, the first part on Sobolev spaces, and the second part on the theory of interpolation spaces, and in the fall of 2001, I wrote lecture notes for an introduction to kinetic theory; of course, it is my plan to finish and review these lecture notes to make them more widely available by publishing them.

I decided at that time to add some information that one rarely finds in courses of mathematics, something about the people who have participated in the creation of the knowledge related to the subject of the course. I had the privilege to study in Paris in the late 1960s, to have great teachers like Laurent SCHWARTZ and Jacques-Louis LIONS, and to have met many famous mathematicians. This has given me a different view of mathematics than the one that comes from reading books and articles, which I find too dry, and I have tried to give a little more life to my story by telling something about the actors; for those mathematicians whom I have met, I have used their first names in the text, and I have tried to give some simple biographical data for all people quoted in the text, in order to situate them, both in time and in space. For mathematicians of the past, a large part of this information comes from using *The MacTutor History of Mathematics archive* (http://www-history.mcs.st-and.ac.uk/history), for which one should thank J. J. O'CONNOR and E. F. ROBERTSON, from the University of St Andrews in Scotland, UK, but for many other names I searched the Internet, and it is possible that some of my information is incomplete or even inaccurate. My interest in history is not recent, but my interest in the history of mathematics has increased recently, in part from finding the above-mentioned archive, but also as a result of seeing so many ideas badly attributed, and I have tried to learn more about the mathematicians who have introduced some of the ideas which I was taught when I was a student in Paris in the late 1960s, and be as accurate as possible concerning the work of all. I hope that I shall be given the correct information by anyone who finds one of my inaccuracies, and that I shall be forgiven for these unintentional errors.

I was born in France in December 1946 from a Syrian father and a French mother and I left France for political reasons, and since 1987 I have enjoyed the hospitality of an American university, CARNEGIE MELLON University, in Pittsburgh PA, but I am still a French citizen, and I only have resident status in United States. This may explain my interest in mentioning that others have worked in a different country than the one where they were born, and I want to convey the idea that the development of mathematics is an international endeavor, but I am not interested in the precise citizenship of the people mentioned, or if they feel more attached to the country were they were born or the one where they work; for example, I quote Olga OLEINIK as being born in Ukraine and having worked in Moscow, Russia, and obviously Ukraine was not an independent country when she was born, but was when she died; a French friend, Gérard TRONEL, has told me that she did feel more Russian than Ukrainian, but if I have been told that information about her I completely lack information about others. Because some countries have not always existed or have seen their boundaries change by their own expansion or that of other countries, some of my statements are anachronistic, like when I say that Leonardo DA VINCI was Italian, but I do not say that for ARCHIMEDES, who is known to have died at the hand of a Roman soldier, or decide about EUCLID, or AL KHWARIZMI, as it is not known where they were born.

I observe that there have been efficient schools in some areas of mathematics at some places and at some moments in time, and when I was a student in Paris in the late 1960s, Jacques-Louis LIONS had mentioned that Moscow was the only other place comparable to Paris for its concentration of mathematicians. Although the conditions might be less favorable outside important centers, I want to think that a lot of good work could be done elsewhere, and my desire is that my lecture notes may help isolated researchers participate more in the advance of scientific knowledge. A few years ago, an Italian friend, GianPietro DEL PIERO, told me that he had taught for a few months in Somalia, and he mentioned that one student had explained to him that he should not be upset if some of the students fell asleep during his lectures, because the reason was not their lack of interest in the course, but the fact that sometimes they had eaten nothing for a few days. It was by thinking about these courageous students who, despite the enormous difficulties that they encounter in their everyday life, are trying to acquire some precious knowledge about mathematics, that I devised my plan to write lecture notes and make them available to all, wishing that they could arrive freely to isolated students and researchers, working in much more difficult conditions than those having access to a good library, or in contact with good teachers. I hope that publishing this revised version will have the effect that it will reach many libraries scattered around the world, where isolated researchers have access.

I hope that my lack of organizational skills will not bother the readers too much. I consider teaching courses like leading groups of newcomers into countries which are often unknown to them, but not unknown to me, as I have often wandered around; some members of a group who have already read about the region or have been in other expeditions with guides more organized than me might feel disoriented by my choice of places to visit, and indeed I may have forgotten to show a few interesting places, but my goal is to familiarize the readers with the subject and encourage them to acquire an open and scientific point of view, and not to write a definitive account of the subject.

There are results which are repeated, but it is inevitable in a real course that one should often recall results which have already been mentioned. There are also results which are mentioned without proof, and sometimes they are proven later but sometimes they are not, and if no references are given, one should remember that I have been trained as a mathematician, and that my statements without proofs have indeed been proven in a mathematical sense, because if they had not I would have called them conjectures instead;[5] however, I am also human and my memory is not perfect and I may have made mistakes. I think that the right attitude in mathematics is to be able to explain all the statements that one makes, but in a course one has to assume that the reader already has some basic knowledge of mathematics, and some proofs of a more elementary nature are omitted. Here and there I mention a result that I have heard of, but for which I never read a proof or did not make up my own proof, and I usually say so. If many proofs are mine it does not necessarily mean that I was the first to prove the corresponding result, but that I am not aware of a prior proof, maybe because I never read much. Actually, my advisor mentioned to me that it is useful to read only

---

[5] Some people like to talk of pure mathematics versus applied mathematics, but I do not think that such a distinction is accurate, as I mentioned in the introduction of an article for a conference at École Polytechnique (Palaiseau, France) in the fall of 1983, but because that introduction was cut by political censors, it is worth repeating that for what concerns different parts of mathematics there are those which I know, those which I do not know well but think that they could be useful to me, and those which I do not know well but do not see how they could be useful to me, and all this evolves with time, so I finally wonder if it is reasonable to classify mathematics as being pure or applied. I consider myself as an "applied" mathematician, although I give it a French meaning (a mathematician interested in other fields of science), opposed to a British meaning (a specialist of continuum mechanics, allowed to use an incomplete mathematical proof without having to call the result a conjecture), and in French universities, applied mathematicians in the British style are found in departments of mécanique. Probably for the reason of funding, which strangely enough is given more easily to people who pretend to do applied research, some who have studied to become mathematicians practice the art of using words which make naive people wrongly believe that they know continuum mechanics or physics, and I find this attitude potentially dangerous for the university system.

the statement of a theorem and one should read the proof only if one cannot supply one.[6]

My personal mathematical training has been in functional analysis and partial differential equations, starting at École Polytechnique, Paris, France, where I had two great teachers, Laurent SCHWARTZ and Jacques-Louis LIONS. Having studied there in order to become an engineer, but having had to change my orientation once I had been told that such a career required administrative skills (which I lack completely), I opted for doing research in mathematics with an interest in other sciences and I asked Jacques-Louis LIONS to be my advisor, and it was normal that once I had been taught enough on the mathematical side, I would apply my improved understanding to investigating questions of continuum mechanics and physics which I had heard about as a student, and to developing the new mathematical tools which are necessary for that.

In my lectures I also try to teach mathematicians about the defects of the models used, but I want to apologize for some of the words which I use, which may have offended some. I have a great admiration for the achievements of physicists and engineers[7] during the last century, and a lot of the improvements in our lives result from their understanding, which is so different than the type of understanding that mathematicians are trained to achieve. If I write that something that they say does not make any sense, it is not a criticism towards physicists or engineers, who are following the rules of their profession, but it is a challenge to my fellow mathematicians that there is something there that mathematicians ought to clarify. I am grateful to Robert DAUTRAY[8] for offering me a position at Commissariat à l'Énergie Atomique from 1982 to 1987, and for helping me understand more about physics through his advice during these years; he helped me understand what the challenges

---

[6] The MacTutor archive mentions an interesting anecdote in this respect concerning a visit of Antoni ZYGMUND to the University of Buenos Aires, Argentina, in 1948; Alberto CALDERÓN was a student there and he was puzzled by a question that ZYGMUND had asked, and he said that the answer was in ZYGMUND's own book *Trigonometric Series*, but there was disagreement on this point; what had happened was that CALDERÓN had read a statement in the book and supplied his own proof, which was more general that the one written, so it also answered the question that ZYGMUND had just asked, but CALDERÓN had wrongly assumed that ZYGMUND's proof in his book, which he had never checked, was similar to his. Franco BREZZI mentioned to me that Ennio DE GIORGI had once told Claudio BAIOCCHI something similar, that he almost never read a proof, and that he did his own proofs for the interesting theorems but that he did not bother to think about the uninteresting ones.

[7] I am not mentioning biologists and chemists because biology was not part of my studies, and although I have learnt some chemistry, I only hope to understand it in a better way once my program for understanding continuum mechanics and physics has progressed enough.

[8] A good reference for learning classical mathematical tools and their use in problems of engineering or physics is the collection of books that Robert DAUTRAY had persuaded Jacques-Louis LIONS to edit with him, [4–9]

are, and I hope that through my lecture notes more will understand about the challenges, and that should make Science progress.

The support of a few friends gave me the strength to decide to complete the writing of some unfinished lecture notes and to revise those which I had already written, with a view to publishing them to attain a wider audience. I want to express my gratitude to Thérèse BRIFFOD, for her hospitality when I carried out the first revision of this course in August 2002, but also for her help in making me understand better an important question in life, having compassion for those who are in difficulty. I want to express my gratitude to Lucia OSTONI, for her hospitality when I carried out the second revision of this course in July 2004, and the final adjustments to Springer's formatting in December 2004.

I want to thank my good friends Carlo SBORDONE and Franco BREZZI for having proposed to publish my lecture notes in a series of Unione Matematica Italiana, and for having helped me to arrive at the necessary corrections of my original text.

Milano, December 2004

Luc TARTAR
Correspondant de l'Académie des Sciences, Paris
University Professor of Mathematics
Department of Mathematical Sciences
CARNEGIE MELLON University
Pittsburgh, PA 15213-3890, United States of America

Introduction

In teaching a mathematical course where the Navier[9]–Stokes[10,11] equation plays a role, one must mention the pioneering work of Jean LERAY[12,13] in the 1930s. Some of the problems that Jean LERAY left unanswered are still open today,[14] but some improvements were started by Olga LADYZHENSKAYA[15] [16], followed by a few others, like James SERRIN,[16] and my advisor, Jacques-Louis LIONS[17] [19], from whom I learnt the basic principles for the mathematical analysis of these equations in the late 1960s.

---

[9] Claude Louis Marie Henri NAVIER, French mathematician, 1785–1836. He worked in Paris, France.

[10] Sir George Gabriel STOKES, Irish-born mathematician, 1819–1903. He held the Lucasian chair at Cambridge, England, UK.

[11] Henry LUCAS, English clergyman, 1610–1663.

[12] Jean LERAY, French mathematician, 1906–1998. He received the Wolf Prize in 1979. He held a chair (Théorie des équations différentielles et fonctionnelles) at Collège de France, Paris, France.

[13] Ricardo WOLF, German-born (Cuban) diplomat and philanthropist, 1887–1981. The Wolf Foundation was established in 1976 with his wife, Francisca SUBIRANA-WOLF, 1900–1981, *to promote science and art for the benefit of mankind.*

[14] Most problems are much too academic from the point of view of continuum mechanics, because the model used by Jean LERAY is too crude to be meaningful, and the difficulties of the open questions are merely of a technical mathematical nature. Also, Jean LERAY unfortunately called turbulent the weak solutions that he was seeking, and it must be stressed that turbulence is certainly not about regularity or lack of regularity of solutions, nor about letting time go to infinity either.

[15] Olga Aleksandrovna LADYZHENSKAYA, Russian mathematician, 1922–2004. She worked at Russian Academy of Sciences, St Petersburg, Russia.

[16] James B. SERRIN Jr., American mathematician, born in 1926. He works at University of Minnesota Twin Cities, Minneapolis, MN.

[17] Jacques-Louis LIONS, French mathematician, 1928–2001. He received the Japan Prize in 1991. He held a chair (Analyse mathématique des systèmes et de leur contrôle) at Collège de France, Paris, France. I first had him as a teacher at

In the announcement of the course, I had mentioned that I would start by recalling some classical facts about the way to use functional analysis for solving partial differential equations of continuum mechanics, describing some fine properties of Sobolev[18] spaces which are useful, and studying in detail the spaces adapted to questions about incompressible fluids. I had stated then that the goal of the course was to describe some more recent mathematical models used in oceanography, and show how some of them may be solved, and that, of course, I would point out the known defects of these models.[19] I had mentioned that, for the oceanography part – of which I am no specialist – I would follow a book written by one of my collaborators, Roger LEWANDOWSKI[20] [18], who had learnt about some of these questions from recent lectures of Jacques-Louis LIONS. I mentioned that I was going to distribute notes, from a course on partial differential equations that I had taught a few years before, but as I had not written the part that I had taught on the Stokes equation and the Navier–Stokes equation at the time, I was going to make use of the lecture notes [23] from the graduate course that I had taught at University of Wisconsin, Madison WI, in 1974–1975, where I had added small technical improvements from what I had learnt. Finally, I had mentioned that I would write notes for the parts that I never covered in preceding courses.

I am not good at following plans. I started by reading about oceanography in a book by A. E. GILL[21] [15], and I began the course by describing some of the basic principles that I had learnt there. Then I did follow my plan of discussing questions of functional analysis, but I did not use any of the notes that I had written before. When I felt ready to start describing new models, Roger LEWANDOWSKI visited CARNEGIE[22] MELLON[23] University and gave a talk in the Center for Nonlinear Analysis seminar, and I realized that there were some questions concerning the models and some mathematical techniques which I had not described at all, and I changed my plans. I opted for describing the general techniques for nonlinear partial differential equations that I

École Polytechnique in 1966–1967, and I did research under his direction, until my thesis in 1971.

[18] Sergei L'vovich SOBOLEV, Russian mathematician, 1908–1989. He worked in Novosibirsk, Russia, and there is now a SOBOLEV Institute of Mathematics of the Siberian branch of the Russian Academy of Sciences, Novosibirsk, Russia.

[19] It seems to have become my trade mark among mathematicians, that I do not want to lie about the usefulness of models when some of their defects have already been pointed out. This is obviously the way that any scientist is supposed to behave, but in explaining why I have found myself so isolated and stubborn in maintaining that behavior, I have often invoked a question of religious training.

[20] Roger LEWANDOWSKI, French mathematician, born in 1962. He works at Université de Rennes I, Rennes, France.

[21] Adrian Edmund GILL, Australian-born meteorologist and oceanographer, 1937–1986. He worked in Cambridge, England, UK.

[22] Andrew CARNEGIE, Scottish-born businessman and philanthropist, 1835–1919.

[23] Andrew William MELLON, American financier and philanthropist, 1855–1937.

had developed, *homogenization, compensated compactness* and *H-measures*; there are obviously many important situations where they should be useful, and I found it more important to teach them than to analyze in detail some particular models for which I do not feel yet how good they are (which means that I suspect them to be quite wrong). Regularly, I was trying to explain why what I was teaching had some connection with questions about *fluids*.

It goes with my philosophy to *explain the origin of mathematical ideas* when I know about them, and as my ideas are often badly attributed, I like to mention *why and when I had introduced an idea*.

I have also tried to *encourage mathematicians to learn more about continuum mechanics and physics*, listening to the specialists and then trying to put these ideas into a sound mathematical framework. I hope that some of the discussions in these lecture notes will help in this direction.[24]

[This course mentions a few equations from continuum mechanics, and besides the Navier–Stokes equation I shall mention the Maxwell equation, the equation of linearized elasticity, and the wave equation, at least, but I did not always follow the classical notation used in texts of mechanics, writing $a, \mathbf{b}, \mathsf{C}$ for scalars, vectors and tensors, and using the notation $f_{,j}$ for denoting the partial derivative of f with respect to x_j. This course is intended for mathematicians, and even if many results are stated in an informal way, they correspond to theorems whose proofs usually involve functional analysis, and not just differential calculus and linear algebra, which are behind the notation used in mechanics.

It is then important to notice that partial differential equations are not written as pointwise equalities but in the sense of distributions, or more generally in some variational framework and that one deals with elements of function spaces, using operators and various types of convergence. Instead of the notation $\nabla a, \nabla.\mathbf{b}, \nabla \times \mathbf{b}$ used in mechanics, I write $grad\, a, div\, b, curl\, b$ (and I also recall sometimes the framework of differential forms), and I only use $\mathbf{b}$ for a vector-valued function b when the pointwise value is meant, in particular in integrands.

It may seem analogous to the remark known to mathematicians that "the function $f(x)$" is an abuse of language for saying "the function f whose elements in its domain of definition will often be denoted x", but there is something different here. The framework of functional analysis is not just a change of language, because it is crucial for understanding the point of view that I developed in the 1970s for relating what happens at a macroscopic level from the description at a microscopic/mesoscopic level, using convergences of weak type (and not just weak convergences), which is quite a different idea than the game of using ensemble averages, which destroys the physical meaning of the problems considered.]

---

[24] I have gone further in the critical analysis of many principles of continuum mechanics, which I shall present as a different set of lecture notes, as an introduction to kinetic theory, taught in the Fall of 2001.

Detailed Description of Lectures

a.b refers to definition, lemma or theorem # b in lecture # a, while (a.b) refers to equation # b in lecture # a.

<u>Lecture 1, Basic physical laws and units</u>: The hypothesis of incompressibility and the speed of sound in water; salinity; units in the metric system; oceanography/meteorology; energy received from the Sun: the solar constant S; blackbody radiation, Planck's law, surface temperature of the Sun; absorption, albedo, the greenhouse effect; convection of water induced by gravity and temperature, and salinity; how a greenhouse functions.

<u>Lecture 2, Radiation balance of atmosphere</u>: The observed percentages of energy in the radiation balance of the atmosphere; absorption and emission are frequency-dependent effects; the greenhouse with p layers (2.1)–(2.7); thermodynamics of air and water: lapse rate, relative humidity, latent heat; the Inter-Tropical Convergence Zone (ITCZ), the trade winds, cyclones and anticyclones.

<u>Lecture 3, Conservations in ocean and atmosphere</u>: The differences between atmosphere and ocean concerning heat storage; conservation of angular momentum, the trade winds, east–west dominant wings; conservation of salt; Eulerian and Lagrangian points of view; conservation of mass (3.1)–(3.3).

<u>Lecture 4, Sobolev spaces I</u>: Sobolev spaces $W^{1,p}(\Omega)$ (4.1)–(4.2); weak derivatives, theory of distributions; notation H^s for $p = 2$ and $\mathcal{H}$ for Hardy spaces; functions of $W^{1,p}(\Omega)$ have a trace on $\partial\Omega$ if it is smooth; integration by parts in $W^{1,1}(\Omega)$ (4.3); results from ordinary differential equations (4.4)–(4.8); conservation of mass (4.9)–(4.12); regularity of solutions of the Navier–Stokes and Euler equations, Riesz operators and singular integrals, Zygmund space, BMO, $\mathcal{H}^1$.

<u>Lecture 5, Particles and continuum mechanics</u>: Particles and continuum mechanics, distances between molecules; homogenization, microscopic/mesoscopic/macroscopic scales; "real" particles versus macroscopic particles as tools from numerical analysis; Radon measures (5.1), distributions (5.2)–(5.4);

momentum and conservation of mass (5.5)–(5.6); the homogenization problem related to oscillations in the velocity field.

Lecture 6, Conservation of mass and momentum: Euler equation (6.1); priority of Navier over Stokes and of Stokes over Riemann, Rankine and Hugoniot; similarity of the stationary Stokes equation and stationary linearized elasticity; kinetic theory, free transport equation and conservation of mass (6.2)–(6.4); transport equation with Lorentz force (6.5); Boltzmann's equation (6.6)–(6.7); Cauchy stress in kinetic theory (6.8); conservation of momentum (6.10); pressure on the boundary resulting from reflection of particles.

Lecture 7, Conservation of energy: Internal energy in kinetic theory (7.1); relation between internal energy and Cauchy stress in kinetic theory (7.2); heat flux in kinetic theory (7.3); conservation of energy (7.4); various origins of the internal energy; variation of thermodynamic entropy, H-theorem (7.5)–(7.6); local Maxwellian distribution (7.7); the parametrization of allowed collisions (7.8)–(7.9); the form of interaction term $Q(f, f)$ in Boltzmann's equation (7.10); the proof of (7.5): (7.11); letting the mean free path tend to 0; irreversibility, nonnegative character of solutions of Boltzmann's equation (7.12)–(7.13).

Lecture 8, One-dimensional wave equation: Longitudinal, transversal waves; approximating the longitudinal vibration of a string by small masses connected with springs (8.1)–(8.2); the limiting 1-dimensional wave equation (8.3)–(8.5); different scalings of string constants; time periodic solutions; linearization for the increase in length in 1-dimensional transversal waves and 2- or 3-dimensional problems; the linearized elasticity system (8.6)–(8.11); Cauchy's introduction of the stress tensor, by looking at the equilibrium of a small tetrahedron.

Lecture 9, Nonlinear effects, shocks: Beware of linearization; nonlinear string equation (9.1); Poisson's study of barotropic gas dynamics with $p = C \varrho^\gamma$ (9.2); what led Stokes to discover "Rankine–Hugoniot" conditions; Burgers's equation (9.3)–(9.5); characteristic curves and apparition of discontinuities (9.6)–(9.7); equations in the sense of distributions imply jump conditions (9.8)–(9.9); a two-parameter family of weak solutions for Burgers's equation with 0 initial datum (9.10); Lax's condition and Oleinik's condition for selecting admissible discontinuities; Hopf's derivation of Oleinik's condition using "entropies" (9.11)–(9.13), Lax's extension to systems; the equation for entropies of system (9.14) describing the nonlinear string equation (9.15)–(9.17); transonic flows.

Lecture 10, Sobolev spaces II: Description of functional spaces for the study of the Stokes and Navier–Stokes equations, boundedness of Ω, smoothness of $\partial\Omega$; $H^1(\Omega)$ (10.1), characteristic length, $H_0^1(\Omega)$; Poincaré's inequality (10.2); scaling, Poincaré's inequality does not hold for open sets containing arbitrary large balls (10.3)–(10.4); 10.1: Poincaré's inequality holds if Ω is included in a bounded strip (10.5), if $meas \, \Omega < \infty$ (10.11)–(10.12); Schwartz's convention for the Fourier transform (10.6), its action on derivation and multiplication (10.7); Plancherel's formula (10.8); Schwartz's extension of the Fourier transform to temperate distributions (10.9); the Fourier transform is an isometry

Gagliardo and of Nirenberg (16.3)–(16.5); solving (16.1) as fixed point for Φ (16.6), estimates for Φ giving existence and uniqueness of a solution for small data and $N \leq 4$ (16.7)–(16.12), by the Banach fixed point theorem; solving (16.1) as fixed point for Ψ (16.13), estimates for Ψ (16.14)–(16.20); 16.2: existence of a fixed point for a contraction of a closed bounded nonempty convex set in a Hilbert space, monotone operators.

Lecture 17, Fixed point theorems: Existence of a solution of (16.1) for large data by the Schauder fixed point theorem for $N \leq 3$, by the Tykhonov fixed point theorem for $N = 4$; Faedo–Ritz–Galerkin method; existence of Faedo–Ritz–Galerkin approximations (17.1) by the Brouwer fixed point method applied to approximations Ψ_m (17.2), existence for large data for $N \leq 4$ by extraction of weakly converging subsequence and a compactness argument, valid for $N > 4$ in larger functional spaces; properties of the Brouwer topological degree; 17.1: nonexistence of tangent nonvanishing vector fields on S^{2N}; 17.2: nonexistence of a continuous retraction of a bounded open set of R^N onto its boundary; 17.3: Brouwer fixed point theorem.

Lecture 18, Brouwer's topological degree: $J_\varphi(u)$ (18.1); 18.1: the derivative of $J_\varphi(u)$ in the direction v is an integral on $\partial\Omega$ (18.2)–(18.3), vanishing if v vanish on $\partial\Omega$; 18.2: invariance by homotopy, $J_\varphi(u) = J_\varphi(w)$ if there is a homotopy from u to w avoiding $supp(\varphi)$ on $\partial\Omega$; 18.3: $J_\varphi(u)$ can be defined for $u \in C(\overline{\Omega}; R^N)$ avoiding $supp(\varphi)$ on $\partial\Omega$; 18.4: if $J_\varphi(u) \neq 0$ there exists $\mathbf{x} \in \Omega$ such that $\mathbf{u}(\mathbf{x}) \in supp(\varphi)$; proof of 18.1: (18.4)–(18.7); 18.5: definition of degree $deg(u; \Omega, \mathbf{p})$; 18.6: formula for degree if $\mathbf{u}(\mathbf{z}) = \mathbf{p}$ has a finite number of solutions where ∇u is invertible (18.8); Sard's lemma.

Lecture 19, Time-dependent solutions I: Spaces V, H for the Stokes or Navier–Stokes equations (19.1)–(19.2); semi-group theory; abstract ellipticity for $A \in \mathcal{L}(V, V')$ (19.3); 19.1: $u' + Au = f \in L^1(0, T; H) + L^2(0, T; V')$, $u(0) = u_0 \in H$ (19.4)–(19.5), by Faedo–Ritz–Galerkin (19.6); 19.2: properties of $W^{1,1}(0, T)$ and Gronwall's inequality; estimates for (19.6): (19.7)–(19.16); a variant of Gronwall's inequality (19.17)–(19.19), giving estimate (19.20).

Lecture 20, Time-dependent solutions II: Taking the limit in (19.6), (20.1)–(20.3), giving existence in 19.1; an identity for proving uniqueness in 19.1, (20.4); spaces $W_1(0, T)$ and $W(0, T)$ (20.5)–(20.8); properties of $W_1(0, T)$, for proving (20.4); problem with time derivative in Faedo–Ritz–Galerkin, and special choice for a basis; regularization effect when the initial datum is not in the right space; backward uniqueness in the case $A^T = A$, Agmon–Nirenberg result of log-convexity for $|u(t)|$.

Lecture 21, Time-dependent solutions III: Problem in the definition of H in (19.2); problem with the "pressure" in the nonstationary Stokes equation (21.1)–(21.7); 21.1: regularity in space when $A^T = A$, $u_0 \in V$, $f \in L^2(0, T; H)$, regularizing effect for $u_0 \in H$, $\sqrt{t}\,f \in L^2(0, T; H)$; problem of identifying H' with H; estimate for the "pressure" in the case $\Omega = R^N$ (21.8)–(21.11); avoiding cutting the transport operator into two terms (21.12)–(21.14); the nonlinear term (21.15) and its estimate in dimension 2, 3, 4 (21.16)–(21.17).

Lecture 22, Uniqueness in 2 dimensions: Cutting the transport term into two terms works for $N = 2$; 21.1: uniqueness for the abstract Navier–Stokes equation for $N = 2$ (21.1)–(21.6); a quasilinear diffusion equation (21.7), with the Artola uniqueness result (21.8)–(21.11).

Lecture 23, Traces: $H(div; \Omega)$ (23.1); space is local, $C^\infty(\overline{\Omega}; R^N)$ dense if $\partial\Omega$ smooth; formula defining the normal trace $u.\nu$ (23.2), in dual of traces of $H^1(\Omega)$ (23.3); interpretation in terms of differential forms, $H(curl; \Omega)$ (23.4); $H^s(R^N)$ (23.5); for $s > 1/2$, restriction on $x_N = 0$ is defined on $H^s(R^N)$, and the trace space is $H^{s-(1/2)}(R^{N-1})$ (23.6)–(23.10); 23.1: orthogonal of H in $L^2(\Omega; R^N)$ is the space $\{grad(p) \mid p \in H^1(\Omega)\}$, if injection of $H^1(\Omega)$ into $L^2(\Omega)$ is compact; 23.2: if $meas\,\Omega < \infty$ and $X(\Omega) = L^2(\Omega)$ then V is dense in H; discussion of $X(\Omega) = L^2(\Omega)$ if $\partial\Omega$ is smooth, and how to change the definitions of the spaces if the boundary is not smooth enough; Faedo–Ritz–Galerkin method for existence of Navier–Stokes equation for $N = 3$ (23.11)–(23.12); singular solutions of the stationary Stokes equation in corners (23.13)–(23.18).

Lecture 24, Using compactness: 24.1: J.-L. Lions's lemma (24.1); 24.2: u_n bounded in $L^p(0,T;E_1)$ and convergent in $L^p(0,T;E_3)$ imply u_n convergent in $L^p(0,T;E_2)$ if injection of E_1 into E_2 is compact (24.2); 24.3: u_n bounded in $L^{p_1}(0,T;E_1)$ and convergent in $L^{p_3}(0,T;E_3)$ gives u_n convergent in $L^{p_2}(0,T;E_2)$ if interpolation inequality holds; hypothesis of reflexivity; 24.4: u_n bounded in $L^p(0,T;E)$ and $||\tau_h u_n - u_n||_{L^p(0,T;E)} \leq M\,|h|^\eta$ imply u_n bounded in $L^q(0,T;E)$; 24.5: u_n bounded in $L^p(0,T;E_1)$ and $||\tau_h u_n - u_n||_{L^p(0,T;E_3)} \leq M\,|h|^\eta$ imply u_n compact in $L^p(0,T;E_2)$ if injection of E_1 into E_2 is compact; application to extracting subsequences from Faedo–Ritz–Galerkin approximation with special basis for the Navier–Stokes equation and $N \leq 3$.

Lecture 25, Existence of smooth solutions: 25.1: If $N = 2$ and Ω smooth enough, $u_0 \in V$ and $f \in L^2((0,T) \times \Omega; R^2)$ then regularity of the linear case holds (25.1)–(25.2); can one improve bounds using interpolation inequalities; 25.2: if $N = 3$ and Ω smooth enough, $u_0 \in V$ and $f \in L^2((0,T) \times \Omega; R^3)$ then there exists $T_c \in (0,T]$ and a solution with the regularity of the linear case for $t \in (0,T_c)$ (25.3)–(25.4); 25.3: if $N = 3$ and Ω smooth enough, $|u_0|\,||u_0||$ small and $f = 0$ then a global solution with the regularity of the linear case exists for $t \in (0,\infty)$ (25.5)–(25.7); the case $f \neq 0$ (25.8); extending an idea of Foias for showing $u \in L^1(0,T;L^\infty(\Omega; R^3))$ for $N = 3$ (25.9)–(25.12).

Lecture 26, Semilinear models: Reynolds number, scaling of norms, the problems that norms give global information and not local information; a different approach shown on models of kinetic theory, the 2-dimensional Maxwell model (26.1), Broadwell model (26.2); using functional spaces with physical meaning; a special class of semilinear models (26.3)–(26.4) and why I had introduced it; 26.1: spaces $V_c \subset W_c$ and L^1 estimate in (x,t) for $u\,v$ (26.5)–(26.7); extension of the idea, compensated integrability.

Lecture 27, Size of singular sets: Leray's self-similar solutions (27.1); the question of estimating the Hausdorff dimension of singular sets; a bound for the

1/2 Hausdorff dimension in t (27.2); different scaling in $(\mathbf{x}, t)$ and the equation for "pressure" (27.3); maximal functions (27.4), Hedberg's program of proving local inequalities using maximal functions (27.5), application to pointwise estimates for the heat equation (27.6)–(27.10).

Lecture 28, Local estimates, compensated integrability: Hedberg's truncation method, a proof of F.-C. Liu's inequality using Hedberg's approach (28.1)–(28.2), a Hedberg type version of the Gagliardo–Nirenberg inequality (28.3); a result of compensated integrability improving Wente by estimates based on interpolation and Lorentz spaces.

Lecture 29, Coriolis force: Equations in a moving frame and Coriolis force (29.1)–(29.3); analogy, Lorentz force, incompressible fluid motion, nonlinearity as $\mathbf{u} \times curl(-\mathbf{u}) + grad(|\mathbf{u}|^2/2)$ (29.4)–(29.8), conservation of helicity.

Lecture 30, Equation for the vorticity: Equation for vorticity, for $N = 2$ and for $N = 3$ (30.1)–(30.6).

Lecture 31, Boundary conditions in linearized elasticity: Other boundary conditions for linearized elasticity, Neumann condition (31.1) and compatibility conditions (31.2)–(31.3); studying linearized rigid displacements (31.4); other type of boundary conditions; traction at the boundary for a Newtonian fluid (31.5)–(31.6).

Lecture 32, Turbulence, homogenization: Microstructures in turbulent flows; the defect of probabilistic postulates; homogenization.

Lecture 33, G-convergence and H-convergence: Weak convergence, linear partial differential equations in theory of distributions; conservation of mass using differential forms; G-convergence and H-convergence; exterior calculus, differential forms, exterior derivative, Poincaré lemma; weak convergence as a way to relate mesoscopic and macroscopic levels, analogy between proofs in H-convergence and the way some physical quantities are measured and other physical quantities are identified; 33.1: div-curl lemma, its relation with differential forms.

Lecture 34, One-dimensional homogenization, Young measures: 1-dimensional homogenization by div-curl lemma; the G-convergence and H-convergence approaches; effective coefficients cannot be computed in terms of Young measures in dimension $N \geq 2$, physicists' formulas are approximations; importance of both balance equations and constitutive relations; 34.1: Young measures.

Lecture 35, Nonlocal effects I: Turbulence as an homogenization problem for a first order transport operator (35.1); memory effects appearing by homogenization; a model problem with a memory effect in its effective equation (35.2)–(35.3), proof by the Laplace transform (35.4)–(35.9); irreversibility without probabilistic framework; a transport problem with a nonlocal effect in (x, t) in its effective equation (35.10)–(35.15).

Lecture 36, Nonlocal effects II: Frequency-dependent coefficients in Maxwell's equation (36.1), principle of causality, pseudo-differential operators; the model problem with time dependent coefficients (36.1)–(36.8), by a perturbation expansion approach; "analogies" with Feynman diagrams and Padé approximants.

Lecture 37, A model problem: A model problem with a term $\mathbf{u} \times curl(\mathbf{v}_n)$ added to the stationary Stokes equation (37.1)–(37.2), the derivation of the effective equation (37.3)–(37.14), by methods from H-convergence; an effective term corresponding to a dissipation quadratic in $\mathbf{u}$ and not in $grad\,\mathbf{u}$, which can be computed with H-measures.

Lecture 38, Compensated compactness I: The time dependent analog requires a variant of H-measures; 38.1: the quadratic theorem of compensated compactness (38.1)–(38.4); chronology of discoveries; correction for $U \otimes U$ written as the computation of a convex hull, a formula simplified by introduction of H-measures.

Lecture 39, Compensated compactness II: Constitutive relations (39.1), balance equations (39.2), question about how to treat nonlinear elasticity (39.3); H-measures can handle variable coefficients; how compensated compactness constrains Young measures (39.4); examples: compactness, convexity, monotonicity, Maxwell's equation; proof of necessary conditions.

Lecture 40, Differential forms: Maxwell's equation expressed with differential forms (40.1)–(40.6); 40.1: generalization of div-curl lemma for p-forms and q-forms; generalizations to Jacobians, special case of exact forms (40.7); 40.2: one cannot use the weak topology in the general div-curl lemma; other necessary conditions; how helicity appears in the framework of differential forms, analogy between Lorentz force and the equations for fluid flows.

Lecture 41, The compensated compactness method: 41.1: case when the characteristic set is the zero set of a nondegenerate quadratic form; the question of making the list of interesting quantities in nonlinear elasticity (41.1); wave equation (41.2), conservation of energy (41.3), where the energy goes, equipartition of energy; use of entropies for Burgers's equation for passing to the limit for weakly converging sequences (41.4)–(41.12), entropy condition (41.13) and Murat's lemma.

Lecture 42, H-measures and variants: Wigner transform, avoiding using one characteristic length, the hints for H-measures; definitions for H-measures (42.1)–(42.4); constructing the right "pseudo-differential" calculus (42.5)–(42.12); localization principle (42.13)–(42.14); small-amplitude homogenization (42.15)–(42.18); propagation equations for H-measures (42.19)–(42.27); the variant with one characteristic length, semi-classical measures of P. Gérard (42.28).

Biographical data: Basic biographical information for people whose name is associated with something mentioned in the lecture notes.

Contents

Basic physical laws and units

The goal of the course is to teach questions of partial differential equations (PDE) for variants of the Navier–Stokes equation occurring in oceanography. My first plan was to follow a book by Roger LEWANDOWSKI [18], but after looking at his bibliography for textbooks on oceanography, I selected a book by A. E. GILL [15], and I decided to first follow this book as an introduction to questions of oceanography. It has the advantage of starting with simpler models, but it also provides a way to learn about the magnitude of different effects. It is indeed an important aspect that mathematicians interested in continuum mechanics or physics must learn, as they too often play with partial differential equations without much knowledge of a few obvious facts concerning the "real world".[1,2,3]

An example of this fact concerns *incompressibility*, an assumption that is often made by mathematicians, and which has the nonphysical effect that some perturbations may travel at infinite speed. Although water looks difficult to compress, the *speed of sound* is only of the order of 1.5 km s^{-1}, and for oceans which are thousands of kilometers wide the information does take some time to travel across. The speed of sound actually varies with temperature and pressure; for a *practical salinity* $S = 35$ (i.e. 35 grams of salt per kilogram of

---

[1] In one of his books of interviews, *Surely you're joking, Mr. Feynman!*, FEYNMAN tells that while he enjoyed playing the drums in the samba schools in Rio de Janeiro, Brazil, he also taught some courses at a nearby university. He had been puzzled by the behavior of some graduate students, who never did the exercises that he proposed in order to teach them about orders of magnitude of various physical effects; they learnt physics as if it was a foreign language, so that they could repeat correctly all the definitions but seemed to have no idea that these concerned properties of the real world.

[2] Richard Phillips FEYNMAN, American physicist, 1918–1988. He received the Nobel Prize in Physics in 1965. He worked at Caltech (California Institute of Technology), Pasadena CA.

[3] Alfred NOBEL, Swedish industrialist and philanthropist, 1833–1896. He created a fund to be used as awards for people whose work most benefited humanity.

water), from 1,439.7 m s^{-1} at $-2°$C to 1,547.6 m s^{-1} at 31°C at atmospheric pressure, while at 6,000 m deep it varies from 1,542.6 m s^{-1} at $-2°$C to 1,560.2 m s^{-1} at $+2°$C (and it is not very practical to think of higher temperatures at this depth).

I shall mostly use the *metric system*, not only because it is the one that I learnt in France, but also because it is more natural, and physicists tend to use it anyway. The usual multiplicative prefixes are deca $= 10$, hecto $= 100$, kilo $= 1,000$, mega $= 10^6$, giga $= 10^9$, tera $= 10^{12}$, peta $= 10^{15}$, exa $= 10^{18}$. The usual divisive prefixes are deci $= 10^{-1}$, centi $= 10^{-2}$, milli $= 10^{-3}$, micro $= 10^{-6}$, nano $= 10^{-9}$, pico $= 10^{-12}$, femto $= 10^{-15}$, atto $= 10^{-18}$.

The unit of length is the *meter* m (a kilometer (km) is 1,000 m, a nautical mile is 1.85318 km; a mile is 1.60932 km). The unit of surface is the *square meter* (m^2) (a square kilometer (km^2) is 10^6 m^2; an acre is 4,046.67 m^2). The unit of volume is the *cubic meter* (m^3).

The unit of mass is the *kilogram* (kg) (a kilogram is 1,000 grams, a metric ton is 1,000 kg; a pound is 0.453 kg). The unit of density is the *kilogram per cubic meter* (kg m^{-3}). Water has a density of approximately 1,000, while air has a density of approximately 1.29.

The unit of time is the *second* (s) (a minute is 60 s, an hour 3,600 s, a day 86,400 s, a year 3.1558×10^7 s).

The unit of velocity is the *meter per second* (m s^{-1}) (a km hr^{-1} is 0.2777 m s^{-1}, a knot, i.e. a nautical mile per hour, is 0.51477 m s^{-1}). The unit of acceleration is the *meter per second square* (m s^{-2}). Acceleration under gravity is approximately 9.8 m s^{-2}.

The unit of force is the *newton*[4] (N), i.e. 1 kg m s^{-2}. The unit of pressure is the *pascal*[5] Pa, i.e. 1 newton m^{-2} (a *bar* is 10^5 Pa). Atmospheric pressure is about one bar, and the pressure increases by about one bar each time one goes down 10 meters in the ocean.

The unit of energy is the *joule*[6] (J), i.e. 1 kg m^2 s^{-2} (one calorie is 4.184 J; a *calorie* is about the amount of energy that one needs for increasing the temperature of a gram of water by one degree, at usual temperatures). The unit of power is the *watt*[7] (W), i.e. 1 kg m^2 s^{-3} (a kilowatt (kW) is 1,000 W).

The unit of temperature is the *degree Celsius*[8] °C $= 1$ kelvin[9] (K); the temperature in degree Celsius is the absolute temperature -273.15.

---

[4] Sir Isaac NEWTON, English mathematician, 1642–1727. He held the Lucasian chair at Cambridge, England, UK.

[5] Blaise PASCAL, French mathematician, 1623–1662. He worked in Paris, France.

[6] James Prescot JOULE, English physicist, 1818–1889. He worked in Salford, England, UK.

[7] James WATT, Scottish engineer, 1736–1819. He worked in Glasgow, Scotland, UK.

[8] Anders CELSIUS, Swedish astronomer, 1701–1744. He worked in Uppsala, Sweden.

[9] William THOMSON, Irish-born physicist, 1824–1907. He was made baron KELVIN of Largs in 1892, and thereafter known as Lord KELVIN. He worked in Glasgow, Scotland, UK.

Oceanography is concerned with both air and sea and deals mostly with large scale motions, as well as the temperature and salinity of water in the oceans.

Meteorology is the study of motion, temperature, moisture content and pressure in the atmosphere, but one discovers quickly the important role played by the oceans, and the exchanges between air and sea appear to be crucial.

Everything starts with the Sun. The Earth moves at an average distance of 140 million kilometers from the Sun, and turns on itself with the inclination of its axis being responsible for the seasons. The average energy flux from the Sun at the mean radius of the Earth, called the *solar constant S*, has the value $S = 1.368$ kW m^{-2}. This means that if one was collecting all the energy from the Sun on a panel of one square meter, without any reflection, the panel being oriented perpendicularly to the direction from the Sun, one would get a power of 1.368 kW (a hair drier works between 1 and 1.5 kW, I believe). This energy corresponds to the *blackbody* radiation at a temperature of about 6,000°C, the surface temperature of the Sun (what happens inside the Sun does not matter for us). The blackbody radiation of a body at *absolute temperature T* is given by *Planck's*[10] *law*, which describes the repartition of energy per unit volume of the "photons" having frequency near ν, $u(\nu)d\nu = \frac{8\pi h \nu^3}{c^3(e^{h\nu/kT}-1)} d\nu$, where h is *Planck's constant*, approximately 6.62×10^{-34} J s, c is the speed of light, approximately 3×10^8 m s^{-1}, and k is *Boltzmann's*[11] *constant*, approximately 1.38×10^{-23} J K^{-1}. A surface at absolute temperature T emits energy in all directions, and the power radiated by a surface dS in the solid angle $d^2\omega$ making an angle θ with the normal to the surface, between frequency ν and $\nu + d\nu$ is $\delta W = u(\nu) c \cos\theta \, dS \frac{d^2\omega}{4\pi} d\nu$ (computed as the energy of the photons contained in a cylinder based on dS with length $c\,dt$, and therefore having volume $dS\,c\,dt\cos\theta$), the dependence on $\cos\theta$ is *Lambert's*[12] *law*. The total power radiated in the half space, obtained by integrating in directions and frequencies is $E = \sigma T^4$, where the *Stefan*[13] *constant* σ is approximately 5.67×10^{-8} w m^{-2} K^{-4}. For the Sun, a large part of the energy is in the *visible spectrum*, between 0.4 and 0.8 μ (μ = micron = micrometer), therefore in very short wavelengths. *Absorption* by the atmosphere, water vapor, carbon dioxide, etc. , is specific and varies a lot with frequency.

---

[10] Max Karl Ernst Ludwig PLANCK, German physicist, 1858–1947. He received the Nobel Prize in Physics in 1918. He worked in Berlin, Germany. There is a Max PLANCK Society for the Advancement of the Sciences, which promotes research in many institutes, mostly in Germany.

[11] Ludwig BOLTZMANN, Austrian physicist, 1844–1906. He worked in Graz and Vienna, Austria, and also in Leipzig, Germany.

[12] Johann Heinrich LAMBERT, French-born mathematician, 1728–1777. He worked in Berlin, Germany.

[13] Josef STEFAN, Austrian physicist, 1835–1893. He worked in Vienna, Austria.

The average power received from the Sun per unit surface of the Earth is then $\frac{S}{4\pi} = 344$ W m^{-2}, but there is an *albedo* effect with an average coefficient $\overline{\alpha} = 0.3$, which denotes the proportion of the energy which is reflected immediately; in average, the ground receives then about 240 W m^{-2}, and 100 W m^{-2} is reflected back into space. The albedo is actually varying and depends a lot upon the cloud cover (Venus, covered by clouds, as an albedo of 0.6, while Mars, which has no clouds, has an albedo of 0.15). The albedo of land is about 0.15, and goes up to 0.2–0.3 for deserts, while land covered by snow or ice has an albedo of 0.6. Most of the oceans below latitude 40 have an albedo below 0.1, but the average is between 0.15 and 0.3.

If there was a purely radiative equilibrium between the energy received from the Sun and the energy radiated by the Earth $((1 - \overline{\alpha})\frac{S}{4\pi} = \sigma T_g^4)$, one would observe that the temperature of the ground would only be about 270 K at the equator, 170 K near the north pole and 150 K near the south pole.

It is the fluid cover, water and air, that makes a huge difference from these cold predictions. First, radiation may be absorbed in the atmosphere, and the gases present, water vapor, carbon dioxide, etc. , are important, creating a *greenhouse* effect. Second, there are *convection effects* which take the warm waters from the equator to the polar regions. HALLEY[14,15] had already understood the opposite effects of the Sun and of gravity in 1686: the Sun creates a horizontal variation, heating the waters near the equator much more than near the poles, while gravity likes a vertical variation, drawing the cold waters to the bottom and making the warm waters rise to the surface; the competition between these two effects creates convection. MARSIGLI[16] had already studied in 1679 the existence of an undercurrent in the depth of the Bosphorus, created by difference in salinity between the Mediterranean and the Black Sea (see Lecture 3).

A greenhouse functions from the idea that glass is transparent to the short wavelengths in the solar radiation (which is why glass looks transparent to us, as we can only see wavelengths between 0.4 and 0.8 μ), while it is not transparent to the long wavelengths corresponding to the energy radiated from the ground. From Planck's law one deduces that the frequency ν_m, where u(ν) is maximum, is given by *Wien's*[17] *law* $\nu_m = \frac{2.82\,kT}{h}$h, which shows that ν_m is linear in T; therefore if the Sun at 6,000 K has its maximum around 0.6 μ, a body at 300 K has its maximum at wavelengths 20 times larger, around 12 μ. If glass absorbs most of these long wavelengths, then it gets hotter (if it was absorbing all the radiation from the ground, it would reach the temperature

---

[14] Edmund HALLEY, English mathematician, 1656–1742. He held the Savilian chair at Oxford, England, UK, and he became the second Astronomer Royal.

[15] Sir Henry SAVILE, English scholar, 1549–1622. He worked in Oxford and in Eton, England, UK.

[16] Count Luigi Ferdinando MARSIGLI, Italian soldier and scientist, 1658–1730.

[17] Wilhelm Carl Werner Otto Fritz Franz WIEN, German physicist, 1864–1928. He received the Nobel Prize in Physics in 1911. He worked in München (Munich), Germany.

that the ground would have in absence of glass), and it emits its own radiation, both up and down and therefore the ground gets back a part of the energy that it had radiated away. If I is the downward flux coming from the Sun, which is in short wavelengths, U the upward flux corresponding to the temperature T_g of the ground, e the proportion of U absorbed by the glass and B the flux emitted in each direction (up or down) by the glass, then both U and B correspond to long wavelengths, and the equilibrium equation, under the hypothesis that the glass does not absorb any part of I, is $eU = 2B$, and $I = (1 - e)U + B$, i.e. $I = \left(1 - \frac{e}{2}\right)U$, and therefore $\sigma T_g^4 = U = \frac{I}{1-(e/2)}$, and T_g is higher by up to 19% in the case $e = 1$, which gives $\sigma T_g^4 = 2I$, as $2^{1/4}$ is around 1.19.

[Taught on Monday January 11, 1999. The course met on Mondays, Wednesdays, and Fridays.]

2

Radiation balance of atmosphere

According to GILL [15], an estimate of the radiation balance for the atmosphere is as follows. In order to work with percentages, let us assume that 100 units of incoming solar radiation arrive at the top of the atmosphere, all this energy being in the short wavelengths. Thirty units will be sent back to space, corresponding to an albedo of 0.3, but these 30 units are decomposed into 6 units backscattered by air, 20 units reflected by clouds and 4 units reflected by the ground surface. In the 70 units which are not reflected, 16 are absorbed in the atmosphere by water, dust and ozone, 3 are absorbed by the clouds and the remaining 51 are absorbed by land and oceans.

The surface also absorbs an estimated 98 units of long-wave radiation, sent back from the atmosphere (this effect is explained as in the way the greenhouse functions, rendering it even more efficient by a few additional layers of glass, as shown below). The net surface emission in the longer wavelengths of the infrared, excess of upward over downward radiation, is 21 units, the remaining upward flux of 30 units being by convection. The temperature of the ground corresponds then to $51 + 98 = 149$ units of radiated energy flux, instead of the 70 units emitted at the top of the atmosphere.

From the 21 units of excess infrared radiation from the surface, 15 are absorbed by water and carbon dioxide and 6 units end up in space. Space also receives 38 units emitted by water and carbon dioxide as well as 26 units emitted by clouds, so $6 + 38 + 26 = 70$ units of infrared radiation are sent to space, equilibrating the 70 units of solar radiation which had not been reflected (the albedo is only the fraction of solar radiation which is sent back immediately). From the 30 units used in convection, 23 are used by latent heat for creating vapor from water and 7 correspond to sensible heat flux, used to warm the atmosphere directly.

The balance for the atmosphere is then 16 units absorbed in solar radiation, 15 units absorbed in infrared radiation and 7 units received as heat, corresponding to the 38 units emitted.

The balance for the clouds is 3 units absorbed in solar radiation and 23 units invested in latent heat, corresponding to the 26 units emitted.

The processes of absorption and emission of radiation are not simple. It is worth noticing that they are highly frequency dependent: roughly speaking, there are frequencies which make a particular type of molecule vibrate and if radiation containing these frequencies goes through a gas containing these molecules, some of the energy at these frequencies will be absorbed by the gas. Conversely, a gas containing these molecules may emit spontaneously at those frequencies which it can absorb.

The generalized greenhouse having p well-separated layers of glass which completely absorb the low frequencies but are transparent to the high frequencies is easy to compute: if I is received from solar radiation, $U = B_0$ is emitted by the surface and B_i is emitted on both sides of the glass layer #i, counting from the surface, then the balance equations are

$$I = B_p; \ B_{i+1} + B_{i-1} = 2B_i \text{ for } i = 1, \ldots, p-1, \tag{2.1}$$

whose solution is

$$B_i = (p+1-i)I \text{ for } i = 0, \ldots, p, \text{ i.e. } U = (p+1)I. \tag{2.2}$$

Let us consider now the more general situation where the glass layer #i absorbs a proportion e_i of low-frequency radiation but is completely transparent to high-frequency radiation. Let B_i be the flux emitted on both sides by the glass layer #i, for $i = 1, \ldots, p$, but now let $U = A_0$ and let A_i denote the ascending flux just above glass layer #i (this flux includes the ascending B_i) for $i = 1, \ldots, p$, so that $A_p = I$; similarly, let D_i denote the descending flux just below glass layer #i (this flux includes the descending B_i) for $i = 1, \ldots, p$, and for convenience let $D_{p+1} = 0$. The balance equations are .

$$\begin{aligned}
e_i(A_{i-1} + D_{i+1}) &= 2B_i, i = 1, \ldots, p \\
A_i &= (1 - e_i)A_{i-1} + B_i, i = 1, \ldots, p \\
D_i &= (1 - e_i)D_{i+1} + B_i, i = 1, \ldots, p,
\end{aligned} \tag{2.3}$$

and eliminating B_i gives

$$\begin{aligned}
A_i &= \left(1 - \tfrac{e_i}{2}\right)A_{i-1} + \tfrac{e_i}{2}D_{i+1}, i = 1, \ldots, p \\
D_i &= \left(1 - \tfrac{e_i}{2}\right)D_{i+1} + \tfrac{e_i}{2}A_{i-1}, i = 1, \ldots, p,
\end{aligned} \tag{2.4}$$

which by adding gives $A_i + D_i = A_{i-1} + D_{i+1}$ for $i = 1, \ldots, p$, which is easy to see directly ($A_{i-1} + D_{i+1}$ is the amount received by the glass layer #i, while $A_i + D_i$ is the amount transmitted by the glass layer #i). Therefore $A_i - D_{i+1}$ is independent of $i = 0, \ldots, p$, and using the value for $i = p$ gives

$$A_i - D_{i+1} = I \text{ for } i = 0, \ldots, p, \tag{2.5}$$

from which one deduces $A_i = \left(1 - \tfrac{e_i}{2}\right)A_{i-1} + \tfrac{e_i}{2}(A_i - I), \ i = 1, \ldots, p,$ or

$$A_{i-1} = A_i + \frac{\frac{e_i}{2}}{1 - \frac{e_i}{2}} I, i = 1, \ldots, p, \tag{2.6}$$

and finally

$$U = \left(1 + \sum_{i=1}^{p} \frac{e_i}{2 - e_i}\right) I. \tag{2.7}$$

One deduces the case of a continuous absorbing media: if the layer between z and $z + dz$ absorbs a proportion $f(z)\, dz$ of low-frequency radiation and is transparent to high-frequency radiation, one finds $U = \left(1 + \frac{1}{2} \int_0^\infty f(z)\, dz\right) I$.

Of course, the radiative balance described is not entirely radiative as it relies on observed distribution of water vapor, responsible for a large part of the absorption, and one certainly needs to understand a little more about the *thermodynamics of air and water* in order to explain quantitatively the effects of convection, but a qualitative explanation is possible.

If there was no water vapor in the air and no other mechanism for absorbing radiation in the atmosphere, the atmosphere would stay cold and a larger amount of radiation would arrive at the surface: the surface would get warmer and the air in contact with the ground would also become warmer by conduction. Warm air is lighter than cold air, and therefore it rises; the pressure decreases when one goes up (the origin of atmospheric pressure is mostly the weight of the air above our heads), and when pressure decreases a gas expands and its temperature decreases, so the crucial problem is to compare the decrease in temperature due to expansion and the decrease in temperature due to altitude.

The *lapse rate* denotes the rate at which the temperature of the atmosphere decreases with height; the *dry adiabatic lapse rate* denotes the rate at which the temperature (of dry air) decreases because of expansion, and it is about 10 K km^{-1}: as long as the lapse rate is greater than the adiabatic lapse rate, warm air goes up, and this starts convection; convection carries heat upward, and therefore diminishes the lapse rate.

Of course, the real situation gets complicated by the fact that the atmosphere contains water vapor. If it contains a small amount of water vapor, convection will still occur when the dry adiabatic lapse rate is exceeded. However, air at a given temperature and pressure can only hold a certain amount of water vapor, and the amount of water vapor relative to this saturation value is called the *relative humidity*; when the relative humidity reaches 100%, water droplets condense in clouds, releasing the latent heat which had been required for creating the water vapor near the ground. The *latent heat of vaporization* $L_v(\text{T})$ is given by the formula $L_v(\text{T}) \simeq 2.5008 \times 10^6 - 2.3 \times 10^3 t$ J kg^{-1}, where t denotes the temperature in degrees Celsius (so at boiling temperature at *atmospheric pressure*, $t = 100$, one needs 542 calories for vaporizing one gram of water). Latent heat represents more than 75% (23 units/30 units) of the heat transferred by convection. For saturated air, one must use the *moist adiabatic lapse rate*, which depends on temperature and pressure: in the lower atmosphere it is about 4 K km^{-1} at 20°C and 5 K km^{-1} at 10°C. Because the amount of water to produce saturation decreases with altitude, saturated air stays saturated when it goes up and the moist adiabatic lapse rate is used, but air cannot stay saturated when it goes down and the dry adiabatic lapse rate is used.

The preceding qualitative analysis was concerned with vertical differences in temperature, but there are also horizontal differences, as the Sun warms the equatorial region more than the polar regions. Although HALLEY (1686) had proposed the model that warm air rises near the equator and tropics and falls at higher latitudes, it appears that the rising motion is concentrated in a narrow band called the *Inter-Tropical Convergence Zone* (ITCZ), usually found between 5° and 10° to the north of the equator (the *trade winds*, created by Coriolis[1] forces, push air from the tropics towards the equator); the regions of descending air are dry and include desert regions found between latitudes 20° and 30°. In mid-latitudes, because of the rotation of the Earth, the motion produced by horizontal density gradients is mainly east–west and there is little meridional circulation; however, large disturbances are created, *cyclones and anti-cyclones*, which are very efficient at transporting energy towards the poles.

It seems that ocean and atmosphere are equally important in transporting energy, the atmosphere being most important at 50° N and the ocean most important at 20° N, but the error in the estimate of the ocean transport at 20° N could be as high as 77%!

[Taught on Wednesday January 13, 1999.]

---

[1] Gaspard Gustave CORIOLIS, French mathematician, 1792–1843. He worked in Paris, France.

3

Conservations in ocean and atmosphere

According to GILL [15], the fraction α of solar radiation reflected by the ocean is a function of the angle of incidence and of the surface roughness; at latitudes below 30° it is less than 0.1, but it increases with latitude because of the angle of incidence of rays. Unlike the atmosphere, the ocean absorbs solar radiation rapidly; 80% is absorbed in the top 10 m, and the absorption rate is even greater in coastal areas, where a lot of suspended material exists.

As long-wave radiation is rapidly absorbed in the atmosphere because of the presence of water vapor, it is absorbed very rapidly by the ocean, and absorption and emission occur in a very thin layer, less than 1 mm thick.

Water is about 800 times denser than air (at atmospheric pressure), 1,025 kg m^{-3} compared with 1.2–1.3 kg m^{-3}, and due to the strength of the gravitational restoring force, there is not much mixing, and transfer of properties between the two media takes place near the interface. The atmospheric pressure (1 bar) corresponds to the weight of the atmosphere, but it is just the weight of 10 m of water; the total mass of the ocean is 270 times the total mass of the atmosphere. The *specific heat* (heat capacity per unit mass) of water is 4 times that of air, and therefore the top 2.5 m of ocean has the same heat capacity than the whole atmosphere above (10^7 J m^{-2} K): raising the temperature of the atmosphere by 1 K can be done by lowering the temperature of 2.5 m of ocean by 1 K, or that of 25 m by 0.1 K. Heat can be stored in latent form, and the same amount of heat can be used to evaporate 4 mm of water, or to melt 30 mm of ice (the *evaporation rate* in the tropics is of order of 4 mm per day). Because of this ability to store heat, the ocean surface temperature changes by much smaller amounts than the land surface, which cannot store much heat. The excess heat gained in summer is not transported to the other hemisphere where there is winter, but is stored in the surface layers (about 100 m) and returned to the atmosphere in the same hemisphere in the next winter.

So much for heat, now let us consider the balance of momentum and angular momentum. How are the winds produced and what determines their distribution?

HALLEY (1686) had tried to explain the trade winds, which blow from the tropics to the equator (NE to SW in the northern hemisphere and from SE to NW in the southern hemisphere), but his idea only explains the meridional circulation in the Inter-Tropical Convergence Zone described before, which is not called after him now, but after HADLEY,[1] who gave a better explanation in 1735. If there was no friction, the equator being longer by 2,083 miles than the tropics (at latitude 23° 27', i.e. around 23.5°), he argued that air at rest at the tropics would acquire a westward motion of 2.083 miles day^{-1} when transported to the equator, but as the observed velocity is not as high as this velocity of almost 140 km hr^{-1}, he argued that there was some friction and air had also been given a correcting eastward push from the surface of Earth. He also argued that there must exist opposite winds somewhere in order to compensate the trade winds: this is related to the *conservation of angular momentum*, but it is not valid for the northward/southward component (the moment of a vector parallel to the axis is 0). If the average *eastward force* (or rate of transfer of eastward momentum) per unit area acting on the surface at latitude φ is $\tau^x(\varphi)$, then the average *torque* (or rate of transfer of angular momentum) per unit area about the axis is $a\,\tau^x(\varphi)\cos\varphi$, where a is the radius of the Earth; the area of the strip between latitudes φ and $\varphi + d\varphi$ is $2\pi\, a^2 \cos\varphi\, d\varphi$, so the torque on this strip is $2\pi\, a^3 \tau^x(\varphi)\cos^2\varphi\, d\varphi$, and the balance of angular momentum for the Earth is then $\int_{-\pi/2}^{+\pi/2} \tau^x(\varphi)\cos^2\varphi\, d\varphi = 0$. The force of the atmosphere on the underlying surface is exerted in two ways: one is the force exerted on irregularities in the surface associated with the pressure differences across the irregularities, and the second is by viscous stresses. The irregularities on which forces are exerted may vary in size from mountain ranges down to trees, blades of grass and ocean surface waves. When the irregularities are small enough (as is the case over the ocean), the associated force per unit area added to the viscous stress is called the surface stress or *wind stress*. There is a westward stress in the trade wind zones (latitudes below 30°) and therefore an eastward stress is required at higher latitudes, and one does observe westerly (i.e. eastward) winds at those latitudes. In France there is a dominant wind from the west (I was taught that this is why industrial plants were first built on the east side of Paris so that the wind would push the smoke away from the city; of course the expensive residential areas then developed in the west side of the city!). In the southern hemisphere the wind is indeed from the West, but it is extremely strong, probably because there is no land there to slow it down, and the sailors traveling in latitudes 40° S and 50° S have coined the names *Roaring Forties* and *Furious Fifties*.

---

[1] George HADLEY, English meteorologist, 1685–1768. He worked in London, England, UK.

Winds are produced in the atmosphere, a result of the radiative forcing, which creates horizontal and vertical gradients, and it is difficult to understand these effects without writing partial differential equations models, but winds are of the order of 10 m s^{-1}. Winds transfer momentum to the ocean, producing currents, but the exact process is not so simple as a shear flow near the interface becomes unstable and turbulent eddies are formed (so there are gusts of wind), and one needs to average over time (a few minutes for points a few meters above the ground): the mean stress τ is equal to the mean value of $\varrho\, u\, w$, where u and w are the horizontal and vertical components of the velocity, and ϱ is the density. If one measures u at some level, one can induce by dimensional analysis that $\tau = c_D \varrho\, u^2$, where c_D is a dimensionless parameter called the *drag coefficient*, which depends upon the roughness of the interface and the lapse rate. The drag coefficient c_D for the ocean surface is found to increase with wind speed: for low speeds it is around 1.1×10^{-3}, but for speeds between 6 m s^{-1} and 22 m s^{-1} one often uses the relation $10^3 c_D = 0.61 + 0.063\, u$.

There are, however, other formulas derived, and we shall only remember that various effects have to be modeled near the interface and that finding correct boundary conditions is important.

We start now deriving the basic partial differential equations which describe the variations in space and time of the various physical quantities. We begin by investigating the equation of *conservation of mass*, which is certainly true away from the interface, but not near the interface, where water is lost by *evaporation* and gained by *precipitation*.

Notice that the *conservation of salt* is also important, and salinity increases because of evaporation and decreases because of precipitation. As was first noticed by MARSIGLI in 1681 the difference in salinity between the Black Sea and the Mediterranean is responsible for a deep undercurrent in the Bosphorus, the lighter, less salty waters from the Black Sea flow on the surface towards the Mediterranean, while the heavier, more salty waters from the Mediterranean flow below towards the Black Sea.

Although various coordinate systems are used, it is useful to use a Cartesian[2] system of coordinates (with an orthonormal basis) in order to derive more easily the equations. Position is denoted by $\mathbf{x} = (x_1, x_2, x_3)$, time by t, velocity by $\mathbf{u} = (u_1, u_2, u_3)$, density by ϱ. In the *Lagrangian*[3,4] *point of view*, one refers quantities to the initial position $\boldsymbol{\xi}$ of the particle, i.e. the components

---

[2] René DESCARTES, French-born mathematician and philosopher, 1596–1650. After extensive travels in Europe, he settled near Amsterdam, The Netherlands. Université of Paris 5, France, is named after René DESCARTES.

[3] Giuseppe Lodovico LAGRANGIA (Joseph Louis LAGRANGE), Italian-born mathematician, 1736–1813. He was made count by NAPOLÉON in 1808. He worked in Paris, France.

[4] Napoléon BONAPARTE, French general, 1769–1821. He proclaimed himself emperor, under the name NAPOLÉON I, 1804–1814 (and 100 days in 1815).

x_i are expressed in terms of $\boldsymbol{\xi}$ and t, and so $v_i = \frac{\partial x_i}{\partial t}$. In the *Eulerian*[5,6] *point of view*, one refers quantities to the actual position $\mathbf{x}$ of the particle and t, and this is the point of view that we shall consider, and we shall see that conservation of mass takes the form

$$\frac{\partial \varrho}{\partial t} + \sum_{i=1}^{3} \frac{\partial (\varrho \, u_i)}{\partial x_i} = 0, \tag{3.1}$$

the second part being abbreviated as $div(\varrho \, u)$.

A classical way of deriving this equation, formal because one assumes that everything is smooth enough, is as follows. One considers a set of material points occupying the domain $\omega(0)$ at time 0 and $\omega(t)$ at time t, and one writes that $\int_{\omega(t)} \varrho \, d\mathbf{x}$ is independent of t. The variation on the intersection of $\omega(t)$ and $\omega(t + \delta t)$ is estimated as $\delta t \int_{\omega(t)} \frac{\partial \varrho}{\partial t} \, d\mathbf{x}$. The part on $\omega(t + \delta t) \setminus \omega(t)$ is estimated by a surface integral on $\partial \omega(t)$: as a point $\mathbf{x}$ on the surface moves a distance of about $\mathbf{u}(\mathbf{x}, t) \delta t$, it is as if the surface was pushed in the direction of the exterior normal $\boldsymbol{\nu}$ by an amount $\mathbf{u}.\boldsymbol{\nu} \, \delta t$, and therefore the second part is estimated as the surface integral $\delta t \int_{\partial \omega(t)} \varrho \, \mathbf{u}.\boldsymbol{\nu} \, d\mathbf{x}'$, and the conservation of mass then takes the form

$$\int_{\omega(t)} \frac{\partial \varrho}{\partial t} \, d\mathbf{x} + \int_{\partial \omega(t)} \varrho \, \mathbf{u}.\boldsymbol{\nu} \, d\mathbf{x}' = 0. \tag{3.2}$$

Then one transforms the boundary integral $\int_{\partial \omega(t)} \varrho \, \mathbf{u}.\boldsymbol{\nu} \, d\mathbf{x}'$, using Green's[7] formula, into $\int_{\omega(t)} div(\varrho \, u) \, d\mathbf{x}$, which gives

$$\int_{\omega(t)} \left(\frac{\partial \varrho}{\partial t} + div(\varrho \, u) \right) d\mathbf{x} = 0, \tag{3.3}$$

and varying $\omega(t)$ gives the result.

Although this type of derivation is common practice among physicists, it is useful for mathematicians to think about the hypotheses needed for carrying out the various steps. One may also try to derive that same basic equation in other ways.

[Taught on Friday January 15, 1999.]

---

[5] Leonhard EULER, Swiss-born mathematician, 1707–1783. He worked in St Petersburg, Russia.

[6] EULER had actually introduced both the Eulerian and the "Lagrangian" points of view.

[7] George GREEN, English mathematician, 1793–1841. He was a miller and never held any academic position.

4

Sobolev spaces I

The argument using Green's formula, which is a question of integration by parts, does hold in Sobolev spaces.

For an open set Ω of R^N and $1 \leq p \leq \infty$, the Sobolev space $W^{1,p}(\Omega)$ is defined as

$$W^{1,p}(\Omega) = \left\{ u \in L^p(\Omega) \mid \frac{\partial u}{\partial x_j} \in L^p(\Omega) \text{ for } j = 1, \ldots, N \right\}, \qquad (4.1)$$

equipped with the norm[1]

$$||u||_{W^{1,p}(\Omega)} = \left(\int_\Omega \left(\frac{1}{\lambda^p} |u|^p + \sum_{j=1}^N \left| \frac{\partial u}{\partial x_j} \right|^p \right) dx \right)^{1/p}, \qquad (4.2)$$

where λ is a characteristic length (which mathematicians usually take equal to 1, giving then the impression that they add quantities measured in different units without being the least surprised!).

Elements of $L^p(\Omega)$ are classes of Lebesgue[2,3,4]-measurable functions (two functions equal almost everywhere being identified), and in order to restrict functions of $W^{1,p}(\Omega)$ on the boundary $\partial\Omega$ of Ω, which is usually a set of measure zero, one has to be careful.

The derivatives $\frac{\partial u}{\partial x_j}$ are not computed in a classical way, but are weak derivatives, and this idea was introduced by Sergei SOBOLEV, and used by Jean LERAY in the 1930s for defining weak solutions of the Navier–Stokes

---

[1] For $1 \leq p < \infty$; for $p = \infty$ one uses the smallest value of M such that $|u| \leq \lambda M$ and $\left| \frac{\partial u}{\partial x_j} \right| \leq M$ for all j, almost everywhere in Ω.

[2] Henri Léon LEBESGUE, French mathematician, 1875–1941. He held a chair (Mathématiques) at Collège de France, Paris, France.

[3] It was F. RIESZ who introduced the L^p spaces, in honor of LEBESGUE.

[4] Frigyes (Frederic) RIESZ, Hungarian mathematician, 1880–1956. He worked in Budapest, Hungary.

equation (which were shown to exist globally in time by Olga LADYZHENSKA-YA). Jean LERAY had called these weak solutions "turbulent", but although uniqueness of these weak solutions is still an open question in 3 dimensions, few people believe that Jean LERAY's ideas about turbulence are right,[5,6,7,8] and the ideas that KOLMOGOROV[9] introduced after have received more attention (which does not mean that they are right either[10]). For a function $u \in C^1(\Omega)$ and $\varphi \in C_c^1(\Omega)$, the space of C^1 functions with *compact support* in Ω, one has $\int_\Omega \frac{\partial u}{\partial x_j} \varphi \, d\mathbf{x} = - \int_\Omega u \frac{\partial \varphi}{\partial x_j} \, d\mathbf{x}$, and this formula helps in defining the weak derivatives of u: one says that $\frac{\partial u}{\partial x_j} = f_j \in L^p(\Omega)$ if for all $\varphi \in C_c^1(\Omega)$ one has $- \int_\Omega u \frac{\partial \varphi}{\partial x_j} \, d\mathbf{x} = \int_\Omega f_j \varphi \, d\mathbf{x}$. The theory of weak solutions was put into

[5] If there was non-uniqueness of weak solutions, one would expect that some kind of entropy condition should be valid for a uniquely defined *physical solution*, in a similar way to hyperbolic systems of conservation laws; it is possible, however, that to discover the right definition of a physical solution, one should first put back into the model some physical effects which have been neglected, like compressibility and temperature dependence of the viscosity for example, or some adequate viscoelastic effects. Everyone has observed how the viscosity of cooking oil varies when one heats it (which could also be related to changes in chemical bonds, possibly resulting in non-Newtonian effects), and I have heard Bill PRITCHARD point out that he needed to maintain temperature as constant as possible in his experiments on water droplets, precisely because of the important influence of temperature on viscosity; however, as was pointed out by my good friend Edward FRAENKEL, the influence of temperature was not emphasized in classical treatises on hydrodynamics, like that of LAMB.

[6] William Gardiner PRITCHARD, Australian-born mathematician, 1942–1994. He worked at The Pennsylvania State University, University Park, PA.

[7] Ludwig Edward FRAENKEL, German-born mathematician, born in 1927. He works in Bath, England, UK.

[8] Horace LAMB, English mathematician, 1849–1934. He worked in Manchester and Cambridge, England, UK.

[9] Andrei Nikolaevich KOLMOGOROV, Russian mathematician, 1903–1987. He received the Wolf Prize in 1980. He worked in Moscow, Russia.

[10] The ideas of KOLMOGOROV have the same defect as the naive idea that the effective conductivity of a mixture of materials only depends upon the proportions used, which is false in more than one dimension. In the case of small amplitude variations of the conductivities of the materials used, I have introduced H-measures for computing the quadratic correction for the effective conductivity, and in some exceptional cases the correction only depends upon proportions; one could hope that something similar would occur in the case of a weak turbulence created by small oscillations of the velocity, but there are some other difficulties which appear, and the tools for carrying this kind of analysis will be described at the end of this course.

a more general framework in the 1940s by Laurent SCHWARTZ[11,12,13] in his theory of distributions,[14] which helps one understand the *linear*[15,16,17] partial differential equations with *smooth* coefficients. The way to attack the basic linear equations of continuum mechanics, which may have *discontinuous coefficients* because of the presence of interfaces, was developed in the 1960s, mostly along the lines that Sergei SOBOLEV and Jean LERAY had pioneered.

The case $p = 2$ plays a special role and $W^{1,2}(\Omega)$ is also denoted as $H^1(\Omega)$, and generally one defines $H^s(R^N)$ for $s \in R$ by a Fourier[18] transform, and $H^s(\Omega)$ for $s \geq 0$ by *restriction* to Ω (the case $s = 0$ corresponds to $L^2(\Omega)$), and I shall use the notation H^s instead of $W^{s,2}$. One should be aware that some other spaces, which are natural in harmonic analysis, are also denoted in the same way, the Hardy[19] spaces, but I shall write them as $\mathcal{H}$ instead of H when they appear.

For a bounded open set Ω with a Lipschitz[20] boundary, i.e. an open set which is locally on one side of its boundary which has locally an equation $x_N = F(x_1, \ldots, x_{N-1})$ in an orthonormal basis, with F Lipschitz continuous,

---

[11] Laurent SCHWARTZ, French mathematician, 1915–2002. He received the Fields Medal in 1950. He worked at École Polytechnique, Palaiseau, France, and then at Université Paris 7 (Denis DIDEROT), Paris, France. I had him as a teacher in 1965–1966 (when École Polytechnique was still in Paris).

[12] John Charles FIELDS, Canadian mathematician, 1863–1932. He worked in Toronto, Ontario (Canada).

[13] Denis DIDEROT, French philosopher, 1713–1784. He worked in Paris, France. Université Paris 7, Paris, France, is named after Denis DIDEROT.

[14] Laurent SCHWARTZ has described something about his discovery of the concept of distributions in his biography [21].

[15] When I was a student in the late 1960s (in Paris, France), my advisor (Jacques-Louis LIONS) had pointed out the limitations of attacking nonlinear problems using only tools developed for linear situations; the compensated compactness method (which I developed after obtaining compensated compactness theorem 38.1 with François MURAT) was an attempt to develop a truly nonlinear theory, but it must be improved, in ways which have not been tried yet.

[16] François MURAT, French mathematician, born in 1947. He works at CNRS (Centre National de la Recherche Scientifique) at Université Paris VI (Pierre et Marie CURIE), Paris, France.

[17] Pierre CURIE, French physicist, 1859-1906, and his wife Marie SLODOWSKA CURIE, Polish-born physicist, 1867–1934, jointly received the Nobel Prize in Physics in 1903, and she also received the Nobel Prize in Chemistry in 1911. They worked in Paris, France.

[18] Jean-Baptiste Joseph FOURIER, French mathematician, 1768–1830. He was prefect (under Napoléon) in Grenoble, France. The Institut FOURIER is the department of Mathematics of Université de Grenoble I, itself named after Joseph FOURIER.

[19] Godfrey Harold HARDY, English mathematician, 1877–1947. He worked at Cambridge, England, UK (Sadleirian chair, endowed by Lady SADLEIR in 1710).

[20] Rudolf Otto Sigismund LIPSCHITZ, German mathematician, 1832–1903. He worked in Bonn, Germany.

one can define the *trace* of a function in $W^{1,p}(\Omega)$ on the boundary $\partial\Omega$, by an argument of functional analysis (as the usual restriction to a set of measure 0 has no meaning). One first shows that $C^1(\overline{\Omega})$, the space of restrictions to $\overline{\Omega}$ of functions which are C^1 on an open set containing $\overline{\Omega}$, is dense in $W^{1,p}(\Omega)$ for $1 \le p < \infty$. Then one shows that the linear mapping obtained by restricting a function of $C^1(\overline{\Omega})$ to the boundary (giving a Lipschitz continuous function on the boundary) is continuous if one puts on $C^1(\overline{\Omega})$ the norm of $W^{1,p}(\Omega)$ and on its traces the norm of $L^p(\partial\Omega)$ (for the natural $(N-1)$-dimensional measure on the boundary). The characterization of the space of traces, due to Emilio GAGLIARDO,[21] gives indeed $L^1(\partial\Omega)$ for the case $p = 1$ and the obvious $W^{1,\infty}(\partial\Omega)$ for the case $p = \infty$, and for $1 < p < \infty$ an interpolation space of functions having $1 - \frac{1}{p}$ derivatives in $L^p(\partial\Omega)$. However, one does not need the precise characterization of the space of traces for proving the formula of integration by parts (which implies Green's formula)

$$\int_\Omega \frac{\partial u}{\partial x_j}\,d\mathbf{x} = \int_{\partial\Omega} u\,\nu_j\,d\sigma \text{ for all } u \in W^{1,1}(\Omega), \tag{4.3}$$

where $d\sigma$ is the $(N-1)$-dimensional Hausdorff[22] measure and $\boldsymbol{\nu}$ is the exterior normal to $\partial\Omega$, which exists $d\sigma$ almost everywhere.

The idea for proving the estimate of the integral on $\omega(t+\delta t)\setminus\omega(t)$ is shown in a simpler example. The field $\mathbf{u}$ is assumed to be of class C^1; using the hyperplane $x_N = 0$ instead of the boundary $\partial\Omega$, one has to integrate on a strip swept by the points $\mathbf{x}' + s\,\mathbf{u}(\mathbf{x}')$ for $\mathbf{x}' \in R^{N-1}$ (or a piece of the hyperplane) when s varies from 0 to and δt. One observes that if $u_N(\mathbf{x}') \ne 0$, then the mapping $(\mathbf{x}', s) \mapsto \mathbf{x}' + s\,\mathbf{u}(\mathbf{x}')$ is a local diffeomorphism, whose Jacobian[23] is precisely $u_N(\mathbf{x}')$, and if one integrates a uniformly continuous function φ on the strip, the integral is easily seen to be equivalent to $\delta t \int_{R^{N-1}} \varphi(\mathbf{x}', 0)u_N(\mathbf{x}')\,d\mathbf{x}'$.

In the preceding computation, one should have followed the solution of an ordinary differential equation during a time δt and not just followed the tangent (as in EULER's method for approximating solutions of ordinary differential equations). We shall see now a different derivation of the equation expressing the conservation of mass, based on the analysis of ordinary differential equations, and it will also make the connection with the Lagrangian point of view.

Let us assume that $\mathbf{u}(\mathbf{x}, t)$ is of class C^1 in $\mathbf{x}$ and t, then the differential equation

$$\frac{D\mathbf{x}}{Dt} = \mathbf{x}'(t) = \mathbf{u}(\mathbf{x}(t), t) \text{ for } t > 0; \ \mathbf{x}(0) = \boldsymbol{\xi}, \tag{4.4}$$

---

[21] Emilio GAGLIARDO, Italian mathematician, born in 1930. He worked at Università di Pavia, Pavia, Italy.

[22] Felix HAUSDORFF, German mathematician, 1869–1942. He worked in Bonn, Germany.

[23] Carl Gustav Jacob JACOBI, German mathematician, 1804–1851. He worked in Königsberg (then in Germany, now Kaliningrad, Russia) and Berlin, Germany.

describes the position at time t of a material point starting at $\boldsymbol{\xi}$ at time 0; here $\frac{D}{Dt}$ is the partial derivative in t when $\boldsymbol{\xi}$ is fixed, but we reserve $\frac{\partial}{\partial t}$ to denote the partial derivative in t when $\mathbf{x}$ is fixed, i.e. in the Eulerian point of view; $\frac{D}{Dt}$ is called the material derivative. It seems natural to ask for uniqueness of a solution, and the classical condition is the local version of the following global Lipschitz condition:

$$|\mathbf{u}(\mathbf{x},t) - \mathbf{u}(\mathbf{y},t)| \leq \lambda(t)|\mathbf{x} - \mathbf{y}| \text{ for all } \mathbf{x}, \mathbf{y} \text{ and } t \in (0,T), \qquad (4.5)$$

with $\lambda \in L^1(0,T)$. There is a small improvement due to OSGOOD[24,25] which gives uniqueness when one only assumes that

$$|\mathbf{u}(\mathbf{x},t) - \mathbf{u}(\mathbf{y},t)| \leq \omega(|\mathbf{x} - \mathbf{y}|) \text{ for all } \mathbf{x}, \mathbf{y}, \text{ and}$$
the modulus of uniform continuity ω satisfies $\qquad (4.6)$
$\int_0^1 \frac{ds}{\omega(s)} = +\infty.$

This gives $\mathbf{x} = \Phi(\boldsymbol{\xi},t)$, and with $\mathbf{u}$ of class C^1 in $(\mathbf{x},t)$, it is not difficult to prove that $\boldsymbol{\xi} \mapsto \Phi(\boldsymbol{\xi},t)$ is a local diffeomorphism and that the Jacobian matrix $\frac{\partial \Phi}{\partial \boldsymbol{\xi}}$ satisfies the linear differential equation

$$\frac{D\frac{\partial \Phi}{\partial \boldsymbol{\xi}}}{Dt} = \frac{\partial u}{\partial \mathbf{x}}\frac{\partial \Phi}{\partial \boldsymbol{\xi}} \text{ on } (0,T); \quad \frac{\partial \Phi}{\partial \boldsymbol{\xi}}(0) = \mathsf{I}, \qquad (4.7)$$

so that the Jacobian determinant $det\frac{\partial \Phi}{\partial \boldsymbol{\xi}}$ satisfies

$$\frac{D(det\frac{\partial \Phi}{\partial \boldsymbol{\xi}})}{Dt} = Trace\left(\frac{\partial u}{\partial \mathbf{x}}\right)det\frac{\partial \Phi}{\partial \boldsymbol{\xi}} = div\,u\,det\frac{\partial \Phi}{\partial \boldsymbol{\xi}} \text{ on } (0,T); \quad det\frac{\partial \Phi}{\partial \boldsymbol{\xi}}(0) = 1. \qquad (4.8)$$

As $det\frac{\partial \Phi}{\partial \boldsymbol{\xi}}$ represents the increase in volume by the transformation $\boldsymbol{\xi} \mapsto \mathbf{x}(t)$, conservation of mass may then be written as

$$\varrho(\mathbf{x}(t),t)det\frac{\partial \Phi}{\partial \boldsymbol{\xi}} = \varrho(\boldsymbol{\xi},0) \text{ almost everywhere}, \qquad (4.9)$$

and the equation for the Jacobian determinant may therefore be written as

$$\frac{D(\frac{\varrho(\boldsymbol{\xi},0)}{\varrho})}{Dt} = div\,u\,\frac{\varrho(\boldsymbol{\xi},0)}{\varrho}, \qquad (4.10)$$

or equivalently (using $\varrho(\boldsymbol{\xi},0) > 0$)

$$\frac{D\varrho}{Dt} + \varrho\,div\,u = 0, \qquad (4.11)$$

[24] William Fogg OSGOOD, American mathematician, 1864–1943. He worked at HARVARD University, Cambridge, MA.
[25] John HARVARD, English clergyman, 1607–1638.

which is the desired equation as

$$\frac{D}{Dt} = \frac{\partial}{\partial t} + \sum_{j=1}^{N} u_j \frac{\partial}{\partial x_j}. \tag{4.12}$$

It then seems reasonable to admit the derived form of conservation of mass, but the regularity hypotheses invoked for proving it are a little too strong in some situations. For the Navier–Stokes equation, under the assumption that the fluid is incompressible and that the viscosity is independent of temperature (so that one just forgets about the equation of conservation of energy), one knows uniqueness of the solution in 2 dimensions, and the solution is smooth enough if the initial data are smooth enough. However, uniqueness is not known in 3 dimensions, and it is only for sufficiently small smooth data that one knows that the solution stays smooth; the dissipation of energy by viscosity gives directly that $u \in L^2(0, T; H^1(\Omega; R^3))$, and by improving an argument of Ciprian FOIAS,[26] I proved that $u \in L^1(0, T; Z)$, with Z a little smaller that $W^{1,3}(\Omega; R^3)$ (so that $Z \subset C^0(\overline{\Omega}; R^3)$, for example), but that is far from the $W^{1,\infty}$ regularity used for deriving the equation.

For incompressible 2-dimensional Euler equation, there is a global existence (and maybe uniqueness result) due to KATO,[27] I believe. The *vorticity* $\omega = \frac{\partial u_1}{\partial x_2} - \frac{\partial u_2}{\partial x_1}$ is transported by the flow, i.e. satisfies $\frac{D\omega}{Dt} = 0$, and therefore if the initial vorticity is in $L^\infty(R^2)$ it stays in this space. At a given time one has then $curl\, u \in L^\infty$ and $div\, u = 0$, but that does not imply $u \in W^{1,\infty}(R^2, R^2)$ (L^∞ is not a good space for *singular integrals*).

The singular integrals which often appear in linear systems of partial differential equations with constant coefficients in R^N are convolution equations with a kernel which is homogeneous of degree $-N$ and whose integral on the sphere is 0; very often they are polynomials in the *Riesz*[28] *operators* R_j, which are the natural generalization to R^N of the *Hilbert*[29] *transform* in R. Singular integrals act on spaces like $C^{k,\alpha}$ (results proven in the 1920s–1930s by GIRAUD,[30] I believe), and were extended in the 1950s to L^p with $1 < p < \infty$

---

[26] Ciprian Ilie FOIAS, Romanian-born mathematician, born in 1933. He works at Indiana University, Bloomington, IN, and at Texas A&M, College Station, TX. He was my colleague at Université Paris XI (Paris-Sud), Orsay, France, in 1978–1979.

[27] Tosio KATO, Japanese-born mathematician, 1917–1999. He worked at University of California Berkeley, Berkeley, CA.

[28] Marcel RIESZ, Hungarian-born mathematician, 1886–1969. He worked in Lund, Sweden. He was the brother of Frederic RIESZ.

[29] David HILBERT, German mathematician, 1862–1943. He worked in Göttingen, Germany.

[30] Georges Julien Adolphe GIRAUD, French mathematician, 1889–1943?. He worked in Clermont-Ferrand, France.

by Alberto CALDERÓN [31] and Antoni ZYGMUND.[32] For the case $p = \infty$, they do not act on L^∞ but they act on the larger space BMO (bounded mean oscillations), introduced by Fritz JOHN[33,34] and Louis NIRENBERG[35,36] for the different purpose of studying the limiting case for the Sobolev embedding theorem 16.1. For the case $p = 1$, they do not act on L^1 but they act on the smaller Hardy[37,38,39,40] space $\mathcal{H}^1$. Charles FEFFERMAN[41] proved that the dual of $\mathcal{H}^1$ is BMO.

One may then say that $curl\, u \in L^\infty$ and $div\, u = 0$ imply that all the derivatives $\frac{\partial u_i}{\partial x_j}$ belong to BMO, but one may also use another space, the Zygmund space Λ_1, which serves as a replacement for the space of Lipschitz functions (as it is an interpolation space between $C^{0,\alpha}$ and $C^{1,\beta}$, it inherits the property that singular integrals act in a continuous way over it). One may then

[31] Alberto Pedro CALDERÓN, Argentine-born mathematician, 1920–1998. He received the Wolf Prize in 1989. He worked at University of Chicago, Chicago, IL.

[32] Antoni Szczepan ZYGMUND, Polish-born mathematician, 1900–1992. He worked at University of Chicago, Chicago, IL.

[33] Fritz JOHN, German-born mathematician, 1910–1994. He worked at New York University (COURANT Institute of Mathematical Sciences), New York, NY.

[34] Richard COURANT, German-born mathematician, 1888–1972. He worked at New York University, New York, NY. The Department of Mathematics of New York University is now named the COURANT Institute of Mathematical Sciences.

[35] Louis NIRENBERG, Canadian-born mathematician, born in 1925. He received the Crafoord Prize in 1982. He works at New York University (COURANT Institute of Mathematical Sciences), New York, NY.

[36] Holger CRAFOORD, Swedish businessman, 1908–1982. With his wife Anna-Greta CRAFOORD, 1914–1994, he established the prize in 1980 by a donation to the Royal Swedish Academy *to promote basic scientific research in Sweden and in other parts of the world in mathematics and astronomy, geosciences, and biosciences.*

[37] It was F. RIESZ who introduced the $\mathcal{H}^p$ spaces in honor of HARDY, and they were originally defined as holomorphic functions f in the upper half plane $\Im z > 0$ such that $\int_R |f(x + i\,\varepsilon)|^p\, dx$ is bounded as $\varepsilon > 0$ tends to 0. His brother M. RIESZ proved that for $1 < p < \infty$ the trace of the real part of f on the real axis is any function in $L^p(R)$, by showing that the Hilbert transform maps $L^p(R)$ into itself for $1 < p < \infty$ (using arguments which are at the basis of the theory of Interpolation spaces). He introduced later the Riesz operators on R^N, which generalize the Hilbert transform for $N = 1$, and using these operators, STEIN and WEISS later extended the definition of Hardy spaces to real functions on R^N.

[38] Elias M. STEIN, Belgian-born mathematician, born in 1931. He received the Wolf Prize in 1999. He works at Princeton University, Princeton, NJ.

[39] Guido L. WEISS, Italian-born mathematician, born in 1928. He works at WASHINGTON University, St Louis, MO.

[40] George WASHINGTON, American general, 1732–1799. First President of the United States of America, 1789–1797.

[41] Charles Louis FEFFERMAN, American mathematician, born in 1949. He received the Fields Medal in 1978. He works at Princeton University, Princeton, NJ.

say that $curl\, u \in L^\infty$ and $div\, u = 0$ imply that $u \in \Lambda_1$, and that means[42,43,44] that there exists a constant M such that $|\mathbf{u}(\mathbf{x}+\mathbf{h}) + \mathbf{u}(\mathbf{x}-\mathbf{h}) - 2\mathbf{u}(\mathbf{x})| \leq M|\mathbf{h}|$ for all $\mathbf{x}, \mathbf{h}$, and this implies $|\mathbf{u}(\mathbf{x}+\mathbf{h}) - \mathbf{u}(\mathbf{x})| \leq C|\mathbf{h}| \log(|\mathbf{h}|)$ for $|\mathbf{h}|$ small, and OSGOOD's variant applies.

[Taught on Wednesday January 20, 1999. There were no classes on Monday January 18, in commemoration of Martin Luther KING Jr.'s birthday.][45]

---

[42] This is the definition that I heard from Yves MEYER, and I noticed then that it is an interpolation space, for example $(C^{0,\alpha}, C^{1,\beta})_{\theta,\infty}$ with $0 < \alpha, \beta < 1$ and $(1-\theta)\alpha + \theta\beta = 1$. I suppose that Antoni ZYGMUND had introduced this space before there was a general theory of interpolation between Banach spaces.

[43] Yves François MEYER, French mathematician, born in 1939. He works at École Normale Supérieure, Cachan, France. He was my colleague at Université Paris XI (Paris-Sud), Orsay, France, from 1975 to 1979.

[44] Stefan BANACH, Polish mathematician, 1892–1945. He worked in Lwów, then in Poland, now Lvov in Ukraine. There is a Stefan BANACH Centre of the Polish Academy of Sciences in Warsaw, Poland.

[45] Martin Luther KING Jr., American civil rights leader, 1929–1968.

5

Particles and continuum mechanics

I consider now another approach for deriving the equation of conservation of mass, based on the use of "particles".

One uses partial differential equations in continuum mechanics, and one knows that they are not valid at a very small scale, because matter is made of "particles".[1] If one uses the language of physicists, one calls the level where there are particles microscopic and our level macroscopic, and the intermediate levels are called mesoscopic. At usual temperature and pressure, a *mole* of a gas, either monatomic like one of the rare gases (Ne, Ar, Kr, Xe, Rn) or diatomic (like H^2, N^2, O^2) or a mixture (like air = 79% N^2 + 20% O^2 + 1% Ar) occupies 22.4 liters (one liter = 1 dm^3 = 10^{-3} m^3) and contains a number of particles around 6.023×10^{23} (the *Avogadro*[2] *number*). In a crystal interatomic distances are measured in *angströms*[3] (1 Å = 10^{-10} m). In a liquid, a mole of water (18 grams of H^2O) occupies approximately 18 cm^3, but when transformed into vapor it occupies 22,400 cm^3 so that the average distance between molecules in the vapor is a little more than 10 times that in the liquid, and as the average volume around a molecule in the liquid is about 3×10^{-23} cm^3, it corresponds to an average distance of about 3 angströms. Once one tries to describe what happens at the level of particles, one discovers that they do not behave like classical particles, because they are actually

[1] There is a 19th century point of view where particles behave in a classical way and a 20th century point of view where particles behave in a quantized way; actually, particles do not exist and there are only waves, and the relation with particles is partly explained by H-measures and their variants.

[2] Lorenzo Romano Amedeo Carlo AVOGADRO, count of Quaregna and Cerreto, Italian physicist, 1776–1856. He worked in Torino (Turin), Italy.

[3] Andres Jöns ÅNGSTRÖM, Swedish physicist, 1814–1874. He worked in Uppsala, Sweden.

waves (but not necessarily described by Schrödinger's[4] equation, which is only an approximation).

In the general theory of homogenization (intertwined with the compensated compactness theory) that I developed in the 1970s with François MURAT (extending earlier results of the Italian school, Sergio SPAGNOLO[5], Ennio DE GIORGI[6]), there is no hypothesis of periodicity (but in the work of Henri SANCHEZ-PALENCIA[7] and that of Ivo BABUŠKA[8], who coined the term homogenization, there were periodicity assumptions). I plan to continue to use the term Homogenization for describing our general approach, i.e. without any periodicity hypotheses, and it would be natural that some of the others who apply our ideas only to periodic situations would at least acknowledge that they use our approach of H-convergence. In that general theory, one starts with partial differential equations at a level which I qualified in the early 1970s as microscopic (instead of mesoscopic) and one tries to discover which constitutive relations and which balance relations (and therefore which effective equations) one should use at our level, for which I used the term macroscopic. I first heard the term mesoscopic in the early 1990s, and as our microscopic level is obviously a level for which the equations of continuum mechanics apply, it is then one of the various mesoscopic levels that physicists mention. My use was not in opposition with what physicists often do in order to understand an important effect: they select the smallest level where this effect takes place and they seem to neglect the smaller scales, but they actually use an ad hoc theory for summarizing what they know about what happens at smaller scales (or what they believe must happen there); then they try to understand what happens at this particular level in order to derive the effective equations that they will use at the next higher level (which might well be the microscopic level of the atoms or one of these intermediate mesoscopic levels). In the continuum mechanics approach, the lower scales are summarized in the laws of "thermodynamics" which constrain the constitutive relations that one may use, and one of the goals of my program of studying the evolution of microstructures in partial differential equations is to avoid postulating the constitutive laws and instead to show how to deduce them from more basic principles. Although my approach explains in some way why high-frequency

---

[4] Erwin Rudolf Josef Alexander SCHRÖDINGER, Austrian-born physicist, 1887–1961. He received the Nobel Prize in Physics in 1933. He worked in Oxford, England, UK, then Graz, Austria, and Dublin, Ireland.

[5] Sergio SPAGNOLO, Italian mathematician, born in 1941. He works at Università di Pisa, Pisa, Italy.

[6] Ennio DE GIORGI, Italian mathematician, 1928–1996. He received the Wolf Prize in 1990. He worked at Scuola Normale Superiore, Pisa, Italy.

[7] Enrique (Henri) Evariste SANCHEZ-PALENCIA, Spanish-born mathematician, born in 1941. He works at CNRS (Centre National de la Recherche Scientifique) at Université Paris VI (Pierre et Marie CURIE), Paris, France.

[8] Ivo M. BABUŠKA, Czech-born mathematician, born in 1926. He works at University of Texas, Austin, TX.

solutions of some partial differential equations may behave like particles, it still faces a few theoretical obstacles and cannot explain yet how to interpret what physicists say when they use a discrete description, starting from atoms arranged in a crystalline way (or a polycrystalline way with important effects at the grain boundaries), describing defects in the crystalline arrangement and how these defects move around in order to explain the effects of plasticity, which limit the applicability of theories like nonlinear elasticity (Owen RICHMOND[9] described a few years ago his program, very similar to mine, but which dealt precisely with the scales that I cannot explain at the moment, where there are *dislocations* and other *defects*).

The "particles" that I shall now use have nothing to do with the "real" particles that one encounters every few angströms in polycrystalline solids, or a little further apart in liquids or even in gases. The "real" particles are actually concentrated packets of highly oscillating waves, while the particles that I shall use should better be called macroscopic particles; they provide a convenient way for approaching the solutions of partial differential equations, whether smooth or not so smooth, and they are more widely used now in that way as a numerical approach than thirty years ago when I was learning numerical analysis, as in those early days computers were not so powerful.

At the level of describing conservation of mass, the argument that I shall use is quite similar to that which is used in classical mechanics, where a rigid solid is replaced by its center of mass, and a point $\mathbf{M_i}$ of mass m_i moving at velocity $\mathbf{V_i}$ may well represent a body having $\mathbf{M_i}$ as its center of mass, m_i as its total mass, and $m_i \mathbf{V_i}$ as its total (linear) momentum.

From a mathematical point of view, we need to use Dirac[10] masses and more general objects called Radon[11] measures. However, this will not be enough, and in a second step we shall need to introduce much more general objects called distributions, but although we shall not need so much of the theory of distributions due to Laurent SCHWARTZ, I shall give some general definitions.

Physicists describe the Dirac "function" at the point $\mathbf{a} \in R^N$ as the "function" which is 0 outside $\mathbf{a}$, $+\infty$ at $\mathbf{a}$ and has the integral 1; mathematicians are quick to mention that there is no such (Lebesgue-integrable) function, but that remark is not so important since Laurent SCHWARTZ found a mathematical explanation for many (but not for all) strange formulas that physicists had obtained by using these non existent "functions". DIRAC was not the first to use such a "function", and Garret BIRKHOFF[12] mentioned in [1] that

---

[9] Owen RICHMOND, American mathematician, 1928–2001. He worked at ALCOA (Aluminum Company of America), Alcoa Center, PA.

[10] Paul Adrien Maurice DIRAC, English physicist, 1902–1984. He received the Nobel Prize in Physics in 1933. He held the Lucasian chair at Cambridge, England, UK.

[11] Johann RADON, Czech-born mathematician, 1887–1956. He worked in Vienna, Austria.

[12] Garrett BIRKHOFF, American mathematician, 1911–1996. He worked at HARVARD University, Cambridge, MA.

KIRCHHOFF[13] had used such a "function" around 1890, but this is just a notation for the much older idea of a point mass, and it is not for that simple idea that DIRAC should be mentioned, but for the much bolder move that one could use the derivative of such a "function".

Let Ω be an open set of R^N. The space $L^1_{loc}(\Omega)$ (of *locally integrable* functions in Ω) is the space of (classes of) Lebesgue-measurable functions such that for every compact $K \subset \Omega$, $M_K = \int_K |f(x)| \, dx < \infty$ (it is not a Banach space,[14] but a Fréchet[15,16] space). For $f \in L^1_{loc}(\Omega)$ and $\varphi \in C_c(\Omega)$, the space of continuous functions with compact support in Ω, one can define $\int_\Omega f(\mathbf{x})\varphi(\mathbf{x}) \, d\mathbf{x}$ and one has $|\int_\Omega f(\mathbf{x})\varphi(\mathbf{x}) \, d\mathbf{x}| \leq M_K \max_{\mathbf{x} \in K} |\varphi(\mathbf{x})|$ for all functions $\varphi \in C_c(\Omega)$ which have their support in K. A *Radon measure* μ in Ω is a linear form $\varphi \mapsto \langle \mu, \varphi \rangle$ on $C_c(\Omega)$ (whose elements are called *test functions*), satisfying similar bounds, i.e. for every compact $K \subset \Omega$ there exists a constant $C(K)$ such that

$$|\langle \mu, \varphi \rangle| \leq C(K) \max_{\mathbf{x} \in K} |\varphi(\mathbf{x})| \text{ for all functions } \varphi \in C_c(\Omega) \text{ which have their support in } K. \tag{5.1}$$

The Dirac mass at $\mathbf{a} \in \Omega$ is an example of a Radon measure: it corresponds to $\langle \mu, \varphi \rangle = \varphi(\mathbf{a})$ for all $\varphi \in C_c(\Omega)$ (and $C(K) = 1$ if $\mathbf{a} \in K$, $C(K) = 0$ if $\mathbf{a} \notin K$).

There is a topology on $C_c(\Omega)$ for which the dual space is $\mathcal{M}(\Omega)$, the space of all Radon measures in Ω; we shall not need to know that topology of $C_c(\Omega)$, but there is a useful topology on $\mathcal{M}(\Omega)$, the corresponding weak $\star$ topology $\sigma(\mathcal{M}(\Omega), C_c(\Omega))$, also called the *vague* topology. A sequence μ_n converges vaguely to μ_∞ if and only if $\langle \mu_n, \varphi \rangle$ converges to $\langle \mu_\infty, \varphi \rangle$ for all $\varphi \in C_c(\Omega)$; however, this does not define the topology, because that topology is *not metrizable* (but its restrictions to *bounded sets*, suitably defined, are metrizable). For example, if f_n is a bounded sequence in $L^1(\Omega)$ satisfying $\int_K |f_n(\mathbf{x})| \, d\mathbf{x} \to 0$ for every compact K not containing 0 and $\int_\omega f_n(\mathbf{x}) \, d\mathbf{x} \to 1$ for some open set ω containing 0, then the sequence of measures f_n (an abuse of language for $f_n \, d\mathbf{x}$) converges vaguely to δ_0, the Dirac mass at 0. This explains how to handle the Dirac "function" idea, by using functions with support sufficiently concentrated near the point and passing to the limit.

In order to understand what the derivative of a Dirac mass could be, a natural idea is to use a sequence f_n made of smooth functions, and then take the limit of the sequence of derivatives, but that requires the introduction of more general objects, the distributions of Laurent SCHWARTZ (Radon measures will then appear to be distributions of order 0). A *distribution* T in

---

[13] Gustav Robert KIRCHHOFF, German physicist, 1824–1887. He worked in Berlin, Germany.

[14] A notion used before BANACH by F. RIESZ.

[15] Maurice René FRÉCHET, French mathematician, 1878–1973. He worked in Paris, France.

[16] It was FRÉCHET who coined the term Banach space.

an open set $\Omega \subset R^N$ is defined as a linear form $\varphi \mapsto \langle T, \varphi \rangle$ on $C_c^\infty(\Omega)$, the space[17] of infinitely differentiable functions with compact support in Ω, such that for every compact $K \subset \Omega$ there exists a constant $C(K)$ and an integer $m(K) \geq 0$ such that

$$|\langle T, \varphi \rangle| \leq C(K) \max_{x \in K} \max_{|\alpha| \leq m(K)} |D^\alpha \varphi(x)| \tag{5.2}$$
$$\text{for all functions } \varphi \in C_c^\infty(\Omega) \text{ with support in } K.$$

There are plenty of such functions φ, but just one with a nonzero integral has to be constructed explicitly for the theory to be developed, for example $u(x) = exp\left(-\frac{1}{1-|x|^2}\right)$ for $|x| < 1$ and $u(x) = 0$ for $|x| \geq 1$ has for support the closed unit ball. For a multi-index[18,19] $\alpha = (\alpha_1, \ldots, \alpha_N)$, D^α denotes the operator $\left(\frac{\partial}{\partial x_1}\right)^{\alpha_1} \ldots \left(\frac{\partial}{\partial x_n}\right)^{\alpha_N}$; the length of α is $|\alpha| = |\alpha_1| + \ldots + |\alpha_N|$. If $m(K)$ can be taken independent of K, the distribution T is said to be of finite order and the smallest possible value of $m(K)$ is called the *order* of T, so that Radon measures are exactly the distributions of order 0.

By analogy with the formulas for smooth functions, one can multiply a distribution T by a C^∞ function ψ (or by a function of class C^m if the distribution has a finite order $\leq m$), and ψT is defined by

$$\langle \psi T, \varphi \rangle = \langle T, \psi \varphi \rangle \text{ for all } \varphi \in C_c^\infty(\Omega), \tag{5.3}$$

and one can define derivatives $D^\alpha T$ of T by

$$\langle D^\alpha T, \varphi \rangle = (-1)^{|\alpha|} \langle T, D^\alpha \varphi \rangle \text{ for all } \varphi \in C_c^\infty(\Omega). \tag{5.4}$$

For example, if H denotes the Heaviside[20,21] function $H(x) = 0$ for $x < 0$ and $H(x) = 1$ for $x > 0$, then one quickly checks that $\frac{dH}{dx} = \delta_0$. A simple

---

[17] Laurent SCHWARTZ denoted this space by $\mathcal{D}(\Omega)$, and the space of distributions by $\mathcal{D}'(\Omega)$, which is indeed its dual once $\mathcal{D}(\Omega)$ has been given a natural topology. He also denoted by $\mathcal{E}(\Omega)$ the space of infinitely differentiable functions, whose dual $\mathcal{E}'(\Omega)$ is the space of distributions with compact support.

[18] Laurent SCHWARTZ told me that the notation for multi-indices was often wrongly attributed to him, and that it had been introduced by WHITNEY.

[19] Hassler WHITNEY, American mathematician, 1907–1989. He received the Wolf prize in 1982. He worked at the Institute for Advanced Study, Princeton, NJ.

[20] Oliver HEAVISIDE, English physicist, 1850–1925. He worked as a telegrapher, in Denmark and Newcastle upon Tyne, England, UK, and then did research on his own, in Paignton, in Newton Abbot and in Torquay, England, UK.

[21] HEAVISIDE developed an operational calculus, which was given a precise mathematical explanation by Laurent SCHWARTZ using his theory of distributions, but we also owe to him the simplified version of Maxwell's equation using vector calculus!

$paradox$[22,23,24] will show that not all formulas extend to distributions: let u be the sign function ($u = -1 + 2H$), so that $\frac{du}{dx} = 2\delta_0$, and notice that $u^2 = 1$ and $u^3 = u$, but the formula $\frac{d(u^3)}{dx} = 3u^2\frac{du}{dx}$ does not hold as the left side is $2\delta_0$, while the right side is $6\delta_0$; using the Sobolev embedding theorem, one easily shows that the N-dimensional formula $\frac{\partial(u^3)}{\partial x_j} = 3u^2\frac{\partial u}{\partial x_j}$ is valid for $u \in W^{1,p}(R^N)$ for $p \geq \frac{3N}{N+2}$, or more generally for $u \in W^{1,p}(R^N) \cap L^q(R^N)$ with $\frac{2}{q} + \frac{1}{p} \leq 1$, or if u has partial derivatives which are Radon measures (i.e. $u \in BV_{loc}(R^N)$) and u is continuous, while the counter-example shows that it is not always true if $u \in BV(R^N)$ and u^2 is continuous, with u discontinuous.

Let us consider now a finite number of point masses moving around, the particle #i having mass m_i, position $\mathbf{M_i}(t)$ and velocity $\mathbf{V_i}(t) = \frac{d\mathbf{M_i}}{dt}$ at time t. Conservation of mass is expressed by the fact that m_i is independent of t; although two particles can go through the same point at some time, there is no exchange of mass between them during these "collisions". The analog of a smooth density $\varrho(\mathbf{x}, t)$ is the measure μ defined by

$$\langle \mu, \varphi \rangle = \sum_i \int_0^T m_i\varphi(\mathbf{M_i}(t), t)\, dt, \tag{5.5}$$

and the analog of the *mass density* at time t is the measure $\mu_t = \sum_i m_i\delta_{\mathbf{M_i}(t)}$. We introduce then a momentum measure $\boldsymbol{\pi}$ by

[22] A paradox is usually a statement which appears puzzling to some people, who do not see where a mistake was made. The example shown is a reminder that a theorem is proven under some hypotheses, and that the conclusion might be false if one hypothesis is not met. A classical example saw a paradox if a man living in a city said that all inhabitants of that city are liars, but this paradox is easily resolved; a liar was meant to designate someone who never says the truth, and therefore there exists at least one truthful person in that city, while the man speaking is himself a liar, and a paradox only exists for those who do not know how to negate a proposition. Russell's paradox involving the "set of all sets" was more puzzling to mathematicians at the beginning of the 20th century, and it forced logicians to realize that one had never defined what a set is; once a reasonable definition was used, it appeared that the statement of RUSSELL was a proof that the collection of all sets is not itself a set. A more puzzling situation was then discovered by GÖDEL, that a well-known conjecture could neither be proven nor disproved, and even that an undecidable statement exists in most theories (if the integers are available in order to code the analog of a paradoxical statement known to ancient Greeks).

[23] Bertrand Arthur William, third Earl RUSSELL, Welsh mathematician and philosopher, 1972–1970. He received the Nobel Prize in Literature in 1950. He worked in Cambridge, England, UK.

[24] Kurt GÖDEL, Czech-born mathematician, 1906–1978. He worked at the Institute for Advanced Study, Princeton, NJ.

$$\langle \boldsymbol{\pi}, \varphi \rangle = \sum_i \int_0^T m_i \mathbf{V_i}(t) \varphi(\mathbf{M_i}(t), t)\, dt, \tag{5.6}$$

and the analog of the *momentum density* at time t is the measure $\pi_t = \sum_i m_i \mathbf{V_i}(t) \delta_{\mathbf{M_i}(t)}$. Notice that $\mathbf{V_i}$ are vectors, and therefore $\boldsymbol{\pi}$ is a vector-valued measure, and its components will be written as π_j for $j = 1, \dots, N$. Then conservation of mass implies that

$$\frac{\partial \mu}{\partial t} + \sum_{j=1}^3 \frac{\partial \pi_j}{\partial x_j} = 0. \tag{5.7}$$

One has

$$\sum_i m_i \int_0^T \frac{\partial \varphi}{\partial t}(\mathbf{M_i}(t), t)\, dt + \sum_{j=1}^3 \sum_i m_i \int_0^t (V_i)_j(t) \frac{\partial \varphi}{\partial x_j}(\mathbf{M_i}(t), t)\, dt = 0 \tag{5.8}$$

for any test function $\varphi \in C_c^\infty(\Omega \times (0, T))$, i.e.

$$\left\langle \mu, \frac{\partial \varphi}{\partial t} \right\rangle + \sum_{j=1}^3 \left\langle \pi_j, \frac{\partial \varphi}{\partial x_j} \right\rangle = 0; \tag{5.9}$$

this is a consequence of the coefficient of m_i being 0, and indeed this coefficient is

$$\int_0^T \left[\frac{\partial \varphi}{\partial t}(\mathbf{M_i}(t), t)\, dt + \sum_{j=1}^3 (V_i)_j(t) \frac{\partial \varphi}{\partial x_j}(\mathbf{M_i}(t), t) \right] dt, \tag{5.10}$$

and as the bracketed terms are the total derivative with respect to t of $\varphi(\mathbf{M_i}(t), t)$ the integral is indeed 0.

If by a limiting process where the masses of the particles may tend to 0 and their number to ∞, one creates a sequence of measures μ_n converging vaguely to $\varrho(\mathbf{x}, t)\, d\mathbf{x}\, dt$ with the sequence of measures $\boldsymbol{\pi}_\mathbf{n}$ converging vaguely to $\mathbf{q}(\mathbf{x}, t)\, d\mathbf{x}\, dt$, then one obtains the conservation of mass $\frac{\partial \varrho}{\partial t} + div\, q = 0$, and $\mathbf{q}$ represents the density of momentum; the macroscopic velocity is then defined[25] by $\mathbf{u} = \frac{\mathbf{q}}{\varrho}$.

Notice that the physical quantities which are additive are ϱ and $\mathbf{q}$, but not $\mathbf{u}$.

Notice that a particle may leave the domain Ω without any difficulty in the preceding proof, as it stops being taken into account when it goes out of the support of φ and $\varphi(\mathbf{M_i}(t), t) = 0$ before the particle exits. However, particles may also enter Ω without any problem, and the conservation of mass

---

[25] In plasmas, charged particles have radically different masses, as *electrons* are light compared with *ions*, and conservation of charge is written as $\frac{\partial \varrho}{\partial t} + div\, j = 0$, with ϱ being the density of charge and $\mathbf{j}$ the density of current; one does not introduce a macroscopic velocity $\frac{\mathbf{j}}{\varrho}$, which would not be so meaningful.

is only expressed inside Ω: Radon measures or distributions in Ω do not see the boundary $\partial\Omega$, and in order to treat boundary conditions one will have to use various Sobolev spaces and check the meaning of *boundary conditions.*

As mentioned, there is no mathematical difficulty in having different particles go through the same point with different velocities, and therefore we have not been following the Eulerian point of view, but we have discovered that the velocity **u** is actually the *ratio of two averaged quantities,* and in cases where there are *oscillations of the velocity at a small scale* (and this is analogous to the use of the term fluctuations by those who prefer a probabilistic framework), one has a *turbulent flow* and such flows are not well described with such a naive definition of a macroscopic velocity; in such a case it is important to understand which are the important physical quantities and what *effective equations* they satisfy, which is precisely what homogenization[26] is about.

[Taught on Friday January 22, 1999.]

---

[26] In the general sense that I have studied with François MURAT, and not in the restricted sense used by many who only consider situations with one small length scale, often with unrealistic periodic properties.

6

Conservation of mass and momentum

Let us now consider the equations describing the conservation of momentum and the conservation of angular momentum.

EULER is credited with writing the equations for an inviscid (i.e. not viscous) *ideal* fluid, which are

$$\frac{\partial \varrho}{\partial t} + \sum_{j=1}^{3} \frac{\partial(\varrho\, u_j)}{\partial x_j} = 0$$
$$\frac{\partial(\varrho\, u_k)}{\partial t} + \sum_{j=1}^{3} \frac{\partial(\varrho\, u_k\, u_j)}{\partial x_j} + \frac{\partial p}{\partial x_k} = 0 \text{ for all } k,$$

(6.1)

where p is the *pressure*.

The equation for the motion of a viscous fluid is attributed to NAVIER and to STOKES, but STOKES only considered the linearized problem, and so one uses the term Stokes equation when inertial terms are neglected but one uses the term Navier–Stokes equation when they are taken into account, although NAVIER had discovered it alone. It is unfortunate that so many results are not attributed correctly: the shock conditions expressing the conservation of mass and momentum in gas dynamics, now known after RANKINE[1] and HUGONIOT,[2] were actually first derived in 1848 by STOKES, and then rediscovered in 1860 by RIEMANN[3] for an isentropic gas; STOKES is therefore credited with a discovery of NAVIER but forgotten for some of his own discoveries; it could be by his own fault, as when he edited his complete works around 1870 he did not reproduce his derivation of the jump conditions, and he apologized for his mistake, because he had been (wrongly) convinced by

---

[1] William John Macquorn RANKINE, Scottish engineer, 1820–1872. He worked in Glasgow, Scotland, UK.

[2] Pierre Henri HUGONIOT, French engineer, 1851–1887.

[3] Georg Friedrich Bernhard RIEMANN, German mathematician, 1826–1866. He worked in Göttingen, Germany.

THOMSON (not yet Lord KELVIN) and Lord RAYLEIGH[4,5] that his discontinuous solutions were not physical, as they did not conserve energy. It is a quite amazing fact that such great scientists as STOKES, THOMSON and RAYLEIGH did not understand as late as 1870 that heat was a form of energy and that the missing energy had been transformed into heat (CARNOT[6] and WATT did not need partial differential equations in order to understand that).

The form of Stokes's equation is very similar to that of linearized elasticity, which CAUCHY[7,8] had derived, and that involves something more general than pressure, as he had to introduce *stress* (i.e. what we call now the Cauchy stress tensor, which is symmetric, and appears in the Eulerian point of view, while in the Lagrangian point of view a different stress tensor appears which is not usually symmetric, introduced by PIOLA[9] and by KIRCHHOFF). I think that one should consider the Eulerian point of view as the physical one and the Lagrangian point of view as a mathematical one, but in order to discover which equations are valid in the Eulerian point of view one may resort to using particles, so that it looks like a Lagrangian point of view, but one should also remember that EULER had actually considered both points of view.

Pressure might be considered an easy concept, but I do not think that ARCHIMEDES[10,11] knew that the reason why a body receives an upward force from the water in which one tries to submerse it is that the body receives a stronger force from below than from above because the hydrostatic pressure is higher below. Even in the beginning of the 20th century, after people had giggled at the idea of making flying machines that would be heavier than air, it was thought that the reason a plane could fly was that it was sustained from the air below it, while it is more because it is sucked upwards from the air above it if the profile of the wing is well designed, as an important depression is then created above the wing by the flow. The difficulty, of course, was that static questions about pressure had been well understood for some time, while dynamic questions were quite new. For the static question, and some

---

[4] John William STRUTT, third baron RAYLEIGH, English physicist, 1842–1919. He received the Nobel Prize in Physics in 1904. He held the CAVENDISH professorship at Cambridge, England, UK.

[5] Henry CAVENDISH, English chemist and physicist (born in Nice, not yet in France), 1731–1810. He was wealthy and lived in London, England, UK.

[6] Sadi Nicolas Léonard CARNOT, French engineer, 1796–1832. He worked in Paris, France.

[7] Augustin Louis CAUCHY, French mathematician, 1789–1857. He was made a baron by CHARLES X in 1836. He worked in Paris, France.

[8] Charles-Philippe DE BOURBON, 1757–1836, was King of France from 1824 to 1830 under the name CHARLES X.

[9] Gabrio PIOLA, Italian physicist, 1794–1850. He worked in Milano (Milan), Italy.

[10] ARCHIMEDES, mathematician, 287 BCE – 212 BCE. He worked in Siracusa (Syracuse), then a Greek colony, now in Italy.

[11] BCE = Before Common Era.

dynamic effects, it is clear from some of his drawings that DA VINCI[12] had well understood what pressure is, and that should not be so surprising if one remembers that he was first of all a hydraulic engineer. After TORRICELLI[13] had invented the barometer, PASCAL was the first to study the laws governing hydrostatic pressure, and both were remembered when units were chosen, a torr for a pressure of a millimeter of mercury, and a pascal for the rather small pressure of 1 newton per square meter.

D. BERNOULLI[14] had studied the movement of a vibrating string by considering the approximation of many small masses linked by small springs; he apparently only derived the modes of vibration and it was D'ALEMBERT[15] who first wrote the 1-dimensional wave equation.[16] HUYGENS[17] had some insight about the wave nature of light, but it might have been LAPLACE[18,19] or POISSON[20] who first wrote down the 3-dimensional wave equation.

CAUCHY derived the linearized elasticity equation using the same idea of masses with small springs, but he only found a one-parameter family of *isotropic* elastic materials, and it was LAMÉ[21] who introduced the two-parameter family that we use now for the *constitutive relation* (*strain–stress law*) $\sigma_{ij} = 2\mu\,\varepsilon_{ij} + \lambda\,\delta_{ij}\sum_{k=1}^{3}\varepsilon_{kk}$, where $\varepsilon_{ij} = \frac{1}{2}\left(\frac{\partial u_i}{\partial x_j} + \frac{\partial u_j}{\partial x_i}\right)$ for $i,j = 1,2,3$. If one lets λ go to ∞ and μ go to 0, one finds that $\sum_{k=1}^{3}\varepsilon_{kk}$, which is $div\,u$, tends to 0, and that $\lambda\,div\,u$ tends to a limit, giving the law for an inviscid incompressible fluid $\sigma_{ij} = -p\,\delta_{ij}$ for $i,j = 1,2,3$ (but this "pressure" for an

---

[12] Leonardo DA VINCI, Italian artist, engineer and scientist, 1452–1519. He worked in Milano (Milan) and Firenze (Florence), Italy.

[13] Evangelista TORRICELLI, Italian physicist, 1608–1647. He worked in Firenze (Florence), Italy.

[14] Daniel BERNOULLI, Swiss mathematician, 1700–1782. He worked in St Petersburg, Russia, and then in Basel, Switzerland.

[15] Jean LE ROND, dit D'ALEMBERT, French mathematician, 1717–1783. He worked in Paris, France.

[16] I wonder if D'ALEMBERT derived the wave equation along the line of argument that D. BERNOULLI had followed, or if he simply looked for an equation which could have as solutions all functions of the form $f(x - c\,t)$ and $g(x + c\,t)$ with f and g arbitrary, and having found that $\frac{\partial^2 u}{\partial t^2} - c^2\frac{\partial^2 u}{\partial x^2} = 0$ is such an equation, he had then proved that the general solution has the form $f(x - c\,t) + g(x + c\,t)$ with f and g arbitrary.

[17] Christiaan HUYGENS, Dutch astronomer and physicist, 1629–1695. He worked in Paris, France, and The Hague, The Netherlands.

[18] Pierre-Simon DE LAPLACE, French mathematician, 1749–1827. He was made a count by NAPOLÉON in 1806 and a marquis by LOUIS XVIII in 1817. He worked in Paris, France.

[19] Louis Stanislas Xavier DE BOURBON, 1755–1824, was King of France from 1814 to 1824 under the name LOUIS XVIII.

[20] Siméon Denis POISSON, French mathematician, 1781–1840. He worked in Paris, France.

[21] Gabriel LAMÉ, French mathematician, 1795–1970. He worked in Paris, France.

incompressible fluid is not so physical). It is worth noticing that in his Physics course [10], FEYNMAN qualified Euler's equation as describing dry water, and the Navier–Stokes equation as describing wet water.

CAUCHY may have understood the force exerted by a part of an elastic body onto its complement as the resultant of all these tiny forces transmitted through these microscopic springs, but if that description might be found convenient for a solid, it does not look so realistic for a liquid or a gas.

The first explanation of what creates the pressure in a gas might have appeared in the work on *kinetic theory of gases* of BOLTZMANN and MAXWELL[22] (whose name was actually CLERK when he was born, and became CLERK MAXWELL after his father had inherited from an uncle).

In kinetic theory, one considers a gas with so many particles inside that one may take a limit and describe a *density* $f(\mathbf{x}, \mathbf{v}, t)$ for particles near the point $\mathbf{x}$, having their velocity near $\mathbf{v}$ around the time t (in order to simplify, I assume that all particles have the same mass). If these particles are not interacting and are feeling no exterior forces, the evolution of the density is given by the *free transport equation*

$$\frac{\partial f}{\partial t} + \sum_{j=1}^{3} v_j \frac{\partial f}{\partial x_j} = 0, \tag{6.2}$$

the density of *mass* ϱ and of *momentum* $\mathbf{P}$, and the macroscopic *velocity* $\mathbf{u}$ being defined by

$$\begin{aligned}
\varrho(\mathbf{x}, t) &= \int_{\mathbf{v} \in R^3} f(\mathbf{x}, \mathbf{v}, t) \, d\mathbf{v} \\
\mathbf{P}(\mathbf{x}, t) &= \int_{\mathbf{v} \in R^3} v \, f(\mathbf{x}, \mathbf{v}, t) \, d\mathbf{v} \\
\mathbf{u}(\mathbf{x}, t) &= \frac{\mathbf{P}(\mathbf{x}, t)}{\varrho(\mathbf{x}, t)},
\end{aligned} \tag{6.3}$$

so that if one integrates in $\mathbf{v}$ the free transport equation, one obtains the equation of conservation of mass

$$\frac{\partial \varrho}{\partial t} + \sum_{j=1}^{3} \frac{\partial (\varrho \, u_j)}{\partial x_j} = 0. \tag{6.4}$$

Actually, there are exterior forces, depending both on position and velocity. For electrically charged particles, one must take into account the *Lorentz[23] force* $q(\mathbf{E} + \mathbf{v} \times \mathbf{B})$ for a particle with charge q; in oceanography one must take into account gravity and the Coriolis force created by the rotation of the Earth, and the form is similar. If all particles have the same mass m and the same charge q, and one still assumes that particles do not interact, the evolution of the density of charged particles is given by the *transport equation*

---

[22] James CLERK MAXWELL, Scottish physicist, 1831–1879. He held the first CAVENDISH professorship of Physics at Cambridge, England, UK.

[23] Hendrik Antoon LORENTZ, Dutch physicist, 1853–1928. He received the Nobel Prize in Physics in 1902. He worked at Leyden, The Netherlands.

$$\frac{\partial f}{\partial t} + \sum_{j=1}^{3} v_j \frac{\partial f}{\partial x_j} + \sum_{j=1}^{3} \frac{q}{m}\left(E_j(\mathbf{x},t) + \sum_{k,\ell=1}^{3} \varepsilon_{jk\ell} v_k B_\ell(\mathbf{x},t)\right)\frac{\partial f}{\partial v_j} = 0, \quad (6.5)$$

where ε_{ijk} is the completely antisymmetric tensor, so that ε_{ijk} is the signature of the permutation $123 \mapsto ijk$ when indices are different and 0 if two indices are the same; integrating in $\mathbf{v}$ (while assuming that f is 0 for large $\mathbf{v}$, for example) gives the same form of the conservation of mass because $\sum_{k=1}^{3} \varepsilon_{jk\ell}\delta_{jk} = 0$ for $j = 1,2,3$. I shall describe the Coriolis force in Lectures 29 and 40, and forget about these exterior forces now and concentrate on interior forces, due to "collisions" of particles, and discuss *Boltzmann's equation*, which in the absence of exterior forces, has the form

$$\frac{\partial f}{\partial t} + \sum_{j=1}^{3} v_j \frac{\partial f}{\partial x_j} + Q(f,f) = 0, \quad (6.6)$$

where the *collision* term $Q(f,f)$ is somewhat complicated, but for our present purpose we shall only need the fact that collisions conserve mass, momentum and kinetic energy, corresponding to $Q(f,f)$ satisfying

$$\begin{aligned}
&\int_{\mathbf{v}\in R^3} Q(f,f)\,d\mathbf{v} = 0 \\
&\int_{\mathbf{v}\in R^3} v_j Q(f,f)\,d\mathbf{v} = 0 \text{ for all } j \\
&\int_{\mathbf{v}\in R^3} |\mathbf{v}|^2 Q(f,f)\,d\mathbf{v} = 0.
\end{aligned} \quad (6.7)$$

Integrating the Boltzmann equation in $\mathbf{v}$ then gives again the same equation for conservation of mass, and integrating after multiplication by v_i for all i gives the form of the equation of conservation of momentum, and integrating after multiplication by $|\mathbf{v}|^2$ gives the form of the equation of conservation of energy. The form is independent of what Q is, as long as Q satisfies the constraints (6.7), but we shall have to add constitutive relations. Let us define the symmetric *stress tensor* σ by

$$\sigma_{ij}(\mathbf{x},t) = -\int_{\mathbf{v}\in R^3} f(\mathbf{x},\mathbf{v},t)\big(v_i - u_i(\mathbf{x},t)\big)\big(v_j - u_j(\mathbf{x},t)\big)\,d\mathbf{v}. \quad (6.8)$$

Then as $v_i v_j = u_i u_j + u_i(v_j - u_j) + u_j(v_i - u_i) + (v_i - u_i)(v_j - u_j)$, and $\int_{\mathbf{v}\in R^3} f(\mathbf{x},\mathbf{v},t)\big(v_i - u_i(\mathbf{x},t)\big)\,d\mathbf{v} = 0$ by definition of $\mathbf{u}$, one deduces that

$$\int_{\mathbf{v}\in R^3} v_i v_j f(\mathbf{x},\mathbf{v},t)\,d\mathbf{v} = \varrho(\mathbf{x},t)u_i(\mathbf{x},t)u_j(\mathbf{x},t) - \sigma_{ij}(\mathbf{x},t), \quad (6.9)$$

and the equation of conservation of momentum becomes

$$\frac{\partial(\varrho u_i)}{\partial t} + \sum_{j=1}^{3} \frac{\partial(\varrho u_i u_j)}{\partial x_j} - \sum_{j=1}^{3} \frac{\partial \sigma_{ij}}{\partial x_j} = 0 \text{ for all } i. \quad (6.10)$$

In the case where $\sigma_{ij} = -p\,\delta_{ij}$, p is the pressure, which is *nonnegative*, as the definition of σ shows that it is a negative definite tensor (one assumes that all the mass does not move at velocities in a plane, so that f is a nonnegative function with positive total mass). This is acceptable in a gas, but not in a solid where extension is possible, and I shall derive the same equation using the point of view of D. BERNOULLI and CAUCHY based on small springs.

The pressure has a simple explanation if one looks at what happens at the boundary. If the normal to the boundary going inside the gas is ν, the usual law of reflection, called *specular reflection*, is that a particle arriving with velocity $\mathbf{v}$ with $\mathbf{v}.\nu < 0$ is reflected with velocity $\mathbf{w}$ given by $\mathbf{w} = -2\nu\,\mathbf{v}.\nu + \mathbf{v}$, so that $\mathbf{w}.\nu = -\mathbf{v}.\nu > 0$. Each particle bouncing on the wall then receives a momentum in the direction of ν, and the pressure exerted by the gas is precisely the effect that all the particles transmit to the boundary a momentum in the direction $-\nu$ when they collide the boundary. The specular reflection is not exactly true, because the boundary is also made of particles and if a particle from the gas has enough velocity it may enter slightly into the solid, interact with the particles in the solid, and return in a different direction, after a small delay.

[Taught on Monday January 25, 1999.]

7

Conservation of energy

Let us look now at the conservation of energy for Boltzmann's equation, by multiplying by $\frac{|\mathbf{v}|^2}{2}$ and then integrating in $\mathbf{v}$; as before, the precise form of the term $Q(f, f)$ is not important in this process.

One defines the *internal energy per unit of mass* e by the formula

$$\varrho(\mathbf{x}, t)e(\mathbf{x}, t) = \int_{\mathbf{v} \in R^3} \frac{|\mathbf{v} - \mathbf{u}(\mathbf{x}, t)|^2}{2} f(\mathbf{x}, \mathbf{v}, t) \, d\mathbf{v}, \tag{7.1}$$

and there is then an automatic relation

$$\varrho e = -\frac{1}{2} trace(\sigma), \text{ i.e. } \frac{3p}{2} \text{ in the case where } \sigma_{ij} = -p\,\delta_{ij}, \tag{7.2}$$

and this is an obvious defect of the Boltzmann equation, in that it implies a constitutive relation which is not satisfied exactly by real gases. Actually, Boltzmann's equation should only be considered as a *model for rarefied gases*, in agreement with the way the equation was derived, by assuming that two particles only see each other and none of the other particles in the gas.

One also defines the *heat flux* $\mathbf{q}$ by the formula

$$q_j(\mathbf{x}, t) = \int_{\mathbf{v} \in R^3} (v_j - u_j(\mathbf{x}, t)) \frac{|\mathbf{v} - \mathbf{u}(\mathbf{x}, t)|^2}{2} f(\mathbf{x}, \mathbf{v}, t) \, d\mathbf{v} \text{ for all } j. \tag{7.3}$$

Apart from the relation already noticed between ϱe and σ, there is no other automatic relation between the *thermodynamic quantities* $\varrho, \sigma, e, \mathbf{q}$, i.e. quantities pertaining to the gas and which therefore do not change in a *Galilean*[1] *transformation* (consisting in adding a constant velocity to $\mathbf{u}$). As ϱ is the moment of order 0 of f, the moments of order 1 are 0, the moments of order 2 give σ (and ϱe is a particular combination of these moments), and a particular combination of moments of order 3 is $\mathbf{q}$, it can be shown that the only relations between these moments are the nonnegative character of ϱ and $-\sigma$.

---

[1] Galileo GALILEI, Italian mathematician, 1564–1642. He worked in Firenze (Florence), Italy.

For writing the conservation of energy, one has to compute the integral $\int_{\mathbf{v} \in R^3} \frac{|\mathbf{v}|^2}{2} f \, d\mathbf{v}$, and putting $\mathbf{v} = \mathbf{u} + \boldsymbol{\xi}$, one finds that $\frac{|\mathbf{v}|^2}{2} = \frac{|\mathbf{u}|^2}{2} + \mathbf{u}.\boldsymbol{\xi} + \frac{|\boldsymbol{\xi}|^2}{2}$, and therefore, as $\int_{\mathbf{v} \in R^3} \boldsymbol{\xi} f \, d\mathbf{v} = 0$, one finds $\int_{\mathbf{v} \in R^3} \frac{|\mathbf{v}|^2}{2} f \, d\mathbf{v} = \frac{\varrho |\mathbf{u}|^2}{2} + \varrho \, e$. Also, for each j, one has to compute the integrals $\int_{\mathbf{v} \in R^3} \frac{|\mathbf{v}|^2}{2} v_j f \, d\mathbf{v}$, and one finds that $\frac{|\mathbf{v}|^2}{2} v_j = \frac{|\mathbf{u}|^2}{2} u_j + \frac{|\mathbf{u}|^2}{2} \xi_j + \mathbf{u}.\boldsymbol{\xi} \, u_j + \mathbf{u}.\boldsymbol{\xi} \, \xi_j + \frac{|\boldsymbol{\xi}|^2}{2} u_j + \frac{|\boldsymbol{\xi}|^2}{2} \xi_j$, and therefore $\int_{\mathbf{v} \in R^3} \frac{|\mathbf{v}|^2}{2} v_j f \, d\mathbf{v} = \varrho \frac{|\mathbf{u}|^2}{2} u_j + \sum_{k=1}^{3} \sigma_{jk} u_k + \varrho \, e \, u_j + q_j$. The conservation of energy then appears as

$$\frac{\partial \left(\frac{\varrho |\mathbf{u}|^2}{2} + \varrho \, e \right)}{\partial t} + \sum_{j=1}^{3} \frac{\partial \left[\left(\frac{\varrho |\mathbf{u}|^2}{2} + \varrho \, e \right) u_j + \sum_{k=1}^{3} (\sigma_{jk} u_k) + q_j \right]}{\partial x_j} = 0. \qquad (7.4)$$

One sees from the formula $\int_{\mathbf{v} \in R^3} \frac{|\mathbf{v}|^2}{2} f \, d\mathbf{v} = \frac{\varrho |\mathbf{u}|^2}{2} + \varrho \, e$ that $\varrho \, e$ is the part of the kinetic energy which is hidden at a microscopic/mesoscopic level. In Boltzmann's model all energy is kinetic, i.e. comes from translation effects and none of it comes from rotation effects (as it would if the particles in the gas were also rotating), and the internal energy is that part of the kinetic energy which cannot be explained by looking only at the macroscopic quantity $\mathbf{u}$. The *first principle of thermodynamics* asserts that energy is conserved, but one should count all the various forms of energy (in nuclear reactions even mass must be considered a form of energy, with the celebrated Einstein[2] formula $e = m c^2$); for a gas made of molecules (i.e. all real gases apart from the rare gases which are monatomic gases), besides *translation* and *rotation* effects of a molecule considered as rigid, there are also *vibration* effects due to the internal *degrees of freedom* of the molecule.

BOLTZMANN had also noticed his famous H-theorem (I think that he may have chosen H as the capital letter for η, used for *thermodynamic entropy*); it follows from the relation

$$\int_{\mathbf{v} \in R^3} Q(f, f) \log f \, d\mathbf{v} \geq 0, \qquad (7.5)$$

which implies

$$\frac{\partial \left(\int_{\mathbf{v} \in R^3} f \log f \, d\mathbf{v} \right)}{\partial t} + \sum_{j=1}^{3} \frac{\partial \left(\int_{\mathbf{v} \in R^3} v_j f \log f \, d\mathbf{v} \right)}{\partial x_j} \leq 0. \qquad (7.6)$$

It is a consequence of the symmetric form of the collision operator (and the nonnegativity of the kernel), and equality only occurs for local *Maxwellian distribution*, i.e.

---

[2] Albert EINSTEIN, German-born physicist, 1879–1955. He received the Nobel Prize in Physics in 1921. He worked at the Institute for Advanced Study, Princeton, NJ. The Max PLANCK Institute for Gravitational Physics in Potsdam, Germany is named the Albert EINSTEIN Institute.

$$f(\mathbf{x}, \mathbf{v}, t) = \alpha \, exp(-\beta \, |\mathbf{v} - \mathbf{u}|^2) \text{ with } \alpha, \beta, \mathbf{u} \text{ depending only upon } \mathbf{x}, t. \quad (7.7)$$

One has $\beta = \frac{1}{kT}$, where k is the Boltzmann constant and T the absolute temperature, and then $\beta^{3/2} \varrho = \alpha \pi^{3/2}$, so for local Maxwellian distribution, one checks that e is proportional to T, that $\sigma = -p\mathsf{l}$ and $\mathbf{q} = 0$, with p computed as before, etc.

For a real gas, there is an *equation of state* which relates the various thermodynamic quantities, and this equation of state is not necessarily the one that comes out of the (formal) computation for Boltzmann's equation.

The qualitative form of the collision operator is obtained as follows. Two particles with initial velocities $\mathbf{v}$ and $\mathbf{w}$ "collide" and give two particles of velocities $\mathbf{v}'$ and $\mathbf{w}'$ and, as the masses are equal, conservation of momentum and conservation of kinetic energy are equivalent to the relations

$$\mathbf{v} + \mathbf{w} = \mathbf{v}' + \mathbf{w}'$$
$$|\mathbf{v}|^2 + |\mathbf{w}|^2 = |\mathbf{v}'|^2 + |\mathbf{w}'|^2, \quad (7.8)$$

which give $|\mathbf{v} - \mathbf{w}| = |\mathbf{v}' - \mathbf{w}'|$, and putting $\mathbf{v}' = \mathbf{v} + \mathbf{z}$ and $\mathbf{w}' = \mathbf{w} - \mathbf{z}$ gives $(\mathbf{v} - \mathbf{w}).\mathbf{z} + |\mathbf{z}|^2 = 0$, so that by putting $\boldsymbol{\alpha} = \frac{\mathbf{z}}{|\mathbf{z}|}$ (if $\mathbf{z} = 0$ one takes for $\boldsymbol{\alpha}$ any unit vector orthogonal to $\mathbf{v} - \mathbf{w}$), one can parametrize all the solutions by using $\mathbf{w}$ and a *unit vector* $\boldsymbol{\alpha}$:

$$\mathbf{v}' = \mathbf{v} + (\mathbf{w} - \mathbf{v}.\boldsymbol{\alpha})\boldsymbol{\alpha}$$
$$\mathbf{w}' = \mathbf{w} - (\mathbf{w} - \mathbf{v}.\boldsymbol{\alpha})\boldsymbol{\alpha}, \quad (7.9)$$

and if one defines θ by $|\mathbf{v} - \mathbf{w}| \cos \theta = |(\mathbf{v} - \mathbf{w}).\boldsymbol{\alpha}|$, then the *angle of deflection* is 2θ or $\pi - 2\theta$, i.e. in the Galilean frame of the center of mass (moving at velocity $\frac{\mathbf{v} + \mathbf{w}}{2}$) the final velocity direction makes an angle 2θ or $\pi - 2\theta$ with the initial direction. The kernel depends only upon $|\mathbf{v} - \mathbf{w}|$ (twice the *velocity of approach* in the frame of the center of mass), and θ, as in the center of mass there is a symmetry around the direction of the initial velocity. The term $Q(f, f)$ therefore has the form

$$Q(f, f) = \int_{\mathbf{w} \in R^3} \int_{\boldsymbol{\alpha} \in S^2} B(|\mathbf{v} - \mathbf{w}|, \theta) \big(f(\mathbf{v}) f(\mathbf{w}) - f(\mathbf{v}') f(\mathbf{w}') \big) \, d\mathbf{w} \, d\boldsymbol{\alpha}, \quad (7.10)$$

and the kernel B is nonnegative. Because $\theta = \frac{\pi}{2}$ corresponds to $\mathbf{v}' = \mathbf{v}$ and $\mathbf{w}' = \mathbf{w}$ (or $\mathbf{v}' = \mathbf{w}$ and $\mathbf{w}' = \mathbf{v}$, as particles are *indiscernible*), and particle collisions are avoided outside a small effective *scattering cross-section*, $B(|\mathbf{v} - \mathbf{w}|, \theta)$ tends to $+\infty$ as θ tends to $\frac{\pi}{2}$. That makes Boltzmann's equation quite difficult to study mathematically; following Harold GRAD,[3] one usually considers an *angular cut-off*, i.e. one truncates B near $\theta = \frac{\pi}{2}$.

If one notices that the kernel B does not change if one exchanges $\mathbf{v}$ and $\mathbf{w}$, or if one exchanges the roles of $(\mathbf{v}, \mathbf{w})$ and $(\mathbf{v}', \mathbf{w}')$ (which is like reversing

---

[3] Harold GRAD, American mathematician, 1923–1987. He worked at the COURANT Institute of Mathematical Sciences at New York University, New York, NY.

time so the collision of $\mathbf{v}'$ and $\mathbf{w}'$ may produce $\mathbf{v}$ and $\mathbf{w}$), but $f(\mathbf{v})f(\mathbf{w}) - f(\mathbf{v}')f(\mathbf{w}')$ stays the same for the first transformation and changes sign for the second, then one deduces that

$$
\begin{aligned}
&4 \int_{\mathbf{v}\in R^3} Q(f,f) \log f \, d\mathbf{v} = \\
&\iint_{\mathbf{v},\mathbf{w}\in R^3, \alpha\in S^2} B(|\mathbf{v}-\mathbf{w}|,\theta)\big(f(\mathbf{v})f(\mathbf{w}) - f(\mathbf{v}')f(\mathbf{w}')\big) \qquad (7.11)\\
&\big(\log f(\mathbf{v}) + \log f(\mathbf{w}) - \log f(\mathbf{v}') - \log f(\mathbf{w}')\big) \, d\mathbf{v} \, d\mathbf{w} \, d\alpha \geq 0,
\end{aligned}
$$

the last inequality coming from $\log f(\mathbf{v}) + \log f(\mathbf{w}) - \log f(\mathbf{v}') - \log f(\mathbf{w}') = \log f(\mathbf{v})f(\mathbf{w}) - \log f(\mathbf{v}')f(\mathbf{w}')$, and the fact that the logarithm is an increasing function. Equilibrium corresponds to $\int_{\mathbf{v}\in R^3} Q(f,f) \log f \, d\mathbf{v} = 0$, and this is equivalent to $f(\mathbf{v})f(\mathbf{w}) - f(\mathbf{v}')f(\mathbf{w}') = 0$ for all collisions, or $\log f(\mathbf{v}) + \log f(\mathbf{w}) = \log f(\mathbf{v}') + \log f(\mathbf{w}') = 0$ for all collisions, and that is certainly true if $\log f(\mathbf{v}) = a + \mathbf{b}.\mathbf{v} + c\frac{|\mathbf{v}|^2}{2}$ for some $a, \mathbf{b}, c$ independent of $\mathbf{v}$, giving a local Maxwellian distribution. To show that there are no other solutions requires a little care.

In order to find more relations between ϱ, e, σ and $\mathbf{q}$, one usually quotes a formal argument of HILBERT, or one of CHAPMAN[4,5] and ENSKOG,[6] which starts by considering a term in $\frac{1}{\varepsilon}Q(f,f)$, relating ε to the mean free path between collisions. The formal argument of HILBERT consists in assuming that $\mathbf{u} = \mathbf{u_0} + \varepsilon\,\mathbf{u_1} + \ldots$ and identifying the various terms, the term in $\frac{1}{\varepsilon}$ imposing that $\mathbf{u_0}$ is a local Maxwellian, and the next terms giving Euler's equation for an inviscid perfect gas. The argument of CHAPMAN and ENSKOG produces the Navier–Stokes equation with a viscosity of order ε.

As I mentioned before, letting the mean free path between collisions tend to 0 is in contradiction with the assumption that one deals with a rarefied gas in order to compute the kernel. It does not seem reasonable to assume that Boltzmann's equation is valid for dense gases and liquids, one reason being that if "particles" get nearby, then the only way to deal with them is to consider that they are waves, and not classical particles. Actually, as Boltzmann's equation (formally) predicts a perfect gas behavior, and real gases are not perfect gases, either Boltzmann's equation is not satisfied by real gases, or the formal argument of HILBERT is not valid.

From a philosophical point of view, it is rather curious to observe the efforts made to derive Euler's equation or the Navier–Stokes equation from Boltzmann's equation, as if starting with Boltzmann's equation was a flawless assumption. On the contrary, Boltzmann's equation has already postulated some *irreversibility*, and this is seen by the fact that nonnegative initial data create nonnegative solutions for positive time, a property that is lost after

---

[4] Sydney CHAPMAN, English mathematician, 1888–1970. He held the Sedleian chair at Oxford, England, UK.

[5] Sir William, First Baronet SEDLEY, of Southfleet, Kent, English gentleman, 1555–1618.

[6] David ENSKOG, Swedish mathematician, 1884–1947. He worked in Stockholm, Sweden.

time reversal. Formally this is due to the form of the equation:

$$\frac{\partial f}{\partial t} + \sum_{j=1}^{3} v_j \frac{\partial f}{\partial x_j} + f A(f) = B(f), \tag{7.12}$$

with $B(f) \geq 0$ a.e. when $f \geq 0$ a.e.; if A and B were locally Lipschitz continuous one could obtain the solution by the iterative process

$$\frac{\partial f^{(n+1)}}{\partial t} + \sum_{j=1}^{3} v_j \frac{\partial f^{(n+1)}}{\partial x_j} + f^{(n+1)} A(f^{(n)}) = B(f^{(n)}), \tag{7.13}$$

which gives $f^{(n+1)}$ nonnegative when the initial condition and $f^{(n)}$ are nonnegative.

[Taught on Wednesday January 27, 1999.]

8

One-dimensional wave equation

Let us look at the way D. BERNOULLI and D'ALEMBERT were led to discover the 1-dimensional wave equation, and later CAUCHY was led in the same way to discover the equation for linearized elasticity.

The simplest case, concerning its analysis, is that of a 1-dimensional *longitudinal wave*. The motion of a violin string is different, as it is a *transversal wave*: the waves propagate along the string but the displacement is mostly perpendicular to the string. A 1-dimensional longitudinal wave corresponds to the experimental situation of a metallic bar which one hits with a hammer at one end in the direction of the bar. In linearized elasticity, in 2 or 3 dimensions and in an isotropic material, P-waves (pressure waves) are longitudinal waves, while S-waves (shear waves) are transversal waves.[1]

Let us consider the motion of $N - 1$ small masses linked with springs, with the purpose of letting N tend to ∞; let $x_0(t) = 0$, and $x_N(t) = L$, corresponding to the fixed walls where the first and last masses are attached. Let $x_i(t)$, $i = 1, \ldots, N - 1$, be the positions at time t of the mass #i, let m_i be its mass. Let us assume that the springs are at equilibrium if the masses are at their rest point, corresponding to ξ_i for the mass #i (one takes $\xi_0 = 0$ and $\xi_N = L$), and let $\kappa_{i,i+1}$ be the *constant of the spring* connecting mass #i and mass #$(i+1)$ (with #0 and #N designating the walls), i.e. an increase in length of $\Delta > 0$ creates a restoring force $\kappa \Delta$ (and similarly for a compression); of course, this is only realistic if the displacements are small.

The force acting on mass #i by the spring connecting it to mass #$(i+1)$ is $\kappa_{i,i+1}(x_{i+1} - x_i - \xi_{i+1} + \xi_i)$, and it is therefore natural to put $x_i(t) = \xi_i + y_i(t)$, and the equation of motion (Newton's law) for mass #i is then

---

[1] These waves travel at different velocities, and P-waves travel faster than S-waves, a quite useful property with regard to earthquakes as it is the S-waves which are dangerous for buildings, at least those which have not been built using anti-seismic ideas, so that the first tremor which arrives is a less dangerous P-wave, which announces a possibly more dangerous S-wave which follows it.

$$m_i \frac{d^2 y_i}{dt^2} = \kappa_{i,i+1}(y_{i+1} - y_i) - \kappa_{i-1,i}(y_i - y_{i-1}) \text{ for } i = 1, \ldots, N-1, \quad (8.1)$$

and the initial position and velocity of each of the $N-1$ masses must also be given, and as it is a linear differential system there exists a unique solution (global in time). However, we need precise estimates if we want to understand what happens when N tends to ∞.

Before doing that, it is useful to repeat that one reason why the analysis can be done is that one has chosen a linearized problem without saying it expressly: if a spring has size 1 at rest and is elongated of an amount Δ, the restoring force may be of the order of $\kappa \Delta$ if $|\Delta|$ is small, but it makes no sense having Δ go to -1, where the spring is compressed to zero length, or Δ tend to ∞ as no known material can sustain such a deformation without going through permanent plastic deformation before breaking. Of course, these springs are only an idealized classical version of what happens at a microscopic level: *electric forces* may be *attracting or repulsing* and both occur in an ionic crystal, like salt (NaCl), but forces that bind a metallic crystal are all similar and it is more than the nearby neighbors which play a role in the stability of the crystalline arrangement (at least in liquids, one sometimes invokes Lennard-Jones[2] potentials, which have a long-range attraction potential in $1/r^6$ and a short-range repulsion potential in $1/r^{12}$). Actually, crystals are not very good as elastic materials, because they cannot support much strain and they may change their microstructure to that of a polycrystal if forces applied are too large; the idealized description of linearized elasticity with small springs could have seemed reasonable to CAUCHY (and I am not even sure if that is the way he thought), but it is known to contradict our actual knowledge.

However, one may look at this description in another way, and consider it a *numerical analysis point of view*. Indeed, if one uses *finite difference schemes* or *finite elements* (where finite is in opposition to infinitesimal and not to infinite), it is quite natural to replace the wave equation by the system that I have written, and interpret it as moving masses linked by springs, and replace the equation of linearized elasticity by a system very similar in nature. In the numerical analysis point of view, mathematicians start from the partial differential equations and want to show that the finite-dimensional approximation chosen does approach the solution as the mesh size tends to 0 (while engineers might not even write down the partial differential equations and may only use the finite-dimensional description). We shall first follow this point of view, neglecting nonlinearities by pretending that they are small,[3] and after we shall try to take them into account. The system written has an *invariant*,

---

[2] Sir John Edward LENNARD-JONES, British chemist, 1894–1954. He worked in Cambridge, England, UK.

[3] Although neglecting a term by arguing that it is small is a common practice among physicists or engineers, it is a possibly dangerous practice and mathematicians are warned to be cautious. The first reason is that one should not mistake the properties of an equation that one has written with something which has been

which is the total energy: multiplying equation #i by $\frac{dy_i}{dt}$ and summing in i, one obtains

$$\frac{d}{dt}\left(\sum_{j=1}^{N-1}\frac{m_j}{2}\left|\frac{dy_j}{dt}\right|^2 + \sum_{j=0}^{N-1}\frac{\kappa_{j,j+1}}{2}|y_{j+1}-y_j|^2\right)=0, \qquad (8.2)$$

the first sum being the kinetic energy, and the second sum being the potential energy, i.e. the energy stored inside the springs (and there is one more spring than masses). There is an obvious Hamiltonian[4] framework[5] behind our equation, but one should be aware of the fact that for quasilinear partial differential equations, a Hamiltonian framework is not always so useful, the main reason being that quantities which are conserved, like energy, may suddenly start converting to a new form like heat, which may not be described by the same equation, and suddenly the "conserved quantity" may start to change!

Then one wants to let N tend to ∞, and all questions of scaling should be done with care, as there are cases where different regimes could be considered, but here the matter is straightforward. One way to guess the right scaling is to consider that the values y_j are extended by interpolation, filling the intervals in the space variable ξ and time t (that is a Lagrangian point of view), defining a function u, and that the kinetic part should look like $\frac{1}{2}\int_0^L \varrho(\xi)\left|\frac{\partial u}{\partial t}\right|^2 d\xi$ and the potential part like $\frac{1}{2}\int_0^L \kappa(\xi)\left|\frac{\partial u}{\partial \xi}\right|^2 d\xi$. For example, taking $\xi_j = \frac{jL}{N}$ and $m_j = \frac{M}{N}$ so that M is the total mass of the springs, and $\kappa_{j,j+1} = N\kappa$, corresponds to a uniform density of mass $\varrho = \frac{M}{L}$ and constant κ, and the equation becomes

---

observed in an experiment, because the equation used might not be an accurate model of the part of reality that the experiment is designed to test; actually, proving rigorously something about the equation which contradicts some observation is the only way to decide that the mathematical model is not adapted. Another danger is that a small term may have large derivatives, so that it is problematic to neglect a term because it is small and then take the derivative of the simplified equation; however, a consequence of the theory of distributions of Laurent SCHWARTZ is that in the case of *linear* partial differential equations this practice is not of any consequence because if a sequence of functions converges to 0 then the sequence of derivatives converges to 0 in a weak topology; however, *this is not always true for nonlinear equations.*

[4] Sir William Rowan HAMILTON, Irish mathematician, 1805–1865. He worked in Dublin, Ireland.

[5] LAGRANGE had already discovered the importance of a "Hamiltonian" framework and the use of *canonical transformations*: in his study of perturbations in celestial mechanics, he needed to see how the orbit of a planet is slowly varying, and instead of position and velocity he used the parameters describing the plane of motion and the ellipse which is followed in this plane, which is the solution in the unperturbed case, and he was surprised to see that in these new coordinates the equations had the same "Hamiltonian" form.

$$M\frac{\partial^2 u}{\partial t^2} - \kappa L^2 \frac{\partial^2 u}{\partial \xi^2} = 0, \tag{8.3}$$

corresponding to a propagation speed

$$c = \sqrt{\frac{\kappa}{M}} L, \tag{8.4}$$

and its solutions are of the form $f(x - ct) + g(x + ct)$, as noticed by D'ALEMBERT. One should add the boundary conditions $u(0, t) = u(L, t) = 0$, and the initial conditions

$$u(\xi, 0) = v(\xi); \quad \frac{\partial u}{\partial t}(\xi, 0) = w(\xi) \text{ a.e. in } (0, L). \tag{8.5}$$

This can be proven using standard results of functional analysis (from any bounded sequence in L^2, one can extract a weakly converging subsequence) and use, to a small extent, of the theory of distributions (for pushing the derivatives to the test functions), but one must be careful that the initial condition is approached in the right way, i.e. $v \in H_0^1(0, L)$ and $w \in L^2(0, L)$ (I shall explain the details of the argument in Lecture 17).

As the units of $\kappa_{j,j+1}$ are mass time^{-2}, and the mass scales naturally in $m_j = M/N$, the scaling $\kappa_{j,j+1} = N\kappa$ corresponds to a characteristic time in $1/N$, which is quite natural for a characteristic length L/N and a *finite propagation speed*, but the argument is circular because the discrete system does not have finite propagation speed (a change of position of the first mass is immediately felt at the last one), and it is only the limiting equation that has the finite propagation speed property. However, if the total energy of the initial data is kept fixed and if one takes $\kappa_{j,j+1} = h(N)$ with $h(N)/N$ tending to ∞, then the solution tends to 0 and all the energy goes into vibration, while if $h(N)/N$ tends to 0 there is only kinetic energy at the limit and no interaction between particles; therefore there is only one good scaling!

It seems that D. BERNOULLI only considered the solutions of the form $y_j(t) = e^{i\omega t} z_j$, which he found to be $z_j = \gamma \sin(\frac{j}{N} m\pi)$, corresponding to $\omega^2 = \frac{4N^2\kappa}{M} \sin^2(\frac{m\pi}{2N})$, which tends to $\frac{K m^2 \pi^2}{M}$ as N tends to ∞, and that is not as precise as deriving the wave equation. Physicists often find information for special solutions oscillating at a unique frequency, and the result may show that no partial differential equation of a given type may create the same kind of relation, but even if one has to write down a *pseudo-differential equation*, it is better to understand what all solutions do; actually in a nonlinear setting one cannot expect to reconstruct the solution easily from the knowledge of special solutions, and even in linear situations it does not help much for understanding what the boundary conditions are (as every function in $L^2(0, 1)$ can be written as an infinite sum of functions vanishing at 0, one must be careful).

If one considers a 2-dimensional or 3-dimensional array of masses linked by springs, or even the transversal vibrations of a string, the first thing one realizes is that without linearization the problem becomes extremely difficult. With linearization, the idea if that if a spring connects points **A** and

B and that these points move of $\delta\mathbf{A}$ and $\delta\mathbf{B}$, which are small compared with the length of AB, then the new length is $|\mathbf{B} - \mathbf{A} + \delta\mathbf{B} - \delta\mathbf{A}| = \sqrt{|\mathbf{B} - \mathbf{A}|^2 + 2(\mathbf{B} - \mathbf{A}.\delta\mathbf{B} - \delta\mathbf{A}) + |\delta\mathbf{B} - \delta\mathbf{A}|^2} = |\mathbf{B} - \mathbf{A}| + \frac{(\mathbf{B}-\mathbf{A}.\delta\mathbf{B}-\delta\mathbf{A})}{|\mathbf{B}-\mathbf{A}|} + o(|\delta\mathbf{B} - \delta\mathbf{A}|)$, and therefore only the displacement parallel to the initial direction of the spring is taken into account. In a 2-dimensional setting, denoting by x and y the space variables, and by u and v the components of the displacement, one sees that a spring parallel to the x axis corresponds to a potential energy involving $|u_x|^2$ (where subscript denotes differentiation), a spring parallel to the y axis corresponds to a potential energy involving $|v_y|^2$, a spring along the first diagonal corresponds to a potential energy involving $|u_x + u_y + v_x + v_y|^2$, and a spring along the second diagonal corresponds to a potential energy involving $|u_x - u_y - v_x + v_y|^2$. One then understands that the notation

$$\varepsilon_{ij} = \frac{1}{2}\left(\frac{\partial u_i}{\partial x_j} + \frac{\partial u_j}{\partial x_i}\right) \text{ for all } i, j, \tag{8.6}$$

helps in writing the limiting equations as

$$\varrho(\boldsymbol{\xi})\frac{\partial^2 u_i}{\partial t^2} - \sum_{j=1}^{3}\frac{\partial \sigma_{ij}}{\partial x_j} = 0 \text{ for all } i, \tag{8.7}$$

where the Cauchy stress σ has the form

$$\sigma_{ij} = \sum_{k,\ell=1}^{3} C_{ijk\ell}(\boldsymbol{\xi})\varepsilon_{k\ell} \text{ for all } i, j. \tag{8.8}$$

Linearization has the defect of mixing up the Eulerian point of view and the Lagrangian point of view, and it is the symmetric Cauchy stress which appears in the Eulerian point of view, while it is the not necessarily symmetric Piola/Kirchhoff stress which appears in the Lagrangian point of view. In the isotropic case, CAUCHY had found the relation $\sigma_{ij} = 2\mu\,\varepsilon_{ij} + \lambda\,\delta_{ij}\sum_{k=1}^{3}\varepsilon_{kk}$ for $i, j = 1, 2, 3$, but with a special relation between μ (the *shear modulus*) and λ (the Lamé parameter[6,7]), as he had $\lambda = \mu$, and it was LAMÉ who then pointed out that there was a 2-dimensional family of isotropic materials. Because the tensors ε and σ are symmetric, there is no loss of generality in assuming that

$$C_{ijk\ell} = C_{jik\ell} \text{ and } C_{ijk\ell} = C_{ij\ell k} \text{ for all } i, j, k, \ell, \tag{8.9}$$

but there is another symmetry relation, for *hyperelastic* materials, i.e. those materials which have a stored energy function (and this symmetry is probably

---

[6] Engineers prefer to use the *Young's modulus* and the *Poisson ratio*, and rarely mention the Lamé parameter, which is natural as they think in terms of stress, which is what elasticity is about, and not in terms of displacements.

[7] Thomas YOUNG, English physicist, 1773–1829. He worked in London, England, UK.

a necessary condition for the evolution problem to be well posed with the finite propagation speed property, assuming that some kind of ellipticity condition is satisfied),

$$C_{ijk\ell} = C_{k\ell ij} \text{ for all } i, j, k, \ell. \tag{8.10}$$

Under this last condition, the conservation of energy becomes

$$\int_{\mathbf{x} \in R^3} \left(\frac{\varrho(\boldsymbol{\xi})}{2} \sum_{i=1}^{3} \left| \frac{\partial u_i}{\partial t} \right|^2 + \frac{1}{2} \sum_{i,j,k,\ell=1}^{3} C_{ijk\ell}(\boldsymbol{\xi}) \varepsilon_{ij} \varepsilon_{k\ell} \right) d\mathbf{x} = constant. \tag{8.11}$$

The purpose of the preceding discussion was to show the form of the equation, and under a hypothesis of *very strong ellipticity* one can show existence and uniqueness for the evolution problem (and the finite propagation speed property[8]), and the convergence of some natural approximation processes, like the one involving small masses and springs.

However, the Cauchy stress should be discussed in the Eulerian framework, and the argument of CAUCHY that there must exist a stress tensor used the equilibrium of a small tetrahedron. He assumed that for a domain ω (with Lipschitz boundary!), the force acting on a small set of the boundary of ω by the exterior of ω is a force proportional to the surface of the element and depending upon the position and the normal to $\partial\omega$ (the exterior of ω receiving an opposite force, so that conservation of momentum is satisfied); writing then the equilibrium of a tetrahedron small enough so that the dependence is only on the normal, the following argument was used to deduce the fact that the dependence with respect to the normal must be linear. For $a_1, a_2, a_3 > 0$ and small, the faces of the tetrahedron are the planes $x_i = 0$ and the face T of equation $\frac{x_1}{a_1} + \frac{x_2}{a_2} + \frac{x_3}{a_3} = 1$ with $x_j \geq 0$. Let $\mathbf{F_i}$ be the force per unit area on the face $x_i = 0$ and let $\mathbf{G}$ be the force per unit area on the face T, then the equilibrium of the tetrahedron is $\frac{a_2 a_3}{2}\mathbf{F_1} + \frac{a_3 a_1}{2}\mathbf{F_2} + \frac{a_1 a_2}{2}\mathbf{F_3} + S\,\mathbf{G} = 0$, where S is the area of the triangle T, but as the normal $\boldsymbol{\nu}$ to T is such that $\nu_j = \frac{C}{a_j}$ for some $C > 0$, and $S = \frac{a_1 a_2}{2\nu_3} = \frac{a_1 a_2 a_3}{2C}$, one finds that $\mathbf{G} = -\nu_1 \mathbf{F_1} - \nu_2 \mathbf{F_2} - \nu_3 \mathbf{F_3}$, and therefore $\mathbf{G}$ is linear with respect to $\boldsymbol{\nu}$.

[Taught on Friday January 29, 1999.]

---

[8] In my 1974–1975 lecture notes [23], I had introduced a proof (which might have been classical) of the finite speed propagation for the scalar wave equation in an inhomogeneous medium. I extended it to the linearized elasticity case a few years ago, but although the finite propagation speed is only expressed in terms of the coefficient of *strong ellipticity*, my proof requires a hypothesis of very strong ellipticity.

9

Nonlinear effects, shocks

Linearization may often seem a reasonable step when some quantities are believed to be small, and a function that one may want to neglect may indeed be small, but the danger comes from the fact that its derivative might not be small. For what concerns hyperbolic equations, which are more or less the partial differential equations for which information travels at finite speed, the difference between the linear and the nonlinear case (actually, the *quasilinear* case) is quite important.

In the case of an *elastic string*, taking into account *large deformations* leads to an equation of the form

$$\frac{\partial^2 w}{\partial t^2} - \frac{\partial}{\partial x}\left(f\left(\frac{\partial w}{\partial x}\right)\right) = 0, \tag{9.1}$$

where w denotes the *vertical displacement*, $\frac{\partial w}{\partial t}$ the (vertical) *velocity*, $\frac{\partial w}{\partial x}$ the *strain*, and $f\left(\frac{\partial w}{\partial x}\right)$ the *stress*, and the function f is no longer affine, but it satisfies $f' > 0$, and $\sqrt{f'}$ appears to be the local *speed of propagation* of perturbations.

The first to study such an equation was POISSON, around 1807, but he was concerned with *gas dynamics*[1,2] in a simplified form, i.e. the system

$$\begin{array}{c} \frac{\partial \varrho}{\partial t} + \frac{\partial (\varrho u)}{\partial x} = 0 \\ \frac{\partial (\varrho u)}{\partial t} + \frac{\partial (\varrho u^2 + p)}{\partial x} = 0, \end{array} \tag{9.2}$$

with p being a nonlinear function of ϱ (a model classed as *barotropic*). NEWTON had apparently computed the *velocity of sound* in air, but his calculation had given a value almost 100 m s^{-1} short of the real velocity (which is

---

[1] I read in an article by SAINT-VENANT that the motivation for POISSON's computation with compressible gases was to estimate the exit velocity of a shell out of the barrel of a gun.

[2] Adhémar Jean Claude BARRÉ DE SAINT-VENANT, French mathematician, 1797–1886. He worked in Paris, France.

a little above 300 m s^{-1} under usual conditions). He had certainly not written the wave equation, but he must have used what he knew about *compressibility of air*, i.e. he had used p as a linear function of ϱ, according to the law of perfect gases $PV = constant$ at constant temperature, which had just been found by BOYLE[3] in 1662 (and MARIOTTE[4] in 1676). The relation $PV = RT$ appeared after the work of GAY LUSSAC[5] in 1802, whose law states that at fixed volume the pressure is proportional to the absolute temperature (although the notion was not yet defined, and appeared just as the temperature in degrees C + 273), and he mentions a law found in 1787 (but not published) by CHARLES,[6] stating that at constant pressure the volume is proportional to the absolute temperature.

POISSON was using a relation[7] $p = c\varrho^\gamma$, which LAPLACE may have suggested, and the thermodynamic interpretation came much later: as the waves are fast phenomena, the mechanical energy has no time to be transformed into heat for processes which happen at the maximum possible speed, and the process is therefore adiabatic ($\delta Q = 0$), or equivalently, isentropic (as $\delta Q = T\,dS$). POISSON's solution was not analytical but had an implicit form, and CHALLIS[8] found in 1848 that POISSON's formula could not be true for all time, which prompted STOKES to explain that profiles were becoming steeper and steeper, until one had to introduce a discontinuity, for which he computed the velocity, by expressing the conservation of mass and the conservation of momentum.

The basic ideas are more easily explained on the inviscid Burgers[9] equation

$$\frac{\partial u}{\partial t} + u\frac{\partial u}{\partial x} = 0, \tag{9.3}$$

which will have to be written in conservation form as

$$\frac{\partial u}{\partial t} + \frac{\partial(\frac{u^2}{2})}{\partial x} = 0, \tag{9.4}$$

because some solutions will not be smooth, although they cannot be general distributions, for which one cannot define u^2. In order to be consistent, u

---

[3] Robert BOYLE, Irish-born physicist, 1627–1691. He worked in London, England, UK.

[4] Edme MARIOTTE, French physicist and clergyman, 1620–1684. He was prior of Saint-Martin-sous-Beaune, near Dijon, France.

[5] Joseph Louis GAY LUSSAC, French physicist, 1778–1840.

[6] Jacques Alexandre César CHARLES, French physicist, 1746–1823.

[7] Thermodynamics tells us that γ is the ratio $\frac{c_p}{c_v}$, where c_p is the heat capacity per unit mass at fixed pressure, and c_v the heat capacity per unit mass at fixed volume; it is about 5/3 for air.

[8] James CHALLIS, English astronomer, 1803–1882. He worked at Cambridge, England, UK.

[9] Johannes Martinus BURGERS, Dutch-born mathematician, 1895–1981. He worked at University of Maryland, College Park, MD.

must have the dimension of a velocity. In 1948, BURGERS had proposed the equation

$$\frac{\partial u}{\partial t} + u\frac{\partial u}{\partial x} - \varepsilon\frac{\partial^2 u}{\partial x^2} = 0, \tag{9.5}$$

as a 1-dimensional model of turbulence, and apart from pointing out immediately that turbulence is something very different, Eberhard HOPF[10] had been able to study the limiting case $\varepsilon \to 0$ by using a nonlinear transformation which changes the equation into the linear heat equation, and that transformation is now known as the Hopf–Cole[11,12] transform,[13,14,15,16] because Julian COLE had found it independently.[17] The work of Peter LAX and of Olga OLEINIK[18] then opened the way for more general cases.

If $a(x,t)$ is Lipschitz continuous in x, the solution of $\frac{\partial u}{\partial t} + a\frac{\partial u}{\partial x} = 0$ for $x \in R$ and $t > 0$ and $u(\cdot,0) = v$ in R is obtained by the method of *characteristic curves*, going back at least to CAUCHY: along the solution of $\frac{dx(t)}{dt} = a(x(t),t)$ and $x(0) = \xi$, the solution u satisfies $\frac{d}{dt}\big(u(x(t),t)\big) = 0$ and so $u(x(t),t) = v(\xi)$. Assuming that the solution u of Burgers's equation is Lipschitz continuous in x for $0 \le t < T$, the characteristic curve is $\frac{dx(t)}{dt} = u(x(t),t)$, and as $u(x(t),t) = v(\xi)$, one finds that the characteristic curve is a line on which u is constant:

---

[10] Eberhard HOPF, German-born mathematician, 1902–1983. He worked at Indiana University, Bloomington, IN.

[11] Julian D. COLE, American mathematician, 1925–1999. He worked at RPI (RENSSELAER Polytechnic Institute), Troy, NY.

[12] Kilean VAN RENSSELAER, Dutch diamond merchant, c. 1580–1644.

[13] I heard Peter LAX mention that equation (9.3) was used before BURGERS. It seems that BATEMAN in 1915 had studied equation (9.5) and the limiting case (9.3) as ε tends to 0, but FORSYTH had introduced (9.5) ten years before, in connection with what is called now the Hopf–Cole transform for linearizing it into the usual heat equation.

[14] Peter D. LAX, Hungarian-born mathematician, born in 1925. He received the Wolf Prize in 1987. He works at New York University (COURANT Institute of Mathematical Sciences), New York, NY.

[15] Harry BATEMAN, English-born mathematician, 1882–1946. He worked in Caltech (California Institute of Technology), Pasadena, CA.

[16] Andrew Russell FORSYTH, Scottish mathematician, 1858–1942. He worked in London, England, UK.

[17] It seems that FORSYTH was working backwards, starting from the heat equation and transforming it by various nonlinear functions. Eberhard HOPF and Julian COLE might have integrated the conservation form of (9.5), so that if U solves $\frac{\partial U}{\partial t} - \varepsilon\frac{\partial^2 U}{\partial x^2} + \frac{1}{2}\big(\frac{\partial u}{\partial x}\big)^2 = 0$, then $u = \frac{\partial U}{\partial x}$ solves (9.5), and it is natural to remark that if $\frac{\partial w}{\partial t} - \varepsilon\frac{\partial^2 w}{\partial x^2} = 0$ then $\varphi(w)$ satisfies $\frac{\partial\varphi(w)}{\partial t} - \varepsilon\frac{\partial^2\varphi(w)}{\partial x^2} = -\varepsilon\,\varphi''(w)\big(\frac{\partial w}{\partial x}\big)^2$, so that one can have $U = \varphi(w)$ if $\varepsilon\,\varphi''(w) = \frac{1}{2}(\varphi')^2$.

[18] Olga Arsen'evna OLEINIK, Ukrainian-born mathematician, 1925–2001. She worked in Moscow, Russia.

$$x(t) = \xi + t\,v(\xi)$$
$$u(x(t), t) = v(\xi). \tag{9.6}$$

In the spirit of the implicit equation found by POISSON, one could write that $\xi = x(t) - t\,u(x(t), t)$, and therefore for a given t, the function $x \mapsto u(x, t)$ solves the implicit equation

$$v\big(x - t\,u(x, t)\big) = u(x, t) \text{ for all } x. \tag{9.7}$$

I think that CHALLIS's argument was to use $v(x) = \sin x$ (I do not know if he had to use POISSON's result in a question on astronomy[19,20,21,22,23,24]). If one believes that u is continuous, the zeros of u must stay at $k\pi$, but looking for the points where $u = 1$ creates a problem, as it gives $v(x - t) = 1$, and therefore $x = t + \frac{\pi}{2} + 2k\pi$, and for $t = \frac{\pi}{2}$ it gives a point where u is known to be 0.

The parametrization $u(x, t) = v(\xi)$ on the line $x = \xi + t\,v(\xi)$ shows the problem more easily: if $\xi < \eta$ but $v(\xi) > v(\eta)$, the lines coming out of ξ and η intersect and there is a conflict between two different values of u at the intersection. More precisely, if v is of class C^1 and $v' \geq 0$, the mapping $\xi \mapsto \xi + t\,v(\xi)$ is a global diffeomorphism from R to R, but in the opposite case, if $-\alpha = \inf_\xi v'(\xi)$ with $\alpha > 0$, and $T_c = \frac{1}{\alpha}$ then the solution is of class C^1 for $0 < t < T_c$, but for some t slightly larger than T_c an intersection of two characteristic lines occur.

The way STOKES resolved the paradox pointed out by CHALLIS was to observe that the solution became discontinuous, so that POISSON's derivation of the implicit equation does not apply after the appearance of a discontinuity. STOKES's computation for discontinuous solutions are easily explained using the theory of distributions that Laurent SCHWARTZ introduced a century after, and a solution of the equation in the sense of distributions is now called a weak solution, but in doing such a computation one must avoid multiplying

---

[19] CHALLIS was the astronomer in Cambridge and is usually mentioned for a negative reason. ADAMS had computed the position of a new planet and transmitted his computations to AIRY, who asked CHALLIS to look for it in the sky, but he failed to see it. Meanwhile LE VERRIER was better served by GALLE, and got all the fame of the discovery, and the right to call it Neptune. Neptune had actually been observed before, by LALANDE in 1795, and in 1613 by GALILEO!

[20] John Couch ADAMS, English astronomer, 1819–1892. He worked in Cambridge, England, UK.

[21] George Biddell AIRY, English mathematician, 1801–1892. He worked in Greenwich, England, UK, as the seventh Astronomer Royal.

[22] Urbain Jean Joseph LE VERRIER, French astronomer, 1811–1877. He worked in Paris, France.

[23] Johann Gottfried GALLE, German astronomer, 1812–1910. He worked in Berlin, Germany.

[24] Joseph-Jérôme LE FRANÇOIS DE LA LANDE, French astronomer, 1732–1807. He worked at Collège de France, Paris, France.

the derivative of u by a function of u, and one uses an equation in conservation form, like

$$\frac{\partial u}{\partial t} + \frac{\partial v}{\partial x} + w = 0, \tag{9.8}$$

where u, v, w are locally integrable functions in an open set Ω of the plane, and in our example v is a function of u and $w = 0$. Let us assume that u and v are in $W^{1,1}$ on both sides of a curve $x = Z(t)$ of class C^1 (or just Lipschitz continuous), and so both u and v have limits on each side of the curve, and the equation is satisfied in each of the subdomains $\Omega_- = \{(x,t) \mid x < Z(t)\}$ and $\Omega_+ = \{(x,t) \mid x > Z(t)\}$. The speed of propagation of the curve is $s = Z'(t)$. Writing the equation in the sense of distributions means that for every $\varphi \in C_c^\infty(\Omega)$ one has $\int_\Omega \left(-u\frac{\partial\varphi}{\partial t} - v\frac{\partial\varphi}{\partial x} + w\varphi\right) dx\, dt = 0$, and decomposing the integral into two parts, one on Ω_- and one on Ω_+, one can integrate by parts each of these integrals, and transform them into integrals on the curve: $\int_{\Omega_-} \cdots = \int_{x=Z(t)} (u\,\nu_t^- + v\,\nu_x^-)\varphi\, d\sigma$ and $\int_{\Omega_+} \cdots = \int_{x=Z(t)} (u\,\nu_t^+ + v\,\nu_x^+)\varphi\, d\sigma$, where ν_t and ν_x are the components of the exterior normal ν along t and x, and of course $\nu^- = -\nu^+$. This leads to the *jump condition*

$$v_+ - v_- = s(u_+ - u_-), \tag{9.9}$$

and such conditions are now called Rankine–Hugoniot conditions, although it would have been more natural to call them after STOKES and RIEMANN, who were the first to derive similar relations; one usually writes (9.9) as $[v] = s\,[u]$, where $[g]$ denotes the jump of g through the discontinuity.

However, that is not the end of the story, because there are too many weak solutions: for example let $a > 0$, and $x_0 \in R$, then the following function u is a weak solution of Burgers's equation with initial data 0:

$$u(x,t) = \begin{cases} 0 \text{ if } x < x_0 - a\,t \\ -2a \text{ if } x_0 - a\,t < x < x_0 \\ 2a \text{ if } x_0 < x < x_0 + a\,t \\ 0 \text{ if } x_0 + a\,t < x. \end{cases} \tag{9.10}$$

The problem is that some of the discontinuities in a weak solution may not be physical.

It is easy to check that for $t \neq 0$ the function $\frac{x}{t}$ is a particular solution of the Burgers equation, and one then deduces that for $\varepsilon > 0$ and the initial data $v(x) = 0$ for $x < 0$, $v(x) = \frac{x}{\varepsilon}$ for $0 < x < \varepsilon$, $v(x) = 1$ for $x > \varepsilon$, the solution (also given by the method of characteristic curves) is $u(x,t) = 0$ for $x < 0$, $u(x,t) = \frac{x}{\varepsilon+t}$ for $0 < x < \varepsilon+t$ and $u(x,t) = 1$ for $x > \varepsilon+t$. If one lets ε tend to 0, one sees that the limit of the initial data is the Heaviside function but the limit of the solutions is not a solution with a shock (discontinuity) but a rarefaction wave $u(x,t) = 0$ for $x < 0$, $u(x,t) = \frac{x}{t}$ for $0 < x < t$ and $u(x,t) = 1$ for $x > t$.

If one then decides to prefer the (unique) locally Lipschitz solution when it exists, and one argues by continuity, one is led to reject all discontinuities for which $u_- < u_+$. There is another qualitative explanation for this selection of *shocks* for Burgers's equation: one needs to have $u_- > s = \frac{u_-+u_+}{2} > u_+$ because what creates the shock is the fact that the information on the left side travels faster than the shock and is catching up on it, while the information on the right side travels slower than the shock and is caught up by the shock, and as one cannot have the analog of a breaking wave on a beach because one looks for a single-valued function, the fast side must help the slow side so that both can move together at an intermediate speed. In the case of an equation with $v = f(u)$ (and $w = 0$ for example), *Lax's condition* is the analog of this remark: $f'(u_-) \geq \frac{f(u_+)-f(u_-)}{u_+-u_-} \geq f'(u_+)$. However, if f is neither convex nor concave, a more complete analysis shows that one must impose *Oleinik's condition*: if $u_- < u_+$, the chord joining $(u_-, f(u_-))$ to $(u_+, f(u_+))$ should be above the graph of f, while if $u_- > u_+$, the chord joining $(u_-, f(u_-))$ to $(u_+, f(u_+))$ should be below the graph of f.

A more mathematical way to discover Oleinik condition was found by Eberhard HOPF and by KRUSHKOV,[25] and then extended to the case of systems by Peter LAX, who coined the term *entropy* for designating some functions which appear in supplementary conservation laws in the case of smooth functions; the choice of name is not so good as it is not directly related to the thermodynamic entropy (I have heard once someone use the term Casimir[26] as equivalent to entropy, but the use of this term by differential geometers seems to be more restrictive). Eberhard HOPF's idea was that if one multiplies the equation

$$\frac{\partial u_\varepsilon}{\partial t} + \frac{\partial f(u_\varepsilon)}{\partial x} - \varepsilon \frac{\partial^2 u_\varepsilon}{\partial x^2} = 0 \tag{9.11}$$

by $\varphi'(u_\varepsilon)$, and if one chooses ψ satisfying $\psi' = f'\varphi'$, one obtains

$$\frac{\partial \varphi(u_\varepsilon)}{\partial t} + \frac{\partial \psi(u_\varepsilon)}{\partial x} - \varepsilon \frac{\partial^2 \varphi(u_\varepsilon)}{\partial x^2} + \varphi''(u_\varepsilon)\left(\frac{\partial u_\varepsilon}{\partial x}\right)^2 = 0, \tag{9.12}$$

and therefore if one knows that u_ε converges almost everywhere to u, then of course u is a weak solution of the equation with $\varepsilon = 0$, but it also satisfies the supplementary conditions (called *entropy conditions* by Peter LAX, or E-conditions by Constantine DAFERMOS[27,28] in [3])

$$\frac{\partial \varphi(u)}{\partial t} + \frac{\partial \psi(u)}{\partial x} \leq 0 \text{ for all convex entropy } \varphi, \tag{9.13}$$

and when one tests these conditions for smooth convex functions which approximate the following special functions φ_k (which KRUSHKOV used later

[25] Stanislav Nikolaevich KRUZHKOV, Russian mathematician, 1936–1997. He worked in Moscow, Russia.

[26] Hendrik Brugt Gerhard CASIMIR, Dutch physicist 1909–2000. He worked in Leyden, The Netherlands.

[27] Constantine M. DAFERMOS, Greek-born mathematician, born in 1941. He works at BROWN University, Providence, RI.

[28] Nicholas BROWN Jr., American merchant, 1769–1841.

for the multidimensional scalar case), $\varphi_k(v) = (v-k)_+$, corresponding to $\psi_k(v) = 0$ for $v < k$ and $\psi_k(v) = f(v) - f(k)$ for $v > k$, one discovers Oleinik's condition. This formalism has the advantage of expressing the additional conditions without imposing that the solution be piecewise smooth and have limits on both sides of the curves of discontinuities; for scalar equations, the solution is indeed unique if one imposes these conditions, but the situation for systems is not so clear.

If one writes the equation for a nonlinear string as a system, u being the strain, v the velocity and $\sigma = f(u)$ the stress, then the system is

$$\frac{\partial u}{\partial t} - \frac{\partial v}{\partial x} = 0$$
$$\frac{\partial v}{\partial t} - \frac{\partial f(u)}{\partial x} = 0, \tag{9.14}$$

and in order to deduce that

$$\frac{\partial \varphi(u,v)}{\partial t} + \frac{\partial \psi(u,v)}{\partial x} = 0 \tag{9.15}$$

for all smooth solutions, one requires

$$\frac{\partial \psi(u,v)}{\partial v} = -\frac{\partial \varphi(u,v)}{\partial u}$$
$$\frac{\partial \psi(u,v)}{\partial u} = -f'(u)\frac{\partial \varphi(u,v)}{\partial v}, \tag{9.16}$$

and therefore φ must satisfy the compatibility condition

$$\frac{\partial^2 \varphi(u,v)}{\partial u^2} = f'(u)\frac{\partial^2 \varphi(u,v)}{\partial v^2}, \tag{9.17}$$

Air is quite compressible, and the speed of sound at atmospheric pressure is a little above the velocity of commercial planes. However, the velocity depends upon temperature and pressure, and the shape of the wings of commercial planes has been designed so that the flow of air creates a depression above the wing and in this lower pressure the speed of sound is less than the speed of the plane, and such *transonic flows*[29,30,31,32,33] are therefore important for practical applications.

For water, the velocities involved are always a small fraction of the speed of sound, which is about 1.5 km s^{-1}, but one must remember that the approximation of incompressibility has the unrealistic consequence that a perturbation at one point can be felt immediately very far from it, which is nonphysical. The "pressure" appearing in incompressible Navier–Stokes equation, for example, should not be mistaken for the real pressure.

One should be aware then of the limitations of most of the approximations used.

[Taught on Monday February 1, 1999.]

---

[29] Flows around airfoils were first studied for two-dimensional flows (and therefore unrealistic infinite cylindrical wings), using methods of complex variables for transforming conformally the exterior of the wing onto simpler domains; one particular difficulty is the *Kutta–Joukowski condition* created by the lack of regularity at the trailing edge, which I think is related to *Stokes's paradox* for ideal (inviscid) fluids. Transonic flows have been extensively studied by GARABEDIAN and by Cathleen MORAWETZ.

[30] Martin Wilhelm KUTTA, German mathematician, 1867–1944. He worked in Stuttgart, Germany.

[31] Nikolai Egorovich ZHUKOVSKY, Russian mathematician, 1847–1921. He worked in Moscow, Russia.

[32] Paul Roesel GARABEDIAN, American mathematician. He worked at New York University (COURANT Institute of Mathematical Sciences), New York, NY.

[33] Cathleen SYNGE MORAWETZ, Canadian-born mathematician, born in 1923. She worked at New York University (COURANT Institute of Mathematical Sciences), New York, NY. Her father, John Lighton SYNGE, Irish mathematician, 1897–1995, was from 1946 to 1948 the Head of the Mathematics Department at the CARNEGIE Institute ot Technology in Pittsburgh, later to become CARNEGIE MELLON University.

10

Sobolev spaces II

I now start introducing the basic functional spaces that will be used in the proofs of existence of solutions to some partial differential equations related to questions about fluids.

One of the basic functional spaces that we shall use is $H_0^1(\Omega)$. Ω will usually be an open subset of R^N, and of course one should think of $N = 3$, but there are sometimes problems which are naturally posed in 1 or 2 dimensions only, and mathematicians like to be general and they study problems in R^N without any constraint imposed on N and they want to discover if the dimension matters. Mathematicians also study partial differential equations on *manifolds*, and sometimes it corresponds to a real question (after all there are some global questions about oceans and it is important to avoid being stuck in technical questions related to the parametrization of the surface of the Earth), but some are not so realistic (*periodicity hypotheses* for example are a good way to avoid being bothered by what happens on the boundary, but should be considered only as a preliminary step, because boundaries play a crucial role for real fluids).

Most of the open sets that one encounters are bounded (and anyway the radius of the Earth is only about 6,370 km), but mathematicians do analyze questions in unbounded sets, because it is sometimes useful to consider explicit solutions which may be computed more easily in the whole space, for example; these solutions in the entire space may be the limit of the solutions obtained when one lets the boundary go to infinity, and they may then be good approximations when one is far from the boundary. The Fourier transform is a very important technical tool for studying the properties of some functional spaces, but it applies only to functions defined on the entire space. As we shall see, some properties of functions inside Ω are similar to what happens for the whole space, while some properties of functions near $\partial\Omega$ are similar to what happens near the boundary of a half space.

The *smoothness of the boundary* is often not important, for example for the velocity field because the viscosity imposes that the velocity must be 0 on the boundary, but there are important questions for which one should look in a more precise way at what happens near the boundary, for the "pressure" for example: there may exist *thin boundary layers* near places where the boundary is not very smooth, and one problem is then to understand what are good effective boundary conditions to use outside the boundary layer.

The Sobolev space $H^1(\Omega)$ is the space of (equivalence classes of measurable) functions in $L^2(\Omega)$, whose first derivatives are all in $L^2(\Omega)$, with a norm

$$||u||_{H^1(\Omega)} = \left(\int_\Omega \left(\frac{|u|^2}{L^2} + |grad\, u|^2 \right) d\mathbf{x} \right)^{1/2}, \qquad (10.1)$$

where L is a *characteristic length* (which mathematicians usually take equal to 1!).

$H_0^1(\Omega)$ is then defined as the closure of $C_c^\infty(\Omega)$ in $H^1(\Omega)$. In the case where Ω has a compact boundary which is locally defined by a Lipschitz equation, Ω being only on one side of the boundary, one can show that $C^\infty(\overline{\Omega})$ is dense in $H^1(\Omega)$, and that the restriction to the boundary extends into a linear continuous mapping from $H^1(\Omega)$ into $L^2(\partial\Omega)$ called the trace; then $H_0^1(\Omega)$ is exactly the subspace of $H^1(\Omega)$ of functions which have trace 0 on the boundary.

An important property is the Poincaré[1] inequality (not unrelated to oceanography, as I was told that POINCARÉ had introduced that inequality in his studies of the *tides*), which is

$$\int_\Omega |u|^2 \, d\mathbf{x} \le C \int_\Omega |grad\, u|^2 \, d\mathbf{x}. \qquad (10.2)$$

Of course, the constant C has dimension length squared, and one cannot expect to have Poincaré's inequality in domains like R^N where there is no characteristic length (one does have $H_0^1(R^N) = H^1(R^N)$). The way to express this argument about units in a mathematical way is to replace $u(\mathbf{x})$ by $u(\lambda\,\mathbf{x})$: if Poincaré's inequality was true in R^N then one would have

$$\int_{\mathbf{x}\in R^N} |u(\lambda\,\mathbf{x})|^2 \, d\mathbf{x} \le C \int_{\mathbf{x}\in R^N} \left| \lambda\,grad\big(u(\lambda\,\mathbf{x})\big) \right|^2 d\mathbf{x}, \qquad (10.3)$$

but using the change of variable $\mathbf{y} = \lambda\,\mathbf{x}$ gives

$$|\lambda|^{-N} \int_{\mathbf{y}\in R^N} |u|^2 \, d\mathbf{y} \le C\,|\lambda|^{2-N} \int_{\mathbf{y}\in R^N} |grad\, u|^2 \, d\mathbf{y}, \qquad (10.4)$$

and taking advantage of the fact that different powers of λ appear on both sides (which is what is meant by the statement that $\int_{\mathbf{y}\in R^N} |v|^2 \, d\mathbf{y}$ and

[1] Jules Henri POINCARÉ, French mathematician, 1854–1912. He worked in Paris, France. There is an Institut Henri POINCARÉ (IHP), dedicated to mathematics and theoretical physics, part of Université Paris VI (Pierre et Marie CURIE), Paris, France.

$\int_{\mathbf{y} \in R^N} |grad\, v|^2\, d\mathbf{y}$ are not measured in the same unit), one gets a contradiction by letting λ tend to 0.

The preceding proof actually shows that if Ω is unbounded and contains balls of arbitrarily large size, then Poincaré's inequality does not hold: indeed, if $B(\mathbf{x_m}, r_m) \subset \Omega$ and $r_m \to \infty$, one uses $u(\mathbf{x}) = \varphi(\frac{\mathbf{x} - \mathbf{x_m}}{r_m})$, with $\varphi \in C_c^\infty(B(0,1))$ and if Poincaré's inequality was true one would deduce that $\int_{B(0,1)} |\varphi|^2\, d\mathbf{x} = 0$. Therefore such an open set must be considered to have an infinite characteristic length, but the maximum size of balls contained in Ω is not always the right measure for a characteristic length: if Ω is obtained from R^N by removing all the points with integer coordinates, then Poincaré's inequality does not hold if $N \geq 2$, because one has $H_0^1(\Omega) = H^1(R^N)$ (functions in $H^1(R^N)$ are not necessarily continuous for $N \geq 2$ and therefore points are negligible).

Lemma 10.1. *i) If Ω lies between two parallel hyperplanes separated by a distance D, then Poincaré's inequality holds for $H_0^1(\Omega)$ with the constant D^2/π^2 (which is optimal if Ω occupies all the domain between the two hyperplanes).*

ii) If $meas(\Omega) < \infty$, then Poincaré's inequality holds for $H_0^1(\Omega)$ with a constant $C_* meas(\Omega)^{2/N}$.

Proof of i): As both norms being compared are invariant in an orthogonal transformation, one may suppose that one hyperplane has equation $x_N = 0$ and the other $x_N = D$. Then it is enough to prove that

$$\int_0^D |u(x_N)|^2\, dx_N \leq \frac{D^2}{\pi^2} \int_0^D |u'(x_N)|^2\, dx_N \text{ for all } u \in C_c^\infty(0, D), \quad (10.5)$$

and an integration in $x_1, \ldots, x_{N-1}$ gives the result. If one does not care about the best constant, one notices that $|u(x)|^2 = \left| \int_0^x u'(y)\, dy \right|^2$, which by the Cauchy–(Bunyakovsky)[2]–Schwarz[3] inequality is $\leq x \int_0^x |u'|^2\, dy \leq D \int_0^D |u'|^2\, dy$, and then one integrates in x.

In order to show that D^2/π^2 is the best constant in the above 1-dimensional Poincaré inequality, one develops $0 \leq \int_0^D \left| u'(y) - u(y) \frac{\varphi'(y)}{\varphi(y)} \right|^2 dy$ with $\varphi(y) > 0$ in $(0, D)$, and as $-\int_0^D 2u\, u' \frac{\varphi'}{\varphi}\, dy = \int_0^D |u|^2 \left(\frac{\varphi''}{\varphi} - \frac{|\varphi'|^2}{\varphi^2} \right) dy$, one finds that $\int_0^D |u'|^2\, dy \geq \int_0^D |u|^2 \left(\frac{-\varphi''}{\varphi} \right) dy$, and then one takes $\varphi(y) = \sin\left(\frac{y\pi}{D} \right)$, and the best constant is found by letting u converge to φ. ∎

As the proof of ii) is based on the Fourier transform, I first recall the basic theory, extended by Laurent SCHWARTZ to a particular space of distributions. For a function $f \in L^1(R^N)$, one defines

[2] Viktor Takovlevich BUNYAKOVSKY, Ukrainian-born mathematician, 1804–1889. He worked in St Petersburg, Russia.

[3] Karl Herman Amandus SCHWARZ, German mathematician, 1843–1921. He worked in Berlin, Germany.

$$\mathcal{F}f(\boldsymbol{\xi}) = \int_{\mathbf{x}\in R^N} f(\mathbf{x})e^{-2i\pi(\mathbf{x}.\boldsymbol{\xi})}\,d\mathbf{x}$$
$$\overline{\mathcal{F}}f(\boldsymbol{\xi}) = \int_{\mathbf{x}\in R^N} f(\mathbf{x})e^{+2i\pi(\mathbf{x}.\boldsymbol{\xi})}\,d\mathbf{x}, \qquad (10.6)$$

so that $\overline{\mathcal{F}}f = \overline{\mathcal{F}}\overline{f}$. Notice that I use Laurent SCHWARTZ's *convention*, while specialists of harmonic analysis do not put the factor 2π in the exponential, and so powers of π appear in their form of the inverse and the property for the norm in $L^2(R^N)$. Then one notices that for f a little smoother, one has the two formulas

$$g = \frac{\partial f}{\partial x_j} \text{ implies } \mathcal{F}g(\boldsymbol{\xi}) = 2i\pi\,\xi_j\mathcal{F}f(\boldsymbol{\xi})$$
$$g(\mathbf{x}) = -2i\pi\,x_j f(\mathbf{x}) \text{ implies } \mathcal{F}g = \frac{\partial\mathcal{F}}{\partial\xi_j}. \qquad (10.7)$$

Then Laurent SCHWARTZ introduced the space $\mathcal{S}(R^N)$ of C^∞ functions φ such that $x^\alpha D^\beta\varphi \in L^\infty$ for every multi-index $\boldsymbol{\alpha},\boldsymbol{\beta}$ (it is a Fréchet space but not a Banach space). This is a natural space where the preceding formulas are true, iterated as many times as one wishes (as Laurent SCHWARTZ once told me, at the time he did that, it was a new idea to introduce a functional space adapted to a given operator). One shows that $\mathcal{F}$ is an isomorphism from $\mathcal{S}(R^N)$ onto itself, with inverse $\overline{\mathcal{F}}$, and one proves Plancherel's[4] formula

$$\int_{\mathbf{x}\in R^N} f(\mathbf{x})\mathcal{F}g(\mathbf{x})\,d\mathbf{x} = \int_{\boldsymbol{\xi}\in R^N} \mathcal{F}f(\boldsymbol{\xi})g(\boldsymbol{\xi})\,d\boldsymbol{\xi} \text{ for } f,g \in \mathcal{S}(R^N) \qquad (10.8)$$

(which implies the same formula for $\overline{\mathcal{F}}$), which one uses in order to define the Fourier transform of some distributions (which Laurent SCHWARTZ called *temperate distributions*), namely those which are in $\mathcal{S}'(R^N)$, the dual space of $\mathcal{S}(R^N)$, by

$$\langle \mathcal{F}T, \varphi\rangle = \langle T, \mathcal{F}\varphi\rangle \text{ for all } T \in \mathcal{S}'(R^N) \text{ and all } \varphi \in \mathcal{S}(R^N), \qquad (10.9)$$

and one shows that $\mathcal{F}\overline{\mathcal{F}} = \overline{\mathcal{F}}\mathcal{F} = I$ on $\mathcal{S}(R^N)$, and therefore also on $\mathcal{S}'(R^N)$, and it is related to the fact that the Fourier transform is an isometry of $L^2(R^N)$ onto itself, as well as its inverse $\overline{\mathcal{F}}$,

$$\int_{\mathbf{x}\in R^N} |f(\mathbf{x})|^2\,d\mathbf{x} = \int_{\boldsymbol{\xi}\in R^N} |\mathcal{F}f(\boldsymbol{\xi})|^2\,d\boldsymbol{\xi} \text{ for all } f \in L^2(R^N). \qquad (10.10)$$

Proof of ii): For $u \in C_c^\infty(\Omega)$, extended by 0 outside Ω, one wants to bound $\int_{\boldsymbol{\xi}\in R^N} |\mathcal{F}u(\boldsymbol{\xi})|^2\,d\boldsymbol{\xi}$ in terms of $\int_{\boldsymbol{\xi}\in R^N} 4\pi^2|\boldsymbol{\xi}|^2|\mathcal{F}u(\boldsymbol{\xi})|^2\,d\boldsymbol{\xi}$.

One has $|\mathcal{F}u(\boldsymbol{\xi})| \leq \int_\Omega |u(\mathbf{x})e^{-2i\pi(\mathbf{x}.\boldsymbol{\xi})}|\,d\mathbf{x} \leq meas(\Omega)^{1/2}||u||_{L^2(R^N)}$ for $|\boldsymbol{\xi}| \leq \varrho$, which gives

$$\int_{|\boldsymbol{\xi}|\leq\varrho} |\mathcal{F}u(\boldsymbol{\xi})|^2\,d\boldsymbol{\xi} \leq meas\big(B(0,1)\big)\varrho^N meas(\Omega)\int_{\mathbf{x}\in R^N} |u|^2\,d\mathbf{x}, \qquad (10.11)$$

---

[4] Michel PLANCHEREL, Swiss mathematician, 1885–1967. He worked at ETH (Eidgenössische Technische Hochschule) in Zürich, Switzerland.

and for $|\boldsymbol{\xi}| \geq \varrho$, one uses

$$\int_{|\boldsymbol{\xi}| \geq \varrho} |\mathcal{F}u(\boldsymbol{\xi})|^2 \, d\boldsymbol{\xi} \leq \frac{1}{\varrho^2} \int_{|\boldsymbol{\xi}| \geq \varrho} |\boldsymbol{\xi}|^2 |\mathcal{F}u(\boldsymbol{\xi})|^2 \, d\boldsymbol{\xi} \leq$$
$$\leq \frac{1}{4\pi^2 \varrho^2} \int_{\boldsymbol{\xi} \in R^N} 4\pi^2 |\boldsymbol{\xi}|^2 |\mathcal{F}u(\boldsymbol{\xi})|^2 \, d\boldsymbol{\xi}. \tag{10.12}$$

Choosing the best ϱ (given by $meas\big(B(0,1)\big) \varrho^N meas(\Omega) = \frac{2}{N+2}$) and adding these two inequalities gives the result.

The best constant (probably a result of Giorgio TALENTI,[5] using techniques of radial decreasing rearrangement), is obtained when Ω is a ball and involves the first zero of a Bessel[6] function.∎

In the rough proof of i), only the fact that u was zero on one hyperplane was used, and the best constant for functions which are only zero at 0 is four times larger than for functions which are zero at 0 and at D: it is obtained by taking $\varphi(y) = \sin \frac{y\pi}{2D}$ so that $\varphi'(D) = 0$.

Using my *equivalence lemma* 13.3, one can show that if the injection from $H^1(\Omega)$ into $L^2(\Omega)$ is compact, then Poincaré's inequality holds for every closed subspace of $H^1(\Omega)$ which does not contain the constant function 1 (assuming that Ω is *connected*, of course). The compactness assumption rules out many unbounded domains: if there exists $\alpha > 0$ such that $B(\mathbf{x_m}, \alpha) \subset \Omega$ and $|\mathbf{x_m}| \to \infty$ then the injection is not compact. Indeed, if $\varphi \in C_c^\infty\big(B(0,\alpha)\big)$ is not 0, then $u_n(\mathbf{x}) = \varphi(\mathbf{x} - \mathbf{x_m})$ is a sequence which converges to 0 in $H^1(\Omega)$ weak (because its support tends to infinity), but which does not converge in $L^2(\Omega)$ strong. The compactness assumption holds for bounded domains (or domains with finite measure) under some smoothness hypothesis concerning the boundary.

Our first incursion among models for fluids will be to consider the stationary incompressible Stokes equation. It would be difficult to prove much on the question of starting with the compressible case and letting the Mach[7] number[8] tend to 0, and therefore I shall describe another approach, which is to consider this equation as a limit case of the equation of linearized elasticity. I shall not recall the defects of linearized elasticity, and here we are only interested in the similarities at the level of the equations. As was observed experimentally by Dan JOSEPH,[9] fluids do have some elastic properties, but for the moment I do not want to discuss the limitations of the usual viscosity argument.

---

[5] Giorgio G. TALENTI, Italian mathematician, born in 1940. He works at Università di Firenze, Firenze (Florence), Italy.

[6] Friedrich Wilhelm BESSEL, German mathematician, 1784–1846. He worked in Königsberg, then in Germany, now Kaliningrad, Russia.

[7] Ernst MACH, Czech-born physicist, 1838–1916. He worked in Vienna, Austria.

[8] The Mach number is the ratio of the velocity to the speed of sound.

[9] Daniel D. JOSEPH, American mathematician, born in 1929. He works at University of Minnesota Twin Cities, Minneapolis, MN.

I shall only consider the isotropic case

$$\sigma_{ij} = 2\mu\,\varepsilon_{ij} + \lambda\,\delta_{ij} \sum_{k=1}^{N} \varepsilon_{kk}, \text{ with } \varepsilon_{ij} = \frac{1}{2}\left(\frac{\partial u_i}{\partial x_j} + \frac{\partial u_j}{\partial x_i}\right) \text{ for } i, j = 1, \ldots, N,$$

(10.13)

where the shear modulus $\mu > 0$ and the Lamé parameter λ either satisfy the very strong ellipticity condition $2\mu + N\lambda > 0$ if coefficients are variable, or the strong ellipticity condition $2\mu + \lambda > 0$ if the coefficients are constant and one uses Dirichlet[10] conditions (for continuous coefficients, one uses Gårding's[11] inequality).

[Taught on Wednesday February 3, 1999.]

---

[10] Johann Peter Gustav LEJEUNE DIRICHLET, German mathematician, 1805–1859. He worked in Berlin and Göttingen, Germany.

[11] Lars GÅRDING, Swedish mathematician, born in 1919. He works at Lund University, Lund, Sweden.

Linearized elasticity

We shall approach the stationary incompressible Stokes equation by first considering stationary linearized elasticity for isotropic materials, i.e. studying the equilibrium equation

$$-\sum_{j=1}^{3} \frac{\partial \sigma_{ij}}{\partial x_j} = f_i \text{ in } \Omega \text{ for } i = 1, 2, 3, \tag{11.1}$$

with the stress–strain relation

$$\sigma_{ij} = 2\mu \, \varepsilon_{ij} + \lambda \, \delta_{ij} \sum_{k=1}^{3} \varepsilon_{kk} \text{ for } i = 1, 2, 3, \tag{11.2}$$

where

$$\varepsilon_{ij} = \frac{1}{2} \left(\frac{\partial u_i}{\partial x_j} + \frac{\partial u_j}{\partial x_i} \right) \text{ for } i, j = 1, 2, 3, \tag{11.3}$$

and **u** is the *displacement*. Notice that apart from gravity it is not very realistic to imagine forces acting directly inside Ω, and for a fluid the gravity forces can be incorporated into the pressure term; of course there are electromagnetic forces for conducting fluids, but then one must couple the equation with Maxwell's equation, or there could be chemical forces, but one needs then to consider a larger system taking into account all the chemical species present.

We shall begin by using Dirichlet boundary conditions, because it is reasonable for a fluid, as the *viscosity* imposes that the fluid must move at the same velocity than the boundary. However, because that condition is obviously not good at the surface of the ocean, we shall have to study other boundary conditions. The use of Dirichlet conditions will also simplify the analysis for the linearized elasticity equation, as one does not need to prove Korn's[1],[2]

---

[1] Arthur KORN, German-born physicist, 1870–1945. He worked at the STEVENS Institute of Technology, Hoboken, NJ.

[2] Edwin Augustus STEVENS, American engineer and philanthropist, 1795–1868.

inequality, because of the following identity, for which I shall give two proofs, very similar, the first one using the Fourier transform, the second one using integration by parts.

Lemma 11.1. *For any open set of R^N, one has*

$$\int_\Omega \sum_{j,k=1}^N |\varepsilon_{jk}|^2 \, dx = \frac{1}{2} \int_\Omega \left(|div \, u|^2 + \sum_{j=1}^N |grad \, u_j|^2 \right) dx \qquad (11.4)$$
for all $u_1, \dots, u_N \in H_0^1(\Omega)$.

First proof: The functions u_j are extended by 0 outside Ω and one uses their Fourier transform.

$$
\begin{aligned}
\int_\Omega \sum_{j,k=1}^N |\varepsilon_{jk}|^2 \, dx &= \int_{x \in R^N} \sum_{j,k=1}^N |\varepsilon_{jk}|^2 \, dx = \\
&= \pi^2 \int_{\xi \in R^N} \sum_{j,k=1}^N |\xi_j \mathcal{F}u_k + \xi_k \mathcal{F}u_j|^2 \, d\xi = \\
&= \pi^2 \int_{\xi \in R^N} \sum_{j,k=1}^N \left(|\xi_j|^2 |\mathcal{F}u_k|^2 + |\xi_k|^2 |\mathcal{F}u_j|^2 + 2\xi_j \xi_k \Re(\mathcal{F}u_k \overline{\mathcal{F}u_j}) \right) d\xi = \\
&= 2\pi^2 \int_{\xi \in R^N} \left(|\xi|^2 \sum_{j=1}^N |\mathcal{F}u_j|^2 + |\sum_{j=1}^N \xi_j \mathcal{F}u_j|^2 \right) d\xi = \\
&= \frac{1}{2} \int_{x \in R^N} \left(\sum_{j=1}^N |grad \, u_j|^2 + |div \, u|^2 \right) dx = \\
&= \frac{1}{2} \int_\Omega \left(\sum_{j=1}^N |grad \, u_j|^2 + |div \, u|^2 \right) dx. \blacksquare
\end{aligned}
$$
$$(11.5)$$

Second proof:

$$
\begin{aligned}
\int_\Omega \sum_{j,k=1}^N |\varepsilon_{jk}|^2 \, dx &= \frac{1}{4} \int_\Omega \sum_{j,k=1}^N \left(\frac{\partial u_j}{\partial x_k} + \frac{\partial u_k}{\partial x_j} \right)^2 dx = \\
&= \frac{1}{2} \int_\Omega \sum_{j=1}^N |grad \, u_j|^2 \, dx + \frac{1}{2} \int_\Omega \left(\sum_{j,k=1}^N \frac{\partial u_j}{\partial x_k} \frac{\partial u_k}{\partial x_j} \right) dx,
\end{aligned}
\qquad (11.6)
$$

and the result is a consequence of the property

$$\int_\Omega \frac{\partial v}{\partial x_j} \frac{\partial w}{\partial x_k} \, dx = \int_{x \in R^N} \frac{\partial v}{\partial x_k} \frac{\partial w}{\partial x_j} \, dx \text{ for every } v, w \in H_0^1(\Omega). \qquad (11.7)$$

This last formula is actually true if v or w belongs to $H_0^1(\Omega)$ and the other belongs to $H^1(\Omega)$, as for any distribution T and $\varphi \in C_c^\infty(\Omega)$ one has

$$\left\langle \frac{\partial T}{\partial x_j}, \frac{\partial \varphi}{\partial x_k} \right\rangle = -\left\langle T, \frac{\partial^2 \varphi}{\partial x_j \partial x_k} \right\rangle = \left\langle \frac{\partial T}{\partial x_k}, \frac{\partial \varphi}{\partial x_j} \right\rangle, \qquad (11.8)$$

and therefore the result is true for $v \in H^1(\Omega)$ and $\varphi \in C_c^\infty(\Omega)$, and by density it remains true for $\varphi \in H_0^1(\Omega)$. $\blacksquare$

The basic existence theorem for variational elliptic problems is the following *Lax–Milgram*[3] *lemma*, or one of its variants.

---

[3] Arthur Norton MILGRAM, American mathematician, 1912–1960. He worked at University of Minnesota, Minneapolis, MN.

Lemma 11.2. *(Lax–Milgram Lemma) Let V be a real Hilbert space, and $A \in \mathcal{L}(V, V')$ satisfying the V-ellipticity condition*

$$\text{there exists } \alpha > 0 \text{ such that } (A u, u) \geq \alpha \, ||u||^2 \text{ for all } u \in V. \qquad (11.9)$$

Then A is an isomorphism from V onto V'.

Proof: As $\alpha \, ||u||^2 \leq (A u, u) \leq ||A u||_* ||u||$, where $||\cdot||_*$ denotes the dual norm on V', one deduces that

$$||A u||_* \geq \alpha \, ||u|| \text{ for all } u \in V, \qquad (11.10)$$

and this means that A is injective (one to one) and has a closed range (if $A u_n \to f$, then $A u_n$ is a Cauchy sequence in V', and therefore u_n is a Cauchy sequence in V, which then converges to some u_∞, and $f = A u_\infty$). As $(A u, u) = (u, A^T u)$, one sees that A^T satisfies the same hypothesis, and therefore A^T is injective, which is equivalent to A having a dense range. The range of A being closed and dense is then equal to V'. Therefore A is a bijection from V to V', and its inverse is automatically continuous by the closed graph theorem, which can be avoided here as $||A u||_* \geq \alpha \, ||u||$ for all $u \in V$ shows directly that $||A^{-1}|| \leq \frac{1}{\alpha}$. ∎

From a practical point of view, an equivalent formulation is to have a continuous bilinear form $a(u, v)$ on $V \times V$, and of course $a(u, v) = (A u, v)$ for every $u, v \in V$; the V-ellipticity condition is $a(u, u) \geq \alpha \, ||u||_2$ for all $u \in V$; the conclusion is that for every linear continuous form $L(v)$ on V, there exists a unique $u \in V$ such that $a(u, v) = L(v)$ for every $v \in V$. One advantage of this formulation is that one does not have to identify what V' is. Another advantage is that the same formulation may be used directly for numerical methods (like *finite elements*): one first creates a family of finite-dimensional spaces $V_h \subset V$, usually made of simple functions on a triangulation of Ω with h related to the mesh size, and one computes $u_h \in V_h$, the unique solution of

$a(u_h, v_h) = L(v_h)$ for all $v_h \in V_h$ by using techniques of linear algebra;[4,5,6,7] the rate of convergence of u_h to the exact solution u is related to the fact that every element of V can be well approximated by sequences from V_h, and this is shown explicitly for smooth functions by estimating an interpolation error (usually the functions in V_h are defined by some values at the vertices of the triangulation, and one must compare a smooth function v to the function in V_h which has the same values than v at the vertices of the triangulation).

For a complex Hilbert space, one may use the hypothesis that a is a sesquilinear form on $V \times V$ such that $\Re(a(u,u)) \geq \alpha \, ||u||^2$ for all $u \in V$, or that there exists θ such that $\Re(e^{i\theta}a(u,u)) \geq \alpha \, ||u||^2$ for all $u \in V$, or even more generally that $|(a(u,u)| \geq \alpha \, ||u||^2$ for all $u \in V$, as this condition implies the preceding one by a result of Eduardo ZARANTONELLO[8] on the numerical range of an operator (the set of $\frac{a(v,v)}{||v||^2}$ for $v \neq 0$ is a convex set of the complex plane).

The same result holds without assuming that A is continuous, as continuity of A can be deduced by using the closed graph theorem. The same result is true for a Banach space V, as one makes it a Hilbert space by using an equivalent norm corresponding to the scalar product $a(u,v) + a(v,u)$ in the real case, or $a(u,v) + \overline{a(v,u)}$ in the complex case.

It is useful to know the following variant of the Lax–Milgram lemma 11.2.

Lemma 11.3. *Let V be a real Hilbert space, and $A \in \mathcal{L}(V, V')$ such that*

there exists $\beta > 0$ such that $(A\,u, u) \geq 0$, and $||A\,u||_* \geq \beta \, ||u||$ for all $u \in V$.
$$(11.11)$$
Then A is an isomorphism from V onto V'.

---

[4] The term *algebra* was derived from the title of a mathematical treatise, *Hisab al-jabr w'al-muqabala*, by AL KHWARIZMI, whose name has been used for coining the term *algorithm*.

[5] Abu Ja'far Muhammad ibn Musa AL KHWARIZMI, mathematician, 780–850. He worked in Baghdad, now in Iraq.

[6] AL KHWARIZMI participated in the work of an institute (Bayt al-Hikmah = House of Wisdom) that the Caliph AL MA'MUN had set up in his capital Baghdad with the goal of translating Greek philosophical and mathematical works into Arabic. I have learnt from my father that translations were made in two steps, Christians translating from Greek to Syriac, and Moslems from Syriac to Arabic. As a consequence of AL MA'MUN's enlightened attitude, some Greek texts have survived only through their translation in Arabic, because in the Occident the Christian scholars were not interested in saving them from oblivion by having copies made (usually by monks), probably because these works had been written by pagans!

[7] Abu al-Abbas Abd Allah AL MA'MUN ibn Harun, 7th Caliph of the Abbasid dynasty, 786–833.

[8] Eduardo H. ZARANTONELLO, Argentine mathematician, born in 1918. He worked in Mendoza, Argentina.

Proof: Let Λ be the canonical isometry from V onto V', i.e. $\Lambda \in \mathcal{L}(V, V')$, satisfies $||\Lambda u||_* = ||u||$ and $(\Lambda u, u) = ||u||^2$ for all $u \in V$. Then for $\varepsilon > 0$, one may apply the Lax–Milgram lemma 11.2 to $A + \varepsilon \Lambda$, with $\alpha = \varepsilon$ and therefore for a given $f \in V'$ there exists a unique $u_\varepsilon \in V$ such that $A u_\varepsilon + \varepsilon \Lambda u_\varepsilon = f$. Taking the scalar product with u_ε gives the first estimate $\varepsilon ||u_\varepsilon|| \leq ||f||_*$, from which one deduces that $||A u_\varepsilon||_* \leq ||f||_* + \varepsilon ||\Lambda u_\varepsilon||_* \leq 2||f||_*$. Therefore $\beta ||u_\varepsilon|| \leq 2||f||_*$, and one deduces then that $A u_\varepsilon \to f$ in V'. In consequence, $A u_{\varepsilon_n}$ is a Cauchy sequence in V' if $\varepsilon_n \to 0$, so that u_{ε_n} is a Cauchy sequence in V, and its limit u_0 satisfies $A u_0 = f$.∎

I do not know if this variant was known before Tyrrell ROCKAFELLAR[9] proved something analogous for monotone operators; the proof above is the one that I immediately imagined with Jean-Claude NÉDÉLEC[10] when we learnt about Tyrrel ROCKAFELLAR's result in the late 1960s.

As we shall see, the existence theorem for linearized elasticity will be a simple application of the Lax–Milgram lemma 11.2 with $V = H_0^1(\Omega; R^N)$, and there will not be too much difficulty in proving an abstract theorem for describing the limit as the Lamé parameter λ tends to infinity: we shall find that $div\, u = 0$ at the limit, and the limit problem will correspond to a situation where the Lax–Milgram lemma 11.2 applies, but for the space $W = \{u \in H_0^1(\Omega; R^N) \mid div\, u = 0\}$. Expressing in a concrete way the partial differential equation that the limit satisfies will formally involve a Lagrange multiplier, the "pressure", but there will be some technical obstacles to overcome before we can assert in what functional space the "pressure" is.

If one could apply the variant, deriving the equation would be straightforward, but checking the hypothesis of the variant will be a technical obstacle equivalent to estimating the "pressure" in the preceding approach.

[Taught on Friday February 5, 1999.]

---

[9] Ralph Tyrrell ROCKAFELLAR, American mathematician, born in 1935. He works at University of Washington, Seattle, WA.

[10] Jean-Claude NÉDÉLEC, French mathematician, born in 1943. He works at École Polytechnique, Palaiseau, France.

12

Ellipticity conditions

In the case of stationary linearized elasticity with variable coefficients (in $L^\infty(\Omega)$), in the general form $\sigma_{ij} = \sum_{k,\ell=1}^{N} C_{ijk\ell}\varepsilon_{k\ell}$ for $i,j = 1,\ldots,N$, one uses the *very strong ellipticity condition*

$$\text{there exists } \alpha > 0 : \sum_{i,j,k,\ell=1}^{N} C_{ijk\ell}(x)\mathsf{M}_{k\ell}\mathsf{M}ij \geq \alpha \sum_{i,j=1}^{N} \mathsf{M}ij^2$$
$$\text{for all symmetric } \mathsf{M}, \text{ a.e. } \mathbf{x} \in \Omega. \tag{12.1}$$

In the case of isotropic materials, i.e. $\sigma_{ij} = 2\mu\,\varepsilon_{ij} + \lambda\,\delta_{ij}\sum_{k=1}^{N}\varepsilon_{kk}$ for $i,j = 1,\ldots,N$, it means that $2\mu\sum_{i,j=1}^{N}\mathsf{M}ij^2 + \lambda(\sum_{k=1}^{N}\mathsf{M}kk)^2 \geq \alpha\sum_{i,j=1}^{N}\mathsf{M}ij^2$, and therefore $2\mu(\mathbf{x}) \geq \alpha > 0$ (if $N \geq 2$) and $2\mu(\mathbf{x}) + N\lambda(\mathbf{x}) \geq \alpha > 0$ a.e. in Ω. Then the Lax–Milgram lemma 11.2 applies: the Hilbert space V is $H_0^1(\Omega; R^N)$, the continuous bilinear form a is given by $a(u,v) = \int_\Omega \sum_{i,j,k,\ell=1}^{N} C_{ijk\ell}\varepsilon_{k\ell}(u)\varepsilon_{ij}(v)\,d\mathbf{x}$, where $\varepsilon_{ij}(v)$ means $\frac{1}{2}\big(\frac{\partial v_i}{\partial x_j} + \frac{\partial v_i}{\partial x_j}\big)$, the linear continuous form L is given by $L(v) = \int_\Omega \sum_{i=1}^{N} f_i v_i\,d\mathbf{x}$, with $f_i \in L^2(\Omega)$ for $i = 1,\ldots,N$, and therefore if Poincaré's inequality holds for $H_0^1(\Omega)$, the very strong ellipticity condition implies the hypothesis of the Lax–Milgram lemma 11.2.

In the case of stationary linearized elasticity with constant coefficients and Dirichlet conditions, one uses the *strong ellipticity condition*

$$\text{there exists } \alpha > 0 : \sum_{i,j,k,\ell=1}^{N} C_{ijk\ell}a_k\xi_\ell a_i\xi_j \geq \alpha\,|\mathbf{a}|^2|\boldsymbol{\xi}|^2$$
$$\text{for all } \mathbf{a}, \boldsymbol{\xi} \in R^N, \text{ a.e. in } \Omega, \tag{12.2}$$

under the symmetry hypothesis $C_{ijk\ell} = C_{jik\ell} = C_{ij\ell k}$. Instead of using a lower bound for the integrand as when the very strong ellipticity condition holds, one integrates in Ω, but because of the Dirichlet conditions it is an integral on R^N, for which one uses the Fourier transform (one could obtain the same result by integration by parts); using the symmetries of the coefficients, one finds $4\pi^2 \int_{\boldsymbol{\xi} \in R^N} \sum_{i,j,k,\ell=1}^{N} C_{ijk\ell}\mathcal{F}u_k\xi_\ell \overline{\mathcal{F}u_i}\xi_j\,d\boldsymbol{\xi}$, and as the hypothesis implies $\Re\big(\sum_{i,j,k,\ell=1}^{N} C_{ijk\ell}a_k\xi_\ell \overline{a_i}\xi_j\big) \geq \alpha\,|\mathbf{a}|^2|\boldsymbol{\xi}|^2$ for all $\mathbf{a} \in C^N$ and all $\boldsymbol{\xi} \in R^N$, the integral (which is real) is bounded below by $4\alpha \pi^2 \int_{\xi \in R^N} |\xi|^2 \sum_{k=1}^{N} |\mathcal{F}u_k|^2 d\xi$, which is $\alpha \int_{\Omega} \sum_{k=1}^{N} |grad\, u_k|^2 dx$. In the case of isotropic materials, the strong ellipticity condition means $2\mu \geq \alpha$ (if $N \geq 2$) and $2\mu + \lambda \geq \alpha$.

As we are interested in letting λ tend to infinity, we assume that $\mu(x) \geq \beta > 0$ a.e. in Ω and λ is a nonnegative constant tending to $+\infty$. The equilibrium equation can be written as

$$a^0(u^\lambda, v) + \lambda b(u^\lambda, v) = L(v) \text{ for all } v \in V = H_0^1(\Omega; R^N), \quad (12.3)$$

where L is a linear continuous form on V, and

$$
\begin{aligned}
a^0(u,v) &= \int_{\Omega} 2\mu(x) \sum_{i,j=1}^{N} \varepsilon_{ij}(u)\varepsilon_{ij}(v)\, dx \\
b(u,v) &= \int_{\Omega} div\, u\, div\, v\, dx,
\end{aligned}
\quad (12.4)
$$

and it has a unique solution $u^\lambda \in H_0^1(\Omega; R^N)$. This type of problem, related to the method of *penalization* or to questions of *singular perturbations*, has been extensively studied by Jacques-Louis LIONS in the late 1960s / early 1970s. Here the abstract treatment is straightforward: as a^0 is V-elliptic and $b(u,u)$ is the square of the norm of $div\, u$, one uses $v = u^\lambda$ and one obtains $\gamma ||u^\lambda||^2 + \lambda |div\, u^\lambda|^2 \leq C\, ||u^\lambda||$ (where $|| \cdot ||$ is the norm in V, and $| \cdot |$ is the norm in $L^2(\Omega)$), which gives

$$
\begin{aligned}
&u^\lambda \text{ bounded in } V, \\
&div\, u^\lambda \to 0 \text{ in } L^2(\Omega),
\end{aligned}
\quad (12.5)
$$

which permits us to extract a subsequence converging in V weak to u^∞; the fact that u^∞ will be characterized as the unique solution of an associated problem will show that all the sequence converges weakly (we shall also show strong convergence). As $div\, u^\infty = 0$, we introduce the new space

$$W = \{u \in H_0^1(\Omega; R^N) \mid div\, u = 0 \text{ a.e. in } \Omega\}, \quad (12.6)$$

and so $u^\infty \in W$, but as $b(u,v) = 0$ for $v \in W$, one has $a^0(u^\lambda, v) = L(v)$ for every $v \in W$, which shows that

$$a^0(u^\infty, w) = L(w) \text{ for all } w \in W, \text{ and } u^\infty \in W, \quad (12.7)$$

which characterizes u^∞ because a^0 is V-elliptic and therefore W-elliptic. In order to show strong convergence, one notices that $a^0(u^\lambda, u^\lambda) \leq L(u^\lambda)$ and therefore

$$\limsup_{\lambda \to \infty} a^0(u^\lambda, u^\lambda) \leq L(u^\infty) = a^0(u^\infty, u^\infty), \quad (12.8)$$

from which one deduces that

$$\limsup_{\lambda \to \infty} a^0(u^\lambda - u^\infty, u^\lambda - u^\infty) = \limsup_{\lambda \to \infty} a^0(u^\lambda, u^\lambda) - a^0(u^\infty, u^\infty) \leq 0, \quad (12.9)$$

and therefore u^λ converges to u^∞ in V strong as $\lambda \to \infty$.

The problem is now to identify what equation u^∞ satisfies, and the difficulty comes from the fact that, because of the constraint $div\, v = 0$, one is not allowed to use arbitrary test functions in $C_c^\infty(\Omega)$ which would give us an equation in the sense of distributions.

One could use the fact that our problem is equivalent to minimizing $a^0(v,v) - 2L(v)$ on the subspace of V defined by an equation $div\, v = 0$, and argue that there will be a Lagrange multiplier $q \in L^2(\Omega)$ such that v minimizes $a^0(v,v) - 2L(v) + 2(q.div\, v)$ without constraints, but the proof that such a Lagrange multiplier exists requires some care.

One could deduce an equation for u^∞ if one knew to what element $\lambda\, div\, u^\lambda$ converges, but for the moment we only know that $\sqrt{\lambda}\, div\, u^\lambda$ is bounded in $L^2(\Omega)$. In a recent discussion with François MURAT, I noticed that if Ω has a Lipschitz boundary, then $\lambda\, div\, u^\lambda$ stays bounded in $L^2(\Omega)$.

In order to prepare for the discussion, we need to become familiar with $H^{-1}(\Omega)$, which is defined as the dual of $H_0^1(\Omega)$; as $C_c^\infty(\Omega)$ is dense in $H_0^1(\Omega)$, its dual is a space of distributions in Ω, which is characterized by

$$H^{-1}(\Omega) = \left\{ f \in \mathcal{D}'(\Omega) \mid f = g_0 + \sum_{j=1}^{N} \frac{\partial g_j}{\partial x_j} \text{ with } g_0, \ldots, g_N \in L^2(\Omega) \right\}.$$

(12.10)

Indeed, let A be the linear continuous mapping from $H_0^1(\Omega)$ into $L^2(\Omega; R^{N+1})$ defined by

$$A\,u = \left(u, \frac{\partial u}{\partial x_1}, \ldots, \frac{\partial u}{\partial x_N} \right).$$

(12.11)

$H_0^1(\Omega)$ is a closed subspace of $H^1(\Omega)$, which is complete (it is a Hilbert space) because if $u_n \in H^1(\Omega)$ is such that $u_n \to v_0$ in $L^2(\Omega)$ and $\frac{\partial u_n}{\partial x_j} \to v_j$ in $L^2(\Omega)$ then $v_j = \frac{\partial v_0}{\partial x_j}$ by using the definition of derivatives in the sense of distributions: for $\varphi \in C_c^\infty(\Omega)$ one has $\int_\Omega \frac{\partial u_n}{\partial x_j} \varphi\, dx = \langle \frac{\partial u_n}{\partial x_j}, \varphi \rangle = -\langle u_n, \frac{\partial \varphi}{\partial x_j} \rangle = -\int_\Omega u_n \frac{\partial \varphi}{\partial x_j}\, dx$, which gives $\int_\Omega v_j \varphi\, dx = -\int_\Omega v_0 \frac{\partial \varphi}{\partial x_j}\, dx$ for all $\varphi \in C_c^\infty(\Omega)$. By definition of the norm of $H_0^1(\Omega)$, $||A\,u||$ is equivalent to $||u||$, and as $H_0^1(\Omega)$ is complete, the range of A is complete and therefore closed in $L^2(\Omega; R^{N+1})$. If L is a linear continuous form on $H_0^1(\Omega)$, it defines a linear continuous form on $R(A)$, which extends (by Hahn[1]–Banach theorem, or by using orthogonal projection on $R(A)$) to a linear continuous form on $L^2(\Omega; R^{N+1})$, which is of the form $(v_0, \ldots, v_N) \mapsto \sum_{k=0}^{N} \int_\Omega h_k v_k\, dx$ with $h_0, \ldots, h_N \in L^2(\Omega)$, and therefore $L(u) = \int_\Omega h_0 u\, dx + \sum_{j=1}^{N} \int_\Omega h_j \frac{\partial u}{\partial x_j}\, dx$. If Poincaré's inequality holds, one can use $A\,u = (\frac{\partial u}{\partial x_1}, \ldots, \frac{\partial u}{\partial x_N})$ instead, and one finds that every $g_0 \in L^2(\Omega)$ can be written as $\sum_{j=1}^{N} \frac{\partial g_j}{\partial x_j}$ for some $g_j \in L^2(\Omega)$ for $j = 1, \ldots, N$. One notices that each derivative $\frac{\partial}{\partial x_j}$ is a linear continuous operator from $L^2(\Omega)$ into $H^{-1}(\Omega)$.

---

[1] Hans HAHN, Austrian mathematician, 1879–1934. He worked in Vienna, Austria.

For $0 \le \lambda < \infty$, using $v \in C_c^\infty(\Omega; R^N)$ and $L(v) = \sum_{i=1}^N \langle f_i, v_i \rangle$ with $f_i \in H^{-1}(\Omega)$ for $i = 1, \ldots, N$, one deduces the equilibrium equation

$$-\sum_{j=1}^N \frac{\partial}{\partial x_j}\left[\mu\left(\frac{\partial u_i^\lambda}{\partial x_j} + \frac{\partial u_j^\lambda}{\partial x_i}\right)\right] - \frac{\partial(\lambda \, div \, u^\lambda)}{\partial x_i} = f_i \text{ for } i = 1, \ldots, N, \quad (12.12)$$

and therefore, as u^λ is bounded in $H_0^1(\Omega)$, one finds that $\frac{\partial(\lambda \, div \, u^\lambda)}{\partial x_i}$ is bounded in $H^{-1}(\Omega)$, so that a subsequence converges to an element $T_i \in H^{-1}(\Omega)$; at this level it is not obvious that $\lambda \, div \, u^\lambda$ stays bounded, even in the space of distributions. The important property of T_i, $i = 1, \ldots, N$, is that one has the equations

$$-\sum_{j=1}^N \frac{\partial}{\partial x_j}\left[\mu\left(\frac{\partial u_i^\infty}{\partial x_j} + \frac{\partial u_j^\infty}{\partial x_i}\right)\right] - T_i = f_i \text{ in } \Omega, \text{ for } i = 1, \ldots, N, \quad (12.13)$$

$$\sum_{i=1}^N \langle T_i, w_i \rangle = 0 \text{ for all } w \in W, \quad (12.14)$$

by taking the limit either for the equations or for the variational formulation.

One has $\frac{\partial T_i}{\partial x_j} = \frac{\partial T_j}{\partial x_i}$ for all i, j, either by noticing that $T_i^\lambda = \frac{\partial(\lambda \, div \, u^\lambda)}{\partial x_i}$ satisfies the same property as it is a gradient, or (if $i \ne j$) by using $w_i = \frac{\partial \psi}{\partial x_j}$, $w_j = -\frac{\partial \psi}{\partial x_i}$ and $w_k = 0$ for $k \ne i$ and $k \ne j$. By a result of Laurent SCHWARTZ, on each open ball (or any *simply connected* open subset ω) of Ω there exists a distribution S such that $T_i = \frac{\partial S}{\partial x_i}$ for $i = 1, \ldots, N$. However, there are more functions in W, which may see the topology of Ω, and a result of DE RHAM[2] asserts that if distributions T_i, $i = 1, \ldots, N$, satisfy $\sum_{i=1}^N \langle T_i, \varphi_i \rangle = 0$ for all $\varphi_i \in C_c^\infty(\Omega)$, $i = 1, \ldots, N$, satisfying $div \, \varphi = 0$, then there exists a distribution S such that $T_i = \frac{\partial S}{\partial x_i}$ for $i = 1, \ldots, N$. If one admits this result, the question will be to prove that if all T_i belong to $H^{-1}(\Omega)$ then S belongs to $L^2(\Omega)$, and that requires some smoothness of the boundary: I noticed a few months ago that the result is not true if the boundary is not Lipschitz continuous, and François MURAT has just mentioned to me a similar counter-example by Giuseppe GEYMONAT[3] and Gianni GILARDI,[4] motivated by showing that Korn's inequality does not always hold if the boundary is not Lipschitz continuous, the connection between the two questions being that they both are related to the space $X(\Omega) = \{u \in H^{-1}(\Omega) \mid \frac{\partial u}{\partial x_i} \in H^{-1}(\Omega), i = 1, \ldots, N\}$ being $L^2(\Omega)$ or not.

[Taught on Monday February 8, 1999.]

---

[2] Georges DE RHAM, Swiss mathematician, 1903–1990. He worked in Genève (Geneva), Switzerland.

[3] Giuseppe GEYMONAT, Italian-born mathematician. He works at Université des Sciences et Techniques de Languedoc (Montpellier II), Montpellier, France.

[4] Gianni GILARDI, Italian mathematician, born in 1947. He works at Università di Pavia, Pavia, Italy.

13

Sobolev spaces III

The idea of considering the space

$$X(\Omega) = \left\{ u \in H^{-1}(\Omega) \mid \frac{\partial u}{\partial x_j} \in H^{-1}(\Omega), j = 1, \ldots, N \right\}, \qquad (13.1)$$

seems due to Jacques-Louis LIONS.

We have found that $T_j \in H^{-1}(\Omega), j = 1, \ldots, N$, satisfies $\sum_{j=1}^{N} \langle T_j, w_j \rangle = 0$ for all $w \in W$, and DE RHAM's result asserts that there exists a distribution S such that $T_j = \frac{\partial S}{\partial x_j}$ for $j = 1, \ldots, N$. Why consider that S itself should belong to $H^{-1}(\Omega)$? That condition is indeed useful, because the information $u \in H^{-1}(\Omega)$ makes the space *local*, i.e. $u \in X(\Omega)$ implies $\psi u \in X(\Omega)$ for every $\psi \in C_c^\infty(R^N)$, and one can then use partitions of unity in order to study the functions of $X(\Omega)$. The approach of Jacques-Louis LIONS seems to have been to use DE RHAM's result and then prove that $S \in L^2(\Omega)$; he quotes an article by Enrico MAGENES[1] and Guido STAMPACCHIA[2] (which I have not read, but Jacques-Louis LIONS told me that they had mentioned there that they were using one of his results).

I present here the approach which I developed in 1974, which does not rely on DE RHAM's result; I only used it for smooth domains, but the result is indeed true for Lipschitz domains: Olga LADYZHENSKAYA may have used this approach before, but certainly Jindrich NEČAS[3,4] did use it before me;

[1] Enrico MAGENES, Italian mathematician, born in 1923. He works at Università di Pavia, Pavia, Italy.

[2] Guido STAMPACCHIA, Italian mathematician, 1922–1978. He worked at Scuola Normale Superiore, Pisa, Italy.

[3] Jindřich NEČAS, Czech-born mathematician, 1929–2002. He worked at CHARLES University, Prague, Czech Republic, and at Northern Illinois University, De Kalb, IL.

[4] CHARLES IV of Luxembourg, 1316–1378, founded the University of Prague in 1348, and was named Holy Roman Emperor in 1355.

GOBERT[5] is mentioned for Korn's inequality, which is related as follows. Korn's inequality is about proving that if $u \in L^2(\Omega; R^N)$ and $\varepsilon_{ij} \in L^2(\Omega)$ for all $i, j = 1, \ldots, N$, then one has $u \in H^1(\Omega; R^N)$; one notices that

$$2\frac{\partial^2 u_i}{\partial x_j \partial x_k} = \frac{\partial}{\partial x_j}\left(\frac{\partial u_i}{\partial x_k} + \frac{\partial u_k}{\partial x_i}\right) - \frac{\partial}{\partial x_i}\left(\frac{\partial u_k}{\partial x_j} + \frac{\partial u_j}{\partial x_k}\right) + \frac{\partial}{\partial x_k}\left(\frac{\partial u_j}{\partial x_i} + \frac{\partial u_i}{\partial x_j}\right), \quad (13.2)$$

and therefore $\frac{\partial^2 u_i}{\partial x_j \partial x_k} \in H^{-1}(\Omega)$; as all $\frac{\partial u_i}{\partial x_j}$ also belong to $H^{-1}(\Omega)$, it shows that all $\frac{\partial u_i}{\partial x_j}$ belong to $X(\Omega)$, and therefore if $X(\Omega) = L^2(\Omega)$, then Korn's inequality holds (Giuseppe GEYMONAT and Gianni GILARDI have shown that Korn's inequality does not hold for some non-Lipschitz domains, and therefore $X(\Omega) \neq L^2(\Omega)$ for the open set Ω that they considered; my remark, that $X(\Omega) \neq L^2(\Omega)$ in many non-Lipschitz domains, seems to provide simpler explicit counter-examples, as I checked with François MURAT).

The result we are interested in is the following:

Lemma 13.1. *Assume that Ω is smooth enough. If $T_i \in H^{-1}(\Omega)$ for $i = 1, \ldots, N$ and $\sum_{i=1}^{N} \langle T_i, w_i \rangle = 0$ for all $w \in W = \{u \in H_0^1(\Omega; R^N) \mid div\, w = 0\}$, then there exists $p \in L^2(\Omega)$ such that $T_i = \frac{\partial p}{\partial x_i}$ for $i = 1, \ldots, N$.*

This result is equivalent to the following:

Lemma 13.2. *Assume that Ω is smooth enough. For $g \in L^2(\Omega)$ satisfying $\int_\Omega g\, dx = 0$, there exists $u \in H_0^1(\Omega; R^N)$ such that $div\, u = g$.*

Indeed, $A = grad$ operates continuously from $L^2(\Omega)$ into $H^{-1}(\Omega; R^N)$, so its transpose $A^T = -div$ operates continuously from $H_0^1(\Omega; R^N)$ into $L^2(\Omega)$. The kernel of A^T is W, and therefore its orthogonal is the closure of $R(A)$, so Lemma 13.1 is equivalent to saying that $R(A)$ is closed. The kernel of A is generated by the constant function 1 (as Ω is always assumed to be connected), and its orthogonal is the closure of $R(A^T)$, so Lemma 13.2 is equivalent to saying that $R(A^T)$ is closed, which is indeed equivalent to the statement that $R(A)$ is closed.

My method, based on the *equivalence lemma* 13.3, appeared to be a generalization of an earlier result of Jaak PEETRE,[6] sufficient for the present situation; it shows that if the injection of $H_0^1(\Omega)$ into $L^2(\Omega)$ is compact (which is the case if $meas(\Omega) < \infty$), and if $X(\Omega) = L^2(\Omega)$ (which requires some smoothness of the boundary), then $R(A)$ is closed.

---

[5] Jules GOBERT.

[6] Jaak PEETRE, Estonian-born mathematician, born in 1935. He worked at Lund University, Lund, Sweden.

The other methods that I have heard of are concerned with solving $div\, u = g$. One, which was mentioned to me by Charles GOULAOUIC,[7] uses some results that he had obtained with Salah BAOUENDI,[8] and consists in solving an equation $-div(\delta\, grad\, v) = g$, where δ is the distance to the boundary, and uses $u = -\delta\, grad\, v$. Their regularity theorem, asserting that $v \in H^1(\Omega)$ and $\delta\, v \in H^2(\Omega)$ was proven using pseudo-differential operators[9,10] which require a C^∞ boundary (in the late 1960s I had derived a simpler proof based on interpolation, which certainly does not require as much smoothness). One should notice that the basic existence result uses the space of functions such that $\sqrt{\delta}\, grad\, v \in L^2(\Omega)$, in which $C_c^\infty(\Omega)$ is dense, and therefore no boundary conditions are added. I think that this approach was analyzed by BOLLEY[11] and CAMUS.[12]

One may construct explicit solutions in R_+^N and estimates using the Calderón–Zygmund theorem; I believe that this was the approach taken by Olga LADYZHENSKAYA; Paolo GALDI[13] reproduces in [12,13] a proof valid for Lipschitz boundaries.

Lemma 13.3. *(Equivalence Lemma) Let E_1 be a Banach space and E_2, E_3 be normed spaces; let $A \in \mathcal{L}(E_1, E_2)$ and $B \in \mathcal{L}(E_1, E_3)$ satisfy the hypotheses*

$$||u||_{E_1} \approx ||A\,u||_{E_2} + ||B\,u||_{E_3} \text{ and } B \text{ is compact.} \qquad (13.3)$$

Then i) the kernel of A is finite dimensional;

ii) the range of A is closed;

iii) there exists a constant K such that if $L \in \mathcal{L}(E_1, F)$ for a normed space F and satisfies $L\,u = 0$ on the kernel of A, then one has the estimate $||L\,u||_F \leq K||L||_{\mathcal{L}(E_1,F)}\,||A\,u||_{E_2}$ for all $u \in E_1$;

iv) if $p(u)$ is a continuous semi-norm on E_1 which is a norm on the kernel of A, then $||u||_{E_1} \approx ||A\,u||_{E_2} + p(u)$.

Proof: i) On $X = ker(A)$, one has $||u||_1 \approx ||B\,u||_3$ and as B is compact one deduces that the closed unit ball of X is compact, which by a theorem of F. RIESZ proves that X is finite dimensional.

[7] Charles GOULAOUIC, French mathematician, –1983. He was my colleague at Université Paris XI (Paris-Sud), Orsay, France, from 1975 to 1979, and then worked at École Polytechnique, Palaiseau, France.

[8] Mohamed Salah BAOUENDI, Tunisian-born mathematician, born in 1937. He works at University of California San Diego, La Jolla, CA.

[9] They were first introduced by Joseph KOHN and Louis NIRENBERG.

[10] Joseph J. KOHN, Czech-born mathematician, born in 1932. He works at Princeton University, Princeton, NJ.

[11] Pierre BOLLEY, French mathematician, born in 1943. He works at Université de Nantes, Nantes, France.

[12] Jacques CAMUS, French mathematician, born in 1942. He works at Université de Rennes I, Rennes, France.

[13] Giovanni Paolo GALDI, Italian-born mathematician, born in 1947. He works at University of Pittsburgh, Pittsburgh, PA.

ii) By the Hahn–Banach theorem, X has a topological supplement Y, so there exists C such that $||A\,u||_{E_2} \geq C\,||u||_{E_1}$ on Y: if the bound was not true there would exist a sequence v_n with norm 1 in Y such that $A\,v_n \to 0$ in E_2, but as $B\,v_n$ belongs to a compact of E_3 one could extract a subsequence v_m such that $B\,v_m$ converges in E_3, and then $A\,v_m$ and $B\,v_m$ being Cauchy sequences, v_m would be a Cauchy sequence, converging to an element v_∞ satisfying the contradictory properties $A\,v_\infty = 0$ and $||v_\infty||_Y = 1$. This inequality shows that $R(A)$ is closed, because if $A\,u_n \to f$, then $A\,u_n$ is a Cauchy sequence, and therefore u_n is a Cauchy sequence, converging to u_∞ and $f = A\,u_\infty$.

iii) A is a bijection from Y onto its image Z, and its inverse D is continuous (so that Z is a Banach space even if E_2 is not complete), and as $u - D\,A\,u$ belongs to $ker(A)$, one has $L\,u = L\,D\,A\,u$ for all $u \in E_1$, and one can take $K = ||D||$.

iv) If $u \mapsto ||A\,u||_{E_2} + p(u)$ was not an equivalent norm on E_1, there would exist a sequence w_n with norm 1 in E_1 such that $||A\,w_n||_{E_2} + p(w_n) \to 0$. As $B\,w_n$ belongs to a compact of E_3 one could extract a subsequence w_m such that $B\,w_m$ converges, and using $A\,w_m \to 0$, one would deduce that w_m is a Cauchy sequence and converges to an element w_∞ satisfying the contradictory properties $A\,w_\infty = 0$, $p(w_\infty) = 0$ and $||w_\infty||_{E_1} = 1$. ∎

The result of Jaak PEETRE assumed reflexive Banach spaces, and was concerned with the special case where B is the injection from E_1 into E_3, and this is usually the case in applications.

In our setting, $E_1 = L^2(\Omega)$, $A = grad$ and $E_2 = H^{-1}(\Omega; R^N)$, B is the injection into $E_3 = H^{-1}(\Omega)$, and the hypothesis is satisfied if $meas(\Omega) < \infty$ (so that the injection from $H^1_0(\Omega)$ into $L^2(\Omega)$ is compact), and if $X(\Omega) = L^2(\Omega)$, proving that $R(A)$ is closed.

Some classical inequalities correspond to E_1 being a subspace of $H^1(\Omega)$, $A = grad$ and $E_2 = L^2(\Omega; R^N)$, B is the injection into $E_3 = L^2(\Omega)$: if the injection from $H^1(\Omega)$ into $L^2(\Omega)$ is compact, Poincaré's inequality holds if (and only if) the constant function 1 does not belong to E_1 (Ω being assumed to be connected), and in the case where $1 \in E_1$ the condition for L is $L(1) = 0$, and the condition for p is $p(1) \neq 0$.

Lemma 13.4. $X(R^N) = L^2(R^N)$.

Proof: Using the Fourier transform. $u \in H^{-1}(R^N)$ is equivalent to $\dfrac{\mathcal{F}u}{\sqrt{a^2+|\xi|^2}} \in L^2(R^N)$ (if the components of $\mathbf{x}$ have the dimension length, a and the components of $\boldsymbol{\xi}$ have dimension length^{-1}). As $\mathcal{F}\frac{\partial u}{\partial x_j} = 2i\pi\xi_j\mathcal{F}u$, one deduces that $u \in X(R^N)$ is equivalent to $\dfrac{(1+\sum_{j=1}^{N}|\xi_j|)\mathcal{F}u}{\sqrt{a^2+|\xi|^2}} \in L^2(R^N)$, and as $\dfrac{1+\sum_{j=1}^{N}|\xi_j|}{\sqrt{a^2+|\xi|^2}}$ is bounded above and below, $X(R^N)$ is $L^2(R^N)$. ∎

[Taught on Wednesday February 10, 1999.]

14

Sobolev spaces IV

I shall describe a few techniques for analyzing functions in various Sobolev spaces, *truncation*, *regularization*, *localization* and *extension*, and although we want to apply them to $X(\Omega)$, it is useful first to start with the simpler case corresponding to $H^1(\Omega)$, but the same results are true for $W^{1,p}(\Omega)$, for $1 \leq p < \infty$.

Truncation: This is used in order to show that functions with compact support are dense, in the case where Ω is not bounded, of course. One chooses $\varphi \in C_c^\infty(R^N)$ such that $\varphi(\mathbf{x}) = 1$ for $|\mathbf{x}| \leq 1$, and for $u \in L^p(\Omega)$ one defines u_n by $u_n(\mathbf{x}) = u(\mathbf{x})\varphi(\frac{\mathbf{x}}{n})$, which has compact support, such that $|u_n(\mathbf{x})| \leq K|u(\mathbf{x})|$ and $u_n(\mathbf{x}) \to u(\mathbf{x})$ a.e. (as $u_n(\mathbf{x}) = u(\mathbf{x})$ for $n \geq |\mathbf{x}|$), and therefore $u_n \to u$ in $L^p(\Omega)$ strong by the *Lebesgue dominated convergence theorem*. If $u \in W^{1,p}(\Omega)$, $\frac{\partial u_n}{\partial x_j}$ is the sum of two terms, the first one being $\frac{\partial u}{\partial x_j}\varphi(\frac{\mathbf{x}}{n})$ which converges to $\frac{\partial u}{\partial x_j}$ by the Lebesgue dominated convergence theorem, and the second being $\frac{u}{n}\frac{\partial \varphi}{\partial x_j}(\frac{\mathbf{x}}{n})$ which has a norm in $L^p(\Omega)$ of the order of $1/n$.

Regularization: This is used in order to approach a given function by smoother functions, but the convolution process which is used for that purpose increases the size of the support, and that creates small problems. On R^N, the convolution product h of two integrable functions f and g, denoted $h = f \star g$, is defined by

$$h(\mathbf{x}) = \int_{\mathbf{y} \in R^N} f(\mathbf{x} - \mathbf{y})g(\mathbf{y}) \, d\mathbf{y} = \int_{\mathbf{y} \in R^N} f(\mathbf{y})g(\mathbf{x} - \mathbf{y}) \, d\mathbf{y}, \qquad (14.1)$$

and by Fubini's[1] theorem one has $||f \star g||_{L^1} \leq ||f||_{L^1}||g||_{L^1}$. The convolution product is commutative and associative on $L^1(R^N)$. The convolution product is well defined if $f \in L^1(R^N)$ and $g \in L^p(R^N)$ and gives $f \star g \in L^p(R^N)$

---

[1] Guido FUBINI, Italian-born mathematician, 1879–1943. He worked in Torino (Turin), Italy, and after 1939 in New York, NY.

with $||f \star g||_{L^p(R^N)} \leq ||f||_{L^1(R^N)}||g||_{L^p(R^N)}$ for $1 \leq p \leq \infty$, as can be easily seen by applying Hölder's[2] inequality (or *Jensen's*[3] *inequality, which says that for every convex function Φ one has $\Phi\left(\int u\,f\,dx\right) \leq \int \Phi(u)\,f\,dx$ if $f \geq 0$ and $\int f\,dx = 1$*). One also has

$$||f \star g||_{L^r(R^N)} \leq C(p,q)||f||_{L^p(R^N)}||g||_{L^q(R^N)}, \text{ if } p,q,r \geq 1 \text{ and } \frac{1}{r} = \frac{1}{p} + \frac{1}{q} - 1,$$
(14.2)

but using Hölder's inequality as before gives $C(p,q) = 1$, which is not the best constant if p, q and r are different from 1.

A regularizing sequence is a sequence $\varrho_n \in C_c^\infty(R^N)$ such that the support of ϱ_n converges to 0, ϱ_n is bounded in $L^1(R^N)$ and $\int_{\mathbf{x} \in R^N} \varrho_n\,d\mathbf{x} \to 1$ (usually one chooses $\varrho_1 \in C_c^\infty(R^N)$ having its support in the closed unit ball, with $\varrho_1 \geq 0$ and $\int_{\mathbf{x} \in R^N} \varrho_1\,d\mathbf{x} = 1$, and then one defines ϱ_n by $\varrho_n(\mathbf{x}) = n^N \varrho_1(n\,\mathbf{x})$). For $\varphi \in C_c(R^N)$, one sees easily by using the uniform continuity of φ that $\varrho_n \star \varphi$ converges to φ uniformly; then for $1 \leq p < \infty$, using the density of $C_c(R^N)$ in $L^p(R^N)$, one deduces that

$$\varrho_n \star g \to g \text{ in } L^p(R^N) \text{ for every } g \in L^p(R^N) \text{ if } 1 \leq p < \infty. \tag{14.3}$$

Because the convolution product commutes with translations (which are actually convolutions with Dirac masses), one sees easily that one has

$$\frac{\partial(f \star g)}{\partial x_j} = \frac{\partial f}{\partial x_j} \star g \text{ for } j = 1, \ldots, N, f \in C_c^1(R^N) \text{ and } g \in L^p(R^N), 1 \leq p \leq \infty,$$
(14.4)

by using the uniform continuity of the partial derivatives of f. One then deduces easily that the same formula holds if $f \in W^{1,p}(R^N)$ and $g \in L^1(R^N)$ for example, by using the definition of derivatives in the sense of distributions and observing that $\int_{\mathbf{x} \in R^N} (f \star g)\varphi\,dx = \int_{\mathbf{x} \in R^N} f(\breve{g} \star \varphi)\,dx$ with $\breve{g}(\mathbf{x}) = g(-\mathbf{x})$ for all $\varphi \in C_c^\infty(R^N)$. If $1 \leq p < \infty$ and $f \in W^{1,p}(R^N)$, one first uses truncation to approach f by functions $f_m \in W^{1,p}(R^N)$ with compact support, and then using regularization one deduces that $\varrho_n \star f_m \in C_c^\infty(R^N)$ and converges to f_m in $W^{1,p}(R^N)$, showing that $C_c^\infty(R^N)$ is dense in $W^{1,p}(R^N)$, i.e. $W_0^{1,p}(R^N) = W^{1,p}(R^N)$, which for $p = 2$ is written $H_0^1(R^N) = H^1(R^N)$.

The convolution product has been extended by Laurent SCHWARTZ to some pairs of distributions, but one must be careful with questions of support. For measurable functions one says that $\mathbf{x} \notin support(f)$ if $f(\mathbf{y}) = 0$ a.e. $\mathbf{y} \in B(\mathbf{x}, r)$ for some $r > 0$, and one deduces that in the classical case considered above, one has

$$support(f \star g) \subset support(f) + support(g). \tag{14.5}$$

---

[2] Otto Ludwig HÖLDER, German mathematician, 1859–1937. He worked in Leipzig, Germany.

[3] Johan Ludwig William Valdemar JENSEN, Danish mathematician, 1859–1925. He never held any academic position, and worked for a telephone company.

For functions in $L^1_{loc}(R)$ with support in $[0,\infty)$ one can define the convolution product, which also has support in $[0,\infty)$, and a theorem of TITCHMARSH[4] states that $f \star g = 0$ if and only if one of the functions f or g is 0, and this was extended to N dimensions by Jacques-Louis LIONS, who has proven (at least for bounded supports) that

$$\overline{conv[support(f \star g)]} = \overline{conv[support(f)]} + \overline{conv[support(g)]}, \qquad (14.6)$$

where $\overline{conv[A]}$ is the closed convex hull of A, but it is rarely necessary to use such a refinement.

The remark on the support of convolution products is used in order to define $\varrho_n \star f$ even though f is not defined everywhere: for example, if $f \in W^{1,p}(R^N_+)$ and ϱ_n has its support in the strip $\alpha_n \leq x_N \leq \beta_n$, then $\varrho_n \star f$ is well defined in $x_N > \beta_n$, i.e. if g is an extension of f to R^N, then the restriction of $\varrho_n \star g$ to the open set $x_N > \beta_n$ is always the same, whatever the extension is. After truncation, let $1 \leq p < \infty$ and let $f \in W^{1,p}(R^N_+)$ with compact support, let ϱ_n be a regularizing sequence with $\beta_n < 0$, and let S denote the operator of restriction to R^N_+; then $S(\varrho_n \star f) \to f$ in $L^p(R^N_+)$, and $\frac{\partial[S(\varrho_n \star f)]}{\partial x_j} = S\left(\varrho_n \star \frac{\partial f}{\partial x_j}\right) \to \frac{\partial f}{\partial x_j}$ in $L^p(R^N_+)$. As $\varrho_n \star f$ is C^∞ for $x_N > \beta_n$, one can multiply it by a C^∞ function of x_N alone which is 1 for $x_N > 0$ and 0 for $x_N < \frac{\beta_n}{2}$, and therefore $S(\varrho_n \star f) = S\varphi_n$ with $\varphi_n \in C^\infty_c(R^N)$, and so $S(\varrho_n \star f) \in C^\infty(\overline{R^N_+})$, which is by definition the space of restrictions to R^N_+ of functions of $C^\infty_c(R^N_+)$; therefore $C^\infty(\overline{R^N_+})$ is dense in $W^{1,p}(R^N_+)$ for $1 \leq p < \infty$.

Localization: In order to show that $C^\infty(\overline{\Omega})$ is dense in $W^{1,p}(\Omega)$ for $1 \leq p < \infty$, one needs some regularity of the boundary of Ω, but the first step is to apply an argument of localization. Assume, for example, that the boundary $\partial\Omega$ of Ω is compact and that around each point of the boundary there is a small open ball in which the boundary has an equation $x_N = F(x_1,\ldots,x_{N-1})$ in some orthonormal basis, with F continuous, and that Ω is on one side of the boundary, say $x_N > F(\mathbf{x}')$, where $\mathbf{x}'$ denotes $(x_1,\ldots,x_{N-1})$. The family of all these open balls is an open covering of $\partial\Omega$ from which one extracts a finite covering, to which one adds Ω, in order to have a finite open covering of $\overline{\Omega}$; let $\omega_j, j = 1,\ldots,J$, be that family of open sets, for which there exists a partition of unity, i.e. functions $\theta_j \in C^\infty_c(\omega_j)$ such that $\sum_{j=1}^J \theta_j = 1$ on $\overline{\Omega}$. One decomposes $u \in W^{1,p}(\Omega)$ as $\sum_{j=1}^J \theta_j u$, and one studies each $\theta_j u$ separately. For an index j corresponding to an open set ω_j which does not intersect the boundary, one uses the techniques developed for R^N, i.e. convolution by a regularizing sequence without paying much attention to its support (as long as it converges to 0), while for the other indices j one uses the techniques developed for R^N_+, i.e. convolution by a regularizing sequence with an adapted

---

[4] Edward Charles TITCHMARSH, English mathematician, 1899–1963. He held the Savilian chair at Oxford, England, UK.

support. The problem is the same as for an open set Ω defined globally by an equation $x_N > F(\mathbf{x}')$, with F uniformly continuous, and for $f \in W^{1,p}(\Omega)$ having compact support. By uniform continuity of F, for every $\varepsilon > 0$ there exists $\delta(\varepsilon) > 0$ such that $|\mathbf{x}' - \mathbf{y}'| \le 2\delta(\varepsilon)$ implies $|F(\mathbf{x}') - F(\mathbf{y}')| \le \varepsilon$ (one may assume that $\delta(\varepsilon) \to 0$ as $\varepsilon \to 0$). One chooses a sequence $\varepsilon_n \to 0$, and one chooses the regularizing sequence ϱ_n with its support in $|\mathbf{x}'| \le \delta(\varepsilon_n)$ and $x_N \le -\varepsilon_n - \delta(\varepsilon_n)$, and then $\varrho_n \star f$ is defined in a domain extending at least a distance $\delta(\varepsilon_n)$ beyond the boundary of Ω, and the method used for R_+^N applies (with S denoting the operator of restriction to Ω).

Of course, there are open sets Ω for which $C^\infty(\overline{\Omega})$ is not dense in $W^{1,p}(\Omega)$ for $1 \le p \le \infty$, the plane R^2 to which one removes the nonnegative x axis, for example: functions of $W^{1,p}(\Omega)$ may have different values on both sides of the removed half axis, while functions of $C^\infty(\overline{\Omega})$ must have the same values on both sides.

Extension: For open sets for which $C^\infty(\overline{\Omega})$ is dense in $W^{1,p}(\Omega)$ for $1 \le p < \infty$, one cannot always construct a linear continuous extension from $W^{1,p}(\Omega)$ into $W^{1,p}(R^N)$, i.e. a map P such that $S_\Omega P u = u$ for all $u \in W^{1,p}(\Omega)$ (S_Ω being the operator of restriction to Ω). One must extend each $\theta_j u$, and one is led to add the requirement that each F is Lipschitz continuous; in the case of one open set of equation $x_N > F(\mathbf{x}')$, one defines the extension P by

$$
\begin{aligned}
P u(\mathbf{x}', x_N) &= u(\mathbf{x}', x_N) \text{ if } x_N > F(\mathbf{x}') \\
P u(\mathbf{x}', x_N) &= u(\mathbf{x}', -x_N + 2F(\mathbf{x}')) \text{ if } x_N < F(\mathbf{x}').
\end{aligned}
\tag{14.7}
$$

For $u \in C^\infty(\overline{\Omega})$, one checks that the norm of Pu is controlled by the norm of u, and the important fact is that Pu is continuous at the boundary of Ω.

The extension of $W^{m,p}(\Omega)$ to $W^{m,p}(R^N)$ for $m \ge 2$ is a little more technical (Alberto CALDERÓN had a proof using the Calderón–Zygmund theorem, and therefore it does not cover the cases $p = 1$ and $p = \infty$, but STEIN has constructed an extension which serves for all $W^{m,p}$ and is valid for $1 \le p \le \infty$), but for the case of $\Omega = R_+^N$, after proving that $C^\infty(\overline{\Omega})$ is dense, one defines P by

$$
\begin{aligned}
P u(\mathbf{x}', x_N) &= u(\mathbf{x}', x_N) \text{ if } x_N > 0 \\
P u(\mathbf{x}', x_N) &= \sum_{j=1}^m a_j u(\mathbf{x}', -b_j x_N) + \text{ if } x_N < 0,
\end{aligned}
\tag{14.8}
$$

with distinct $b_j > 0$ for $j = 1, \ldots, m$. The continuity of Pu (or its partial derivative in x_j for $j < N$) at $x_N = 0$ requires $\sum_{j=1}^m a_j = 1$. The continuity of $\frac{\partial Pu}{\partial x_N}$ at $x_N = 0$ requires $\sum_{j=1}^m a_j b_j = -1$, and more generally, the continuity of $\frac{\partial^k Pu}{\partial x_N^k}$ at $x_N = 0$ requires

$$
\sum_{j=1}^m a_j(-b_j)^k = 1 \text{ for } k = 0, \ldots, m - 1.
\tag{14.9}
$$

As the b_j are distinct, the values of a_j for $j = 1, \ldots, m$ are determined by the system of equations corresponding to the continuity for $k = 0, \ldots, m-1$, which is a Vandermonde[5] system.

The extension property does not hold for some open sets with F only Hölder continuous with exponent $\alpha < 1$. The following counter-example relies on the Sobolev embedding theorem, i.e. $H^1(R^2) \subset L^p(R^2)$ for all $p < \infty$; if there exists a continuous extension P from $H^1(\Omega)$ into $H^1(R^2)$, it implies that $H^1(\Omega) \subset L^p(\Omega)$ for all $p < \infty$, and the counter-example consists in showing a function of $H^1(\Omega)$ which does not belong to all $L^p(\Omega)$. For example, the case $\Omega = \{(x,y) \mid 0 < x < 1, 0 < y < x^\gamma\}$ with $\gamma > 1$, corresponds to F being Hölder continuous with exponent $\frac{1}{\gamma}$, and one looks for a counter-example u of the form $u(x,y) = x^\beta$; then $u \in L^p(\Omega)$ if and only if $p\beta + \gamma > -1$ and $u \in H^1(\Omega)$ if and only if $2(\beta - 1) + \gamma > -1$, and therefore one has $u \in H^1(\Omega)$, but $u \notin L^p(\Omega)$ if one can find β satisfying $\frac{-\gamma+1}{2} < \beta < \frac{-\gamma-1}{p}$, and this is possible for $p > \frac{2(\gamma+1)}{\gamma-1}$.

[Taught on Friday February 12, 1999.]

---

[5] Alexandre Théophile VANDERMONDE, French mathematician, 1735–1796. He worked in Paris, France.

15

Sobolev spaces V

We have seen in Lemma 13.4 that $X(R^N) = L^2(R^N)$, and we consider now the case of $X(R^N_+)$. We have noticed that $X(\Omega)$ is a *local space*, i.e. $u \in X(\Omega)$ implies $\varphi u \in X(\Omega)$ for every $\varphi \in C^\infty_c(R^N)$, because $u \in H^{-1}(\Omega)$ implies $\varphi u \in H^{-1}(\Omega)$, and this kind of property is seen by duality: for $v \in C^\infty_c(\Omega)$, one has $\langle \varphi u, v \rangle = \langle u, \varphi v \rangle$ and therefore the key point is that multiplication by φ is continuous from $H^1_0(\Omega)$ into itself (it is sufficient that $\varphi \in W^{1,\infty}(\Omega)$, or if one uses Sobolev embedding theorem, that $\varphi \in W^{1,N}(\Omega) \cap L^\infty(\Omega)$ for $N \geq 3$, $\varphi \in W^{1,p}(\Omega)$ with $p > 2$ for $N = 2$ and with $p = 2$ for $N = 1$).

After the preceding localization argument, one proves that $C^\infty\big(\overline{R^N_+}\big)$ is dense in $X(R^N_+)$ in a similar way to $H^1(R^N_+)$, but a little care is useful if one wants to avoid using convolution of distributions. Every $u \in H^{-1}(R^N_+)$ can be written $f_0 + \sum_{i=1}^N \frac{\partial f_i}{\partial x_i}$, with $f_0, \ldots, f_N \in L^2(R^N_+)$, and one may consider that the f_j are extended for $x_N < 0$, but the precise extension will not be used; one chooses a regularizing sequence ϱ_n with support in $\alpha_n \leq x_N \leq \beta_n$ with $\beta_n < 0$, and one chooses a C^∞ function ψ_n of x_N alone, which is 1 for $x_N \geq 0$ and 0 for $x_N < \frac{\beta_n}{2}$, and one approaches u by $\psi_n(\varrho_n \star u)$ restricted to R^N_+. If $\varphi \in C^\infty_c(\Omega)$, then $\langle \psi_n(\varrho_n \star u), \varphi \rangle = \langle \varrho_n \star u, \psi_n \varphi \rangle$, but $\psi_n \varphi = \varphi$, and so it is $\langle u, \breve{\varrho}_n \star \varphi \rangle = \langle f_0 + \sum_{i=1}^N \frac{\partial f_i}{\partial x_i}, \breve{\varrho}_n \star \varphi \rangle = \langle \varrho_n \star f_0, \varphi \rangle + \sum_{i=1}^N \langle \frac{\partial(\varrho_n \star f_i)}{\partial x_i}, \varphi \rangle$, and as the restriction of $\varrho_n \star f_j$ to R^N_+ converges strongly to f_j in $L^2(R^N_+)$, one sees that the restriction of $\psi_n(\varrho_n \star u)$ converges to u strongly in $H^{-1}(R^N_+)$. Then one notices that $\langle \frac{\partial[\psi_n(\varrho_n \star u)]}{\partial x_i}, \varphi \rangle = -\langle \varrho_n \star u, \psi_n \frac{\partial \varphi}{\partial x_i} \rangle$, and as $\psi_n = 1$ on the support of φ, it is $\langle \varrho_n \star \frac{\partial u}{\partial x_i}, \varphi \rangle$, and therefore $\frac{\partial[\psi_n(\varrho_n \star u)]}{\partial x_i}$ converges to $\frac{\partial u}{\partial x_i}$ strongly in $H^{-1}(R^N_+)$.

In order to construct an extension of $X(R^N_+)$ into $X(R^N)$, it is simpler to work by duality, and we shall construct an adequate restriction from $H^1(R^N)$ into $H^1_0(R^N_+)$. Let S denote the operator of restriction to R^N_+, which is linear continuous from $H^{-1}(R^N)$ into $H^{-1}(R^N_+)$, as it is the transpose of the extension by 0 (denoted $\tilde{\cdot}$), which is linear continuous from $H^1_0(R^N_+)$ into $H^1(R^N)$. An extension is an operator P such that $SP = identity$, then by transposition

$P^{T\tilde{}} = identity$. We are looking then for an operator from $H^1(R^N)$ into $H_0^1(R_+^N)$ such that if $\varphi \in H_0^1(R_+^N)$ then $P\tilde{\varphi} = \varphi$ on R_+^N. Of course, we define the transposed operator $Q = P^T$ on $C_c^\infty(R^N)$, which is dense in $H^1(R^N)$, the continuity of Q being easy to check when one uses the norms in H^1:

$$Q\,u(\mathbf{x}', x_N) = u(\mathbf{x}', x_N) + \sum_{j=1}^{2} a_j u(\mathbf{x}', -b_j x_N) \text{ for } x_N > 0, \qquad (15.1)$$

with $b_1, b_2 > 0$ and $b_1 \neq b_2$. Q is obviously linear continuous from $H^1(R^N)$ into $H^1(R_+^N)$ and satisfies $Q\tilde{\varphi} = \varphi$ for every $\varphi \in H_0^1(R_+^N)$, but one needs the condition $1 + a_1 + a_2 = 0$ in order to ensure that Q maps $H^1(R^N)$ into $H_0^1(R_+^N)$. For $j < N$, one has $\frac{\partial(Q\varphi)}{\partial x_j} = Q\left(\frac{\partial\varphi}{\partial x_j}\right)$ for all $\varphi \in C_c^\infty(R^N)$, but for $j = N$ we introduce the operator R defined by

$$R\,u(\mathbf{x}', x_N) = u(\mathbf{x}', x_N) - \sum_{j=1}^{2} \frac{a_j}{b_j} u(\mathbf{x}', -b_j x_N) \text{ for } x_N > 0, \qquad (15.2)$$

and one adds the condition $1 - \frac{a_1}{b_1} - \frac{a_2}{b_2} = 0$ (and so a_1 and a_2 are determined), in order to ensure that R maps $H^1(R^N)$ into $H_0^1(R_+^N)$, and one has $\frac{\partial(R\varphi)}{\partial x_N} = Q\left(\frac{\partial\varphi}{\partial x_N}\right)$. Therefore, using $P = Q^T$, if $u \in X(R_+^N)$, one has

$$\begin{aligned}
\langle P\,u, \varphi \rangle &= \langle u, Q\varphi \rangle \\
\left\langle \frac{\partial(P\,u)}{\partial x_i}, \varphi \right\rangle &= -\left\langle P\,u, \frac{\partial\varphi}{\partial x_i} \right\rangle = -\left\langle u, Q\frac{\partial\varphi}{\partial x_i} \right\rangle = -\left\langle u, \frac{\partial(Q\varphi)}{\partial x_i} \right\rangle = \left\langle \frac{\partial u}{\partial x_i}, Q\varphi \right\rangle \\
\left\langle \frac{\partial(P\,u)}{\partial x_N}, \varphi \right\rangle &= -\left\langle P\,u, \frac{\partial\varphi}{\partial x_N} \right\rangle = -\left\langle u, Q\frac{\partial\varphi}{\partial x_N} \right\rangle = -\left\langle u, \frac{\partial(R\varphi)}{\partial x_N} \right\rangle = \left\langle \frac{\partial u}{\partial x_N}, R\varphi \right\rangle,
\end{aligned}$$
$$(15.3)$$

and this proves that $P\,u$ belongs to $X(R^N)$, which is $L^2(R^N)$, and therefore its restriction u belongs to $L^2(R_+^N)$.

I shall discuss the proof that $X(\Omega) = L^2(\Omega)$ for bounded domains with Lipschitz boundary in Lecture 23. I had noticed in the fall[1] that it is not true for domains of the form $\{(x, y) \mid 0 < x < 1, 0 < y < x^2\}$, and therefore that one should not expect the "pressure" to belong to $L^2(\Omega)$ for such domains. François MURAT informed me during his recent visit that it had also been noticed by Giuseppe GEYMONAT and Gianni GILARDI that $X(\Omega) \neq L^2(\Omega)$ for a particular domain that they had constructed for showing that Korn's inequality may fail in non-Lipschitz domains.[2] If $X(\Omega) = L^2(\Omega)$ algebraically, then the norms are equivalent by the closed graph theorem as the injection

---

[1] The course having been taught in the spring of 1999, it was in the fall of 1998 that I had made this observation, and the motivation for having thought about this question was to consider particles in a fluid and wonder about the case where the particles touch, as this may create for the fluid a domain which does not have a Lipschitz boundary.

[2] My proof was not shown in the original notes. It consists in showing that if $u \in H_0^1(\Omega; R^2)$ and $f = div\,u$, then f satisfies some relation that not all functions

of $L^2(\Omega)$ into $X(\Omega)$ is always continuous. One hypothesis of the equivalence lemma 13.3 is that the injection of $L^2(\Omega)$ into $H^{-1}(\Omega)$ is compact, and by transposition it is equivalent to prove that the injection of $H^1_0(\Omega)$ into $L^2(\Omega)$ is compact:

Lemma 15.1. *If $meas(\Omega) < \infty$, then the injection of $H^1_0(\Omega)$ into $L^2(\Omega)$ is compact.*

Proof: We extend functions by 0 outside Ω; we have already noticed that Poincaré's inequality holds for such domains, and we follow a similar proof using the Fourier transform (an argument which I heard attributed to Lars HÖRMANDER,[3] different from the approach which I had learnt from my advisor). We consider a bounded sequence in $H^1_0(\Omega)$ and we want to show that one can extract a subsequence converging strongly in $L^2(\Omega)$. Because $L^2(\Omega)$ is *separable, the weak topology is metrizable on bounded sets*, and one can extract a weakly converging sequence in $L^2(\Omega)$, and by translation, one may assume then that the sequence u_n converges weakly to 0 in $L^2(\Omega)$, that it is bounded in $H^1_0(\Omega)$ and one wants to prove that it converges strongly to 0 in $L^2(\Omega)$. One has $\mathcal{F}u_n(\boldsymbol{\xi}) = \int_{\mathbf{x} \in R^N} u_n(\mathbf{x}) e^{-2i\pi(\mathbf{x}.\boldsymbol{\xi})} \, d\mathbf{x} = \int_\Omega u_n(\mathbf{x}) e^{-2i\pi(\mathbf{x}.\boldsymbol{\xi})} \, d\mathbf{x} \to 0$, as it is the $L^2(\Omega)$ scalar product of u_n with a fixed function which belongs to $L^2(\Omega)$ by the hypothesis $meas(\Omega) < \infty$; on the other hand, one has $|\mathcal{F}u_n(\boldsymbol{\xi})| \leq C$, and by the Lebesgue dominated convergence theorem one has $\int_{|\boldsymbol{\xi}| \leq \varrho} |\mathcal{F}u_n(\boldsymbol{\xi})|^2 \, d\boldsymbol{\xi} \to 0$ for any $\varrho < \infty$. Because u_n is bounded in $H^1_0(\Omega)$, one has $\int_{\boldsymbol{\xi} \in R^N} |\boldsymbol{\xi}|^2 |\mathcal{F}u_n(\boldsymbol{\xi})|^2 \, d\boldsymbol{\xi} \leq C$, and therefore $\int_{|\boldsymbol{\xi}| \geq \varrho} |\mathcal{F}u_n(\boldsymbol{\xi})|^2 \, d\boldsymbol{\xi} \leq C/\varrho^2$; one deduces that $\limsup \int_{\mathbf{x} \in R^N} |u_n - u_m|^2 \, d\mathbf{x} = \limsup \int_{\boldsymbol{\xi} \in R^N} |\mathcal{F}u_n - \mathcal{F}u_m|^2 \, d\boldsymbol{\xi} \leq \limsup \int_{|\boldsymbol{\xi}| \leq \varrho} |\mathcal{F}u_n - \mathcal{F}u_m|^2 \, d\boldsymbol{\xi} + \limsup \int_{|\boldsymbol{\xi}| \geq \varrho} |\mathcal{F}u_n - \mathcal{F}u_m|^2 \, d\boldsymbol{\xi} \leq 4C/\varrho^2$, and therefore $\limsup \int_{\mathbf{x} \in R^N} |u_n - u_m|^2 \, d\mathbf{x} = 0$, so that u_n is a Cauchy sequence and converges strongly (to 0).∎

For a (connected) bounded open set Ω with Lipschitz boundary, the conditions of equivalence lemma 13.3 are satisfied with $E_1 = L^2(\Omega)$, $A = grad$, $E_2 = H^{-1}(\Omega; R^N)$, B the injection of $L^2(\Omega)$ into $E_3 = H^{-1}(\Omega)$. The range of A is closed, and therefore equal to its closure, which is always the space of

in $L^2(\Omega)$ (with integral 0) satisfy. Let $g(x) = \int_0^{x^2} f(x, y) \, dy$; then by Green's formula $\int_0^z g(x) \, dx = \int_0^{z^2} u_1(z, y) \, dy$. By Poincaré's inequality, one has $\left(\int_0^Y h(y) \, dy \right)^2 \leq CY^3 \int_0^Y \left| \frac{\partial h}{\partial y} \right|^2 \, dy$ if $h(0) = h(Y) = 0$, and therefore one has $\int_0^1 \frac{1}{z^6} \left(\int_0^z g(x) \, dx \right)^2 \, dz < \infty$. If f has a singularity near the origin in x^α (with $\alpha < 0$), then $f \in L^2(\Omega)$ if and only if $2\alpha + 2 > -1$, i.e. $\alpha > -\frac{3}{2}$, but in this case g behaves as $x^{\alpha+2}$ and $\int_0^z g(x) \, dx$ behaves as $z^{\alpha+3}$ and the condition to be satisfied is then $2(\alpha + 3) - 6 > -1$, i.e. $\alpha > -\frac{1}{2}$.

[3] Lars HÖRMANDER, Swedish mathematician, born in 1931. He received the Fields Medal in 1962, and the Wolf Prize in 1988. He worked at Lund University, Lund, Sweden.

$T \in H^{-1}(\Omega; R^N)$ orthogonal to the kernel of A^T, i.e. $W = \{u \in H_0^1(\Omega; R^N) \mid div\, u = 0\}$, as $A^T = -div$. Of course, one gets as a corollary that the range of A^T is closed, and therefore equal to its closure, which is always the subspace of $f \in L^2(\Omega)$ with $\int_\Omega f\, dx = 0$, i.e. the orthogonal of the kernel of A, which are the constants.

The equivalence lemma 13.3 says a little more: an equivalent norm on $L^2(\Omega)$ is $||grad\, u||_{H^{-1}(\Omega)} + p(u)$, where p is any semi-norm such that $p(1) \neq 0$, for example $p(u) = |\int_\Omega u\, dx|$. Therefore, on the subspace of $v \in L^2(\Omega)$ with $\int_\Omega v\, dx = 0$, one has $||v||_{L^2(\Omega)} \approx ||grad\, v||_{H^{-1}(\Omega)}$. An application of this result is that one can now prove that in the limit $\lambda \to \infty$, the sequence $-\lambda\, div\, u^\lambda$ converges strongly in $L^2(\Omega)$ to a "pressure", as its integral on Ω is 0 (because $u^\lambda \in H_0^1(\Omega; R^N)$) and its gradient converges strongly in $H^{-1}(\Omega; R^N)$ by the equation, as we have already proven that $u^\lambda \to u^\infty$ strongly in $H_0^1(\Omega; R^N)$.

[Taught on Monday February 15, 1999.]

16

Sobolev embedding theorem

We have considered the stationary Stokes equation by taking advantage of the similarity with the equation of stationary linearized elasticity, but the time-dependent equations are not similar, due to the fact that the unknown u in the Stokes equation is a *velocity*, while for linearized elasticity it is a *displacement*. We have also used a coefficient μ, bounded below but variable with $\mathbf{x}$, and that is not physical in general: in the Lagrangian point of view some parameters may depend upon the initial position $\boldsymbol{\xi}$, but in the Eulerian point of view these properties are transported by the flow, and unlike for linearized elasticity where the Lagrangian and Eulerian point of views have been mixed, one mostly uses the Eulerian point of view for fluids.

There is at least one case where μ may reasonably depend upon $\mathbf{x}$: *Poiseuille*[1] *flows*. If different *nonmiscible* fluids move in an infinite cylinder, there are particular *stationary* solutions (of the Stokes equation as well as of the Navier–Stokes equation, because the nonlinearity vanishes for these solutions), where the velocities are all parallel to the axis of the cylinder, the gradient of the pressure is constant, parallel to the axis, and the velocity satisfies an equation $-div(\mu\,grad\,u) = Constant$ with $u \in H_0^1(\omega)$, where ω is the section of the cylinder. Each choice of μ as a function of $\mathbf{x}$ (bounded below and above) gives rise to a Poiseuille flow, but not all these solutions are stable for the evolution problem; these questions have been investigated by Dan JOSEPH and Michael RENARDY,[2] for *Newtonian fluids* (water is used in pipelines carrying oil, and the stable configuration has the water near the boundary, serving as *lubricant*), or *non-Newtonian fluids* (for extrusion of *molten polymers*).

We shall consider the viscosity μ and the density ϱ_0 to be constant (leaving for later the study of mixtures of fluids). As $\sigma_{ij} = 2\mu\,\varepsilon_{ij} - p\delta_{ij}$ for i,

---

[1] Jean-Louis Marie POISEUILLE, French physician, 1797–1869. He worked in Paris, France.

[2] Michael J. RENARDY, German-born mathematician, born in 1955. He works at Virginia Polytechnic Institute and State University, Blacksburg, VA.

$j = 1, \ldots, N$, one finds that $-\sum_{j=1}^{N} \frac{\partial \sigma_{ij}}{\partial x_j} = -\mu \Delta u_i + \frac{\partial p}{\partial x_i}$, and the stationary Navier–Stokes equation is

$$
\begin{aligned}
&\varrho_0 \sum_{j=1}^{N} u_j \frac{\partial u_i}{\partial x_j} - \mu \Delta u_i + \frac{\partial p}{\partial x_i} = f_i \text{ in } \Omega, \\
&\text{div } u = 0 \text{ in } \Omega, \\
&u \in H_0^1(\Omega; R^N).
\end{aligned}
\tag{16.1}
$$

As $div\, u = 0$, the nonlinear term $\sum_{j=1}^{N} u_j \frac{\partial u_i}{\partial x_j}$ can be written as $\sum_{j=1}^{N} \frac{\partial (u_j u_i)}{\partial x_j}$. Of course, apart from the gravitational force, which can be incorporated into the pressure, there are usually no body forces, and the fluid is usually put into motion because part of the boundary moves, but at this stage we shall only consider homogeneous Dirichlet conditions. It is the *kinematic viscosity* $\nu = \frac{\mu}{\varrho_0}$ which really appears in the equation, and it must be noticed that although water is more viscous than air, the kinematic viscosities are in reverse order, 10^{-6} m^2 s^{-1} for water, and 1.4×10^{-5} m^2 s^{-1} for air (at atmospheric pressure). Contrary to the case of Stokes equation, which is linear and where the value of N does not matter much, the value of N is important for the Navier–Stokes equation, because of the nonlinearity; one way the value of N arises is through the following *Sobolev embedding theorem.*

Theorem 16.1. *(Sobolev embedding theorem) If $1 \le p < N$, then $W^{1,p}(R^N)$ is continuously embedded into $L^{p^*}(R^N)$ with $\frac{1}{p^*} = \frac{1}{p} - \frac{1}{N}$.*

$W^{1,1}(R) \subset C_0(R)$, the space of continuous (bounded) functions tending to 0 at infinity.

For $p = N \ge 2$, $W^{1,N}(R^N) \subset L^q(R^N)$ for all $q \in [N, \infty)$.

For $p > N$, $W^{1,p}(R^N) \subset C^{0,\alpha}(R^N)$ with $\alpha = 1 - \frac{N}{p}$.

The original proof of Sergei SOBOLEV started by using the formula $u = \sum_{i=1}^{N} \frac{\partial u}{\partial x_i} \star \frac{\partial E}{\partial x_i}$, where E is a radial *elementary* solution of Δ; noticing that $\left| \frac{\partial E}{\partial x_i} \right| = O\left(\frac{1}{|x|^{N-1}} \right)$, he extended *Young's*[3] inequality for convolution to the case of convolution with $\frac{1}{|x|^{\alpha}}$, using nonincreasing radial rearrangements. His results have been slightly improved for the case $p < N$ by Jaak PEETRE using the larger family of interpolation spaces known as the Lorentz[4] spaces; for the case $p = N$, Fritz JOHN and Louis NIRENBERG have introduced the space BMO (Bounded Mean Oscillation) and noticed that $grad\, u \in L^N(R^N; R^N)$ implies $u \in BMO(R^N)$, and $u \in BMO(R^N)$ implies that there exists $\varepsilon > 0$ such that $e^{\varepsilon |u|} \in L^1_{loc}(R^N)$, but Neil TRUDINGER[5] has shown that

---

[3] William Henry YOUNG, English-born mathematician, 1863–1942. He worked in Lausanne, Switzerland. He worked on so many problems with his wife, Grace CHISHOLM-YOUNG, English-born mathematician, 1868–1944, that it is difficult to know if any result attributed to him is a joint work with his wife or not.

[4] George Gunther LORENTZ, Russian-born mathematician, born in 1910. He worked at University of Texas, Austin, TX.

[5] Neil S. TRUDINGER, Australian mathematician, born in 1942. He works at Australian National University, Canberra, Australia.

$grad\, u \in L^N(R^N; R^N)$ implies $e^{C\,|u|^{N'}} \in L^1_{loc}(R^N)$ for every $C > 0$, where $N' = \frac{N}{N-1}$, the conjugate exponent of N. Sergei SOBOLEV's method only applies to cases where all the derivatives are in the same functional space, and in the case where the derivatives belong to the same Lorentz space, the improvements are due to Jaak PEETRE for the case $p < N$ and to Haïm BREZIS[6,7] and Stephen WAINGER[8] for the case $p = N$, using a formula of O'NEIL[9] for the nonincreasing rearrangement of a convolution product.

A second method, which applies to cases where the derivatives are in different L^r spaces, has been developed by Emilio GAGLIARDO, and independently by Louis NIRENBERG, and maybe also by Olga LADYZHENSKAYA (there is another method which I have introduced which also applies to the case of derivatives in different Lorentz spaces). I show this method in the example $H^1(R^3) \subset L^6(R^3)$, proving the estimates for $u \in C_c^\infty(R^3)$. One has $|u|^4(\mathbf{x}) = \int_{-\infty}^{x_1} \left(4u^3 \frac{\partial u}{\partial x_1}\right)(t, x_2, x_3)\, dt = -\int_{x_1}^{+\infty} \left(4u^3 \frac{\partial u}{\partial x_1}\right)(t, x_2, x_3)\, dt$, and therefore

$$|u|^4(\mathbf{x}) \le 2 \int_{-\infty}^{+\infty} \left(|u|^3 \left|\frac{\partial u}{\partial x_1}\right|\right)(t, x_2, x_3)\, dt = F_1(x_2, x_3),$$
$$\text{and } \int_{R^2} F_1\, dx_2\, dx_3 \le 2||u||_{L^6}^3 \left\|\frac{\partial u}{\partial x_1}\right\|_{L^2}. \tag{16.2}$$

One has similar inequalities $|u|^4(\mathbf{x}) \le F_2(x_1, x_3)$ and $|u|^4(\mathbf{x}) \le F_3(x_1, x_2)$, and therefore

$$|u|^6(\mathbf{x}) \le G_1(x_2, x_3)G_2(x_1, x_3)G_3(x_1, x_2), \text{ with } G_i = \sqrt{F_i}. \tag{16.3}$$

Using $H(x_1, x_2) = \int_R G_1(x_2, x_3)G_2(x_1, x_3)\, dx_3$ and the Cauchy–Schwarz inequality, one has $H^2(x_1, x_2) \le \int_R G_1^2(x_2, x_3)\, dx_3 \int_R G_2^2(x_1, x_3)\, dx_3$, which implies $\int_{R^2} H^2(x_1, x_2)\, dx_1\, dx_2 \le \int_{R^2} G_1^2(x_2, x_3)\, dx_2\, dx_3 \int_{R^2} G_2^2(x_1, x_3)\, dx_1\, dx_3$; finally, by applying the Cauchy–Schwarz inequality to $H\, G_3$ one obtains

$$\int_{\mathbf{x} \in R^3} |u|^6\, d\mathbf{x} \le ||G_1||_{L^2(R^2)}||G_2||_{L^2(R^2)}||G_3||_{L^2(R^2)}. \tag{16.4}$$

Using inequalities like $||G_1||_{L^2(R^2)}^2 = \int_{R^2} F_1\, dx_2\, dx_3 = 2 \int_{\mathbf{x} \in R^3} |u^3| \left|\frac{\partial u}{\partial x_1}\right| d\mathbf{x} \le 2\left\|\frac{\partial u}{\partial x_1}\right\|_{L^2} \left(\int_{R^3} |u|^6\, d\mathbf{x}\right)^{1/2}$, one deduces that

$$||u||_{L^6} \le 2\left\|\frac{\partial u}{\partial x_1}\right\|_{L^2}^{1/3} \left\|\frac{\partial u}{\partial x_2}\right\|_{L^2}^{1/3} \left\|\frac{\partial u}{\partial x_3}\right\|_{L^2}^{1/3}. \tag{16.5}$$

---

[6] Haïm R. BREZIS, French mathematician, born in 1944. He works at Université Paris VI (Pierre et Marie CURIE), Paris (and it seems at RUTGERS University, Piscataway, NJ.).

[7] Henry RUTGERS, American colonel, 1745–1830.

[8] Stephen WAINGER, American mathematician, born in 1936. He works at University of Wisconsin, Madison, WI.

[9] Richard C. O'NEIL, American mathematician. He works at University of Albany, Albany, NY.

Similar inequalities in R^N are proven by using Hölder's inequality, and showing by induction on N that if G_i is independent of x_i and belongs to L^{N-1} for its $N-1$ variables, then $G_1 \ldots G_N \in L^1(R^N)$ and $||G_1 \ldots G_N||_{L^1} \leq ||G_1||_{L^{N-1}} \ldots ||G_N||_{L^{N-1}}$.

I shall prove the other parts of the theorem, or generalizations, when it will become necessary.

A way to find solutions of the stationary Navier–Stokes equation is to use a *fixed point argument*. There are different ways to define the mapping for which one seeks a fixed point, and let us start by studying Φ defined by

$$u = \Phi(v) \text{ is the solution } u \in W \text{ of } \varrho_0 \sum_{j=1}^{N} \frac{\partial(v_j v_i)}{\partial x_j} - \mu \Delta u_i + \frac{\partial p}{\partial x_i} = f_i \text{ in } \Omega,$$

(16.6)

for $f \in H^{-1}(\Omega; R^N)$, the condition $div\, u = 0$ being included in the definition of the space W, defined in (12.6). A natural condition is to take $v \in L^4(\Omega; R^N)$ so that each term $\frac{\partial(v_j v_i)}{\partial x_j}$ belongs to $H^{-1}(\Omega)$, but as one finds $u \in H_0^1(\Omega; R^N)$, it is only for $N \leq 4$ that one is sure to find $u \in L^4(\Omega; R^N)$. One assumes that Poincaré's inequality holds, so that $||v||_{H_0^1(\Omega)}$ can be changed to $||grad\, v||_{L^2(\Omega)}$, and the estimate for u becomes

$$\mu \sum_{i=1}^{N} ||grad\, u_i||_{L^2}^2 \leq \sum_{i=1}^{N} ||f_i||_{H^{-1}} ||grad\, u_i||_{L^2} + \varrho_0 \sum_{i,j=1}^{N} ||v_j v_i||_{L^2} \left|\left| \frac{\partial u_i}{\partial x_j} \right|\right|_{L^2}.$$

(16.7)

A simpler inequality is obtained if one uses linearity, adding the bound for f when $v = 0$ and the bound for v when $f = 0$, i.e.

$$\mu \left(\sum_{i=1}^{N} ||grad\, u_i||_{L^2}^2 \right)^{1/2} \leq \left(\sum_{i=1}^{N} ||f_i||_{H^{-1}}^2 \right)^{1/2} + \varrho_0 \sum_{j=1}^{N} ||v_j||_{L^4}^2,$$

(16.8)

as $\sum_{i,j=1}^{N} ||v_j v_i||_{L^2}^2 \leq \sum_{i,j=1}^{N} ||v_j||_{L^4}^2 ||v_i||_{L^4}^2 = \left(\sum_{j=1}^{N} ||v_j||_{L^4}^2 \right)^2$. If

$$||v||_{L^4(\Omega)} \leq \gamma ||grad\, v||_{L^2(\Omega)} \text{ for all } v \in H_0^1(\Omega),$$

(16.9)

then putting $X^2 = \sum_{j=1}^{N} ||grad\, v_i||_{L^2}^2$ and $A^2 = \sum_{i=1}^{N} ||f_i||_{H^{-1}}^2$, one obtains $\mu \left(\sum_{i=1}^{N} ||grad\, u_i||_{L^2}^2 \right)^{1/2} \leq A + \varrho_0 \gamma^2 X^2$, and in order to find a ball which is sent by Φ into itself it is enough to find $X_0 > 0$ such that $A + \varrho_0 \gamma^2 X_0^2 \leq \mu X_0$, and therefore

$$\text{if } \left(\sum_{i=1}^{N} ||f_i||_{H^{-1}}^2 \right)^{1/2} \leq \frac{\mu^2}{4\varrho_0 \gamma^2}, \text{ then } \Phi \text{ maps } C \text{ into } C \text{ with}$$
$$C = \left\{ v \in W \mid \left(\sum_{j=1}^{N} ||grad\, v_j||_{L^2}^2 \right)^{1/2} \leq \frac{\mu}{2\varrho_0 \gamma^2} \right\}.$$

(16.10)

In order to check if Φ is a strict contraction on C (or just a contraction, which ensures that Φ has at least one fixed point, not necessarily unique, because

W is a Hilbert space, as shown in Lemma 16.2), one takes $v' \in C$, and the estimate becomes

$$\mu \sum_{i=1}^{N} ||grad(u_i - u_i')||_{L^2}^2 \le \varrho_0 \sum_{i,j=1}^{N} ||v_j v_i - v_j' v_i'||_{L^2} \left|\left|\frac{\partial(u_i - u_i')}{\partial x_j}\right|\right|_{L^2}, \quad (16.11)$$

so

$$\mu\left(\sum_{i=1}^{N} ||grad(u_i - u_i')||_{L^2}^2\right)^{1/2} \le \varrho_0 \left[\sum_{i,j=1}^{N} ||v_j v_i - v_j' v_i'||_{L^2}^2\right]^{1/2}$$
$$\le \varrho_0 \left[\sum_{i,j=1}^{N} \left(||v_j||_{L^4}||v_i - v_i'||_{L^4} + ||v_i'||_{L^4}||v_j - v_j'||_{L^4}\right)^2\right]^{1/2}$$
$$\le \varrho_0 \left[\sum_{i,j=1}^{N} ||v_j||_{L^4}^2||v_i - v_i'||_{L^4}^2\right]^{1/2} + \varrho_0 \left[\sum_{i,j=1}^{N} ||v_i'||_{L^4}^2||v_j - v_j'||_{L^4}^2\right]^{1/2}$$
$$= 2\varrho_0 \left(\sum_{j=1}^{N} ||v_j||_{L^4}^2\right)^{1/2}\left(\sum_{i=1}^{N} ||v_i - v_i'||_{L^4}^2\right)^{1/2} \le \frac{\mu}{\gamma}\left(\sum_{i=1}^{N} ||v_i - v_i'||_{L^4}^2\right)^{1/2}$$
$$\le \mu\left(\sum_{i=1}^{N} ||grad(v_i - v_i')||_{L^2}^2\right)^{1/2},$$

$$(16.12)$$

showing that Φ is a contraction on C; one gets a strict contraction by assuming that $\left(\sum_{i=1}^{N} ||f_i||_{H^{-1}}^2\right)^{1/2} \le K < \frac{\mu^2}{4\varrho_0\gamma^2}$, so that one can lower the bound in the definition of C.

A second method for looking for a fixed point is to use the mapping Ψ defined as

$$u = \Psi(v) \text{ is the solution } u \in W \text{ of } \varrho_0 \sum_{j=1}^{N} v_j\frac{\partial u_i}{\partial x_j} - \mu\,\Delta u_i + \frac{\partial p}{\partial x_i} = f_i \text{ in } \Omega,$$

$$(16.13)$$

which is a linear equation with variable coefficients; it can be treated by the Lax–Milgram lemma 11.2 for $N \le 4$, the variational formulation corresponding to the bilinear form a_v defined by

$$a_v(u, \varphi) = \int_{\Omega}\left(\varrho_0 \sum_{j=1}^{N} v_j\frac{\partial u_i}{\partial x_j}\varphi_i + \mu \sum_{i,j=1}^{N} \frac{\partial u_i}{\partial x_j}\frac{\partial \varphi_i}{\partial x_j}\right) d\mathbf{x}, \quad (16.14)$$

and the important fact is that

$$a_v(u, u) = \mu \sum_{i=1}^{N} ||grad\,u_i||_{L^2}^2 \text{ if } v \in W, \quad (16.15)$$

because $\sum_{j=1}^{N} v_j\frac{\partial u_i}{\partial x_j}u_i = \frac{1}{2}\sum_{j=1}^{N} \frac{\partial(v_j u_i^2)}{\partial x_j} +$ a term containing $div\,v$, so that its integral over Ω is 0. If one defines the trilinear form

$$b(v, u, \varphi) = \sum_{i,j=1}^{N} \int_{\Omega} v_j\frac{\partial u_i}{\partial x_j}\varphi_i\,d\mathbf{x} \text{ for } v, \varphi \in L^4(\Omega; R^N), u \in H^1(\Omega; R^N),$$

$$(16.16)$$

then for $N \le 4$ (in order to have $H_0^1(\Omega) \subset L^4(\Omega)$)

$$b(v, u, \varphi) + b(v, \varphi, u) = -\int_\Omega \left(\sum_{i=1}^N u_i \varphi_i\right) div\, v\, d\mathbf{x} \text{ for } u, v, \varphi \in H_0^1(\Omega; R^N),$$

$$(16.17)$$

so that $b(v, u, u) = 0$ for $u, v \in W$. One finds then a bound for u which is independent of v

$$\mu\left(\sum_{i=1}^N \|grad\, u_i\|_{L^2}^2\right)^{1/2} \le \left(\sum_{i=1}^N \|f_i\|_{H^{-1}}^2\right)^{1/2}, \qquad (16.18)$$

i.e. $\mu X \le A$ in the preceding notations. In order to check if Ψ is a strict contraction on the ball $X \le \frac{A}{\mu}$, one takes v' in this ball, and by subtracting $\mu\, a_0(u, \varphi) + \varrho_0 b(v, u, \varphi) = L(\varphi)$ and $\mu\, a_0(u', \varphi) + \varrho_0 b(v', u', \varphi) = L(\varphi)$ and taking $\varphi = u - u'$, one obtains

$$\mu\, a_0(u - u', u - u') = \varrho_0 b(v', u', u - u') - \varrho_0 b(v, u, u - u')$$
$$= \varrho_0 b(v' - v, u', u - u') - \varrho_0 b(v, u - u', u - u') = -\varrho_0 b(v' - v, u - u', u')$$
$$\le \varrho_0 \sum_{i,j=1}^N \|v_j' - v_j\|_{L^4} \left\|\frac{\partial(u_i - u_i')}{\partial x_j}\right\|_{L^2} \|u_i'\|_{L^4},$$

$$(16.19)$$

and therefore

$$\mu\left(\sum_{i=1}^3 \|grad(u_i - u_i')\|_{L^2}^2\right)^{1/2} \le \varrho_0 \left(\sum_{i,j=1}^3 \|v_j' - v_j\|_{L^4}^2 \|u_i'\|_{L^4}^2\right)^{1/2}$$
$$\le \frac{\varrho_0 \gamma A}{\mu}\left(\sum_{i,j=1}^3 \|v_j' - v_j\|_{L^4}^2\right)^{1/2}. \qquad (16.20)$$

One deduces that if $A < \frac{\mu^2}{\varrho_0 \gamma^2}$ there is a unique solution in W, as Ψ is a strict contraction on the ball $\left(\sum_i \|grad(u_i - u_i')\|_{L^2}^2\right)^{1/2} \le \frac{A}{\mu}$, which contains $\Psi(W)$.

Using the *Schauder*[10] *fixed point theorem*, we shall see that there exists a solution without constraint on A. I conclude with the case contractions mentioned before.

Lemma 16.2. *Let $C \ne \emptyset$ be a closed bounded convex set of a Hilbert space H, and let T be a contraction from C into C; then T has at least one fixed point (the set of fixed points in C is a closed convex set). Let $c_0 \in C$, and for $0 \le \theta < 1$ let $x(\theta)$ be the unique fixed point of $x \mapsto (1 - \theta)c_0 + \theta\, T(x)$, then as $\theta \to 1$, $x(\theta)$ converges strongly to $z(c_0)$, which is the fixed point of T in C, which is the nearest from c_0.*

Proof: As $x \mapsto (1 - \theta)c_0 + \theta\, T(x)$ maps C into C and is a strict contraction, it has a unique fixed point $x(\theta)$ (and $x(0) = c_0$). As $x(\theta)$ is bounded, one can extract a sequence $\theta_n \to 1$ such that $x(\theta_n) \rightharpoonup z$ in H weak (and $z \in C$, as closed convex sets are weakly closed). As $x(\theta_n) = (1 - \theta_n)c_0 + \theta_n T(x(\theta_n))$ and C is bounded, one deduces that $x(\theta_n) - T(x(\theta_n)) \to 0$ strongly, and

---

[10] Juliusz Pawel SCHAUDER, Polish mathematician, 1899–1943. He worked in Lwów, then in Poland, now Lvov in Ukraine.

therefore $T\big(x(\theta_n)\big) \rightharpoonup z$. As $M(x) = x - T(x)$ is monotone continuous, this proves that $M(z) = 0$, i.e. $T(z) = z$, as recalled below. If ξ is another fixed point of T in C, then $\big|T\big(x(\theta_n)\big) - \xi\big|^2 = \big|T\big(x(\theta_n)\big) - T(\xi)\big|^2 \le |x(\theta_n) - \xi|^2$, i.e. ξ is nearer to $T\big(x(\theta_n)\big)$ than to $x(\theta_n)$, but $x(\theta_n)$ being between c_0 and $T\big(x(\theta_n)\big)$, one deduces that $|c_0 - x(\theta_n)| \le |c_0 - \xi|$ and therefore $|c_0 - z| \le |c_0 - \xi|$, so that z is the nearest fixed point to c_0; this shows that all the sequence converges weakly to z, but also that the sequence converges strongly as $\limsup |c_0 - x(\theta_n)| \le |c_0 - z|$.∎

M is said to be *monotone* (from a *topological vector space* E into its dual E') if $\langle M(b) - M(a), b - a \rangle \ge 0$ for all a, b; here $E = E' = H$ and as $M(x) = x - T(x)$, one has $(M(b) - M(a), b - a) = |b - a|^2 - (T(b) - T(a), b - a) \ge |b - a|^2 - |T(b) - T(a)|\,|b - a| \ge 0$. M is said to be *hemi-continuous* if $t \mapsto \langle M(a + t\,c), c \rangle$ is continuous on R for every a, c (in the application above, M is Lipschitz continuous). Assuming that M is monotone hemi-continuous, if $e_n \rightharpoonup f$ in E weak, $M(e_n) \rightharpoonup g$ in E' weak $\star$, and $\limsup \langle M(e_n), e_n \rangle \le \langle f, g \rangle$, then $M(f) = g$. Indeed, as $\limsup \langle M(e_n) - M(a), e_n - a \rangle \ge 0$, one deduces that $\langle g - M(a), f - a \rangle \ge 0$ for all a; taking $a = f - \varepsilon c$ with $\varepsilon > 0$ gives $\langle g - M(f - \varepsilon c), c \rangle \ge 0$ and hemi-continuity gives $\langle g - M(f), c \rangle \ge 0$, so that varying c gives $g = M(f)$. A particular case is that $e_n \rightharpoonup f$ in E weak and $M(e_n) \to g$ in E' strong imply $M(f) = g$.

[In the theory of monotone operators, it is usual to call the preceding argument *Minty's*[11] *trick!* If a function is nonnegative and vanishes at a point, the idea that its derivative at this point must be 0 goes back at least to FERMAT[12] in the 17th century, and Minty's trick is exactly that: the hypothesis of hemi-continuity is what is needed for the function $b \mapsto \langle M(b), b - a \rangle$ to be Gateaux[13]-differentiable at a with derivative $M(a)$.]

[Taught on Wednesday February 17, 1999.]

---

[11] George James MINTY Jr., American mathematician, 1929–1986. He worked at Indiana University, Bloomington, IN.

[12] Pierre DE FERMAT, French mathematician, 1601–1665. He worked (as a judge) in Toulouse, France.

[13] René GATEAUX, French mathematician, 1880–1914.

17

Fixed point theorems

For $N \leq 4$ one can show the existence of at least one solution of the stationary Navier–Stokes equation without assuming that the data f_i have a small norm in $H^{-1}(\Omega)$, by using the *Schauder–Tikhonov*[1] *fixed point theorem* for the mapping Ψ defined in (16.13).

For $N \leq 4$, Ψ is continuous from $L^4(\Omega; R^N)$ into W defined in (12.6), which is a subset of $L^4(\Omega; R^N)$, and therefore Ψ maps W into a bounded set of W.

For $N \leq 3$ and $meas(\Omega) < \infty$, the injection from $H_0^1(\Omega)$ into $L^4(\Omega)$ is compact, because $H_0^1(\Omega) \subset L^6(\Omega)$ and the injection of $H_0^1(\Omega)$ into $L^2(\Omega)$ is compact, and therefore the injection of $H_0^1(\Omega)$ into $L^p(\Omega)$ is compact for $p < 6$; indeed, if u^n is bounded in $L^6(\Omega)$ and converges strongly in $L^2(\Omega)$, then by Hölder's inequality $||u^n - u^m||_{L^p} \leq |u^n - u^m||_{L^2}^{1-\theta} ||u^n - u^m||_{L^6}^{\theta}$, with $\frac{1}{p} = \frac{1-\theta}{2} + \frac{\theta}{6}$, and as $\theta < 1$ and u^n is a Cauchy sequence in $L^2(\Omega)$, one deduces that u^n is a Cauchy sequence in $L^p(\Omega)$.

For $N = 4$, the injection from $H_0^1(\Omega)$ into $L^4(\Omega)$ is not compact. More generally, for $\Omega \subset R^N$ and $p < N$, the injection of $W_0^{1,p}(\Omega)$ into $L^{p^*}(\Omega)$ is not compact. In order to show this, let φ be a nonzero function in $C_c^\infty(R^N)$, and for some $\mathbf{z} \in \Omega$ let u^n be defined by $u_n(\mathbf{x}) = n^{N/p^*} \varphi(n(\mathbf{x} - \mathbf{z}))$, so that for n large enough $u^n \in C_c^\infty(\Omega)$. One checks easily that u^n is bounded in $L^{p^*}(\Omega)$ while $grad\, u^n$ is bounded in $L^p(\Omega)$ (because $1 + N/p^* = N/p$), and therefore if the injection of $W_0^{1,p}(\Omega)$ into $L^{p^*}(\Omega)$ was compact, one could extract from u^n a subsequence converging strongly in $L^{p^*}(\Omega)$, and therefore $|u^n|^{p^*}$ would converge strongly in $L^1(\Omega)$, which is not the case, as $|u^n|^{p^*}$ converges in the sense of distributions (or weakly $\star$ in the sense of measures) to $A\,\delta_z$, with $A = \int_{\mathbf{x} \in R^N} |\varphi|^{p^*} d\mathbf{x}$.

---

[1] Andreï Nikolaevich TIKHONOV, Russian mathematician, 1906–1993. He worked in Moscow, Russia.

For $N \leq 3$, the closed ball of W containing $\Psi(W)$ is compact in $L^4(\Omega; R^N)$, and as Ψ is continuous from $L^4(\Omega; R^N)$ into W, the *Schauder fixed point theorem* asserts that Ψ has at least one fixed point.

For $N = 4$, the closed ball of W containing $\Psi(W)$ is not compact in $L^4(\Omega; R^4)$ if one uses the strong topology, but as a bounded closed convex set of W is compact if one uses the weak topology, the *Tikhonov fixed point theorem* (which extends the Schauder fixed point theorem to *locally convex spaces*) asserts that Ψ has at least one fixed point when Ψ is (sequentially) weakly continuous from W into itself (the weak topology is not metrizable, but its restriction to a bounded set containing $\Psi(W)$ is metrizable). In order to prove the continuity, one takes a sequence v^n converging weakly to v^∞ in W, and as the corresponding solutions $u^n = \Psi(v^n)$ are bounded in W, one can extract a subsequence such that u^m converges weakly to u^∞ in W, and the problem is to show that $u^\infty = \Psi(v^\infty)$ (which ensures that all the sequence converges). For this purpose one uses the equation $\mu\, a(u^m, \varphi) + \varrho_0 b(v^m, u^m, \varphi) = L(\varphi)$ for every $\varphi \in W$, and one notices that if one shows that $b(v^m, u^m, \varphi) \to b(v^\infty, u^\infty, \varphi)$ for all $\varphi \in W$, then one has proven that $u^\infty = \Psi(u^\infty)$. As $b(v^m, u^m, \varphi) = -b(v^m, \varphi, u^m)$, one sees that it is enough to show that $v_j^m u_i^m \rightharpoonup v_j^\infty u_i^\infty$ in $L^2(\Omega)$ weak for all i, j. We know that v_j^m and u_i^m are bounded in $L^4(\Omega)$ and therefore $v_j^m u_i^m$ is bounded in $L^2(\Omega)$ and a subsequence converges to g in $L^2(\Omega)$ weak, but as $v_j^m \to v_j^\infty$ and $u_i^m \to u_i^\infty$ in $L^p(\Omega)$ strong for $2 \leq p < 4$, because the injection of $H_0^1(\Omega)$ into $L^p(\Omega)$ is compact for all $2 \leq p < 4$, one deduces that $v_j^m u_i^m \to v_j^\infty u_i^\infty$ in $L^{p/2}(\Omega)$ strong, and therefore $g = v_j^\infty u_i^\infty$.

I now turn to a method that will only use the *Brouwer*[2] *fixed point theorem* 17.3; it will also provide existence of weak solutions for $N > 4$. It uses a method due to FAEDO,[3] RITZ[4] and GALERKIN.[5]

A *topological space* is *separable* if it contains a countable dense subset; equivalently, a metric space is separable if for every $\varepsilon > 0$ it can be covered by a countable number of balls of radius at most ε (and therefore a subset of a separable space is separable). For any (nonempty) open set Ω of R^N, the spaces $L^p(\Omega)$ are separable for $1 \leq p < \infty$ (but $L^\infty(\Omega)$ is not separable); for $1 \leq p < \infty$, $W_0^{1,p}(\Omega)$ is separable; it suffices to approach functions in $C_c^\infty(\Omega)$ in the norm of $L^p(\Omega)$ or $W_0^{1,p}(\Omega)$ by a family of smooth functions which only depend upon a countable number of parameters: this is the first basic step when one wants to do the numerical analysis of solutions of partial differential equations, and there are traditional ways like *finite elements* for doing it, but from a theoretical point of view one checks density by applying the

---

[2] Luitzen Egbertus Jan BROUWER, Dutch mathematician, 1881–1966. He worked in Amsterdam, The Netherlands.

[3] Alessandro FAEDO, Italian mathematician, 1913–2001. He worked in Pisa, Italy.

[4] Walter RITZ, Swiss-born physicist, 1878–1909. He worked in Göttingen, Germany.

[5] Boris Grigorevich GALERKIN, Russian mathematician, 1871–1945. He worked in St Petersburg, Russia.

Hahn–Banach theorem. If $\theta_n \in C_c^\infty(\Omega)$ is 1 when the distance to the boundary is more than $1/n$ (and such a θ_n can be obtained by convolution with a regularizing sequence applied to the characteristic function of the points where the distance is more than $1/2n$ for example), then one obtains a dense set by considering the family of all functions $\theta_n P$, for all n and all polynomials P (and one has a countable dense subset by taking only polynomials with rational coefficients). Indeed, if f is in the dual of $L^p(\Omega)$ or of $W_0^{1,p}(\Omega)$ and is orthogonal to all $\theta_n P$, then each $\theta_n f$ (which is a distribution with compact support in Ω) is orthogonal to all polynomials, and therefore its Fourier transform has all its derivatives at 0 equal to 0, but it must be 0 because the Fourier transform of a distribution with compact support is an analytic function (which extends to C^N with, at most, exponential growth in the imaginary direction, and the Paley[6]–Wiener[7] theorem, extended by Laurent SCHWARTZ to distributions, characterizes these Fourier transforms); this shows that $\theta_n f = 0$ for all n and therefore $f = 0$. A *Faedo–Ritz–Galerkin basis* of a separable topological vector space E is a countable family $e_1, \ldots, e_n, \ldots$ of linearly independent elements which generate a dense subspace of E.

Let $w_1, \ldots, w_n, \ldots$ be any Faedo–Ritz–Galerkin basis of W, and let W_n be the finite-dimensional subspace generated by $w_1, \ldots, w_n$. One considers the approximate equation

$$\mu\, a(u^m, \varphi) + \varrho_0 b(u^m, u^m, \varphi) = L(\varphi) \text{ for all } \varphi \in W_m \qquad (17.1)$$
$$u^m \in W_m,$$

L being a linear continuous form on W and $N \leq 4$. The existence of a solution u^m follows from the Brouwer fixed point theorem 17.3 applied to Ψ_m, where for $v \in W_m$, $u^m = \Psi_m(v)$ is the unique solution of

$$\mu\, a(u^m, \varphi) + \varrho_0 b(v, u^m, \varphi) = L(\varphi) \text{ for all } \varphi \in W_m \qquad (17.2)$$
$$u^m \in W_m.$$

The fact that u^m is defined in a unique way follows from the Lax–Milgram lemma 11.2 (and $b(v, u, u) = 0$ for all $u, v \in W$), and provides a bound $\mu \|grad\, u^m\|_{L^2} \leq C$ independent of v, and independent of m. Using the same method as for Ψ, one sees that the mapping Ψ_m is Lipschitz continuous, and it must have at least one fixed point by the Brouwer fixed point theorem 17.3, as it maps all W_m into a bounded set of W_m.

As u^m is bounded in W, one can extract a subsequence $u^p \rightharpoonup u^\infty$ in W weak; as soon as $p \geq k$ one may take $\varphi = w_k$ and as before, the critical point of the proof is the convergence $b(u^p, u^p, w_k) = -b(u^p, w_k, u^p) \to -b(u^\infty, w_k, u^\infty) = b(u^\infty, u^\infty, w_k)$, and therefore u^∞ satisfies the desired equation for $\varphi = w_k$, i.e. for φ in a dense subspace of W. Because $N \leq 4$ and

---

[6] Raymond Edward Alan Christopher PALEY, English mathematician, 1907–1933. He worked in Cambridge, England, UK.

[7] Norbert G. WIENER, American mathematician, 1894–1964. He worked at MIT (Massachusetts Institute of Technology), Cambridge, MA.

$\varphi \mapsto b(u, u, \varphi)$ is linear continuous on W for $u \in W$, one then obtains the variational formulation for all $\varphi \in W$ and one has therefore found a solution of the Navier–Stokes equation for $N \leq 4$ (of course, one then goes through the interpretation of the equation, involving the "pressure" in $L^2(\Omega)$, if Ω is smooth enough).

The preceding method actually gives the existence of a solution in the sense of distributions for $N > 4$. The first step is to take a Faedo–Ritz–Galerkin basis made of smooth functions, for example by proving that $W = \{\varphi \in C_c^\infty(\Omega; R^N), | \, div \, \varphi = 0\}$ is dense in W (which is true if Ω is bounded with a Lipschitz boundary). Then $b(v, u, \varphi)$ is trilinear continuous on W_m, so the Lax–Milgram lemma 11.2 applies, defining Ψ_m, which by the Brouwer fixed point theorem 17.3 has a fixed point u^m, and as u^m is bounded in W one extracts a subsequence u^p which converges weakly to u^∞, and one can pass to the limit in $b(u^\infty, u^\infty, w_k)$ because w_k is smooth. The interpretation of the variational formulation then involves a "pressure" in some $L^q(\Omega)$, with q depending on the dimension N.

I want to explain now what is behind the Brouwer fixed point theorem 17.3, i.e. the *Brouwer topological degree* theory (which goes back to around 1910, I believe); as was pointed out to me by Jacques-Louis LIONS, the extension done in the 1930s by LERAY and SCHAUDER to infinite dimensions (and for mappings of the form *identity + compact*) is rarely necessary, as one usually finds more precise results by using a Faedo–Ritz–Galerkin basis, applying the methods of topological degree in finite dimensions and then letting the dimensions tend to infinity.

The idea is that if Ω is a bounded open set of R^N and F is a continuous mapping from $\overline{\Omega}$ into R^N, then one can make an *algebraic count* of the number of solutions of $\mathbf{F}(\mathbf{x}) = \mathbf{p}$ which are in Ω, by only looking at the restriction of F on the boundary $\partial\Omega$, assuming that there is no solution of $\mathbf{F}(\mathbf{x}) = \mathbf{p}$ on $\partial\Omega$. This will extend the trivial case of an interval (a, b) in 1 dimension, where the count is 1 if $F(a) < p < F(b)$, -1 if $F(a) > p > F(b)$, and 0 otherwise; it will also extend the case of a holomorphic mapping F for a domain in C bounded by a smooth Jordan[8] curve Γ, where the number is $\frac{1}{2i\pi} \int_\Gamma \frac{F'(z)}{F(z) - p} \, dz$ (always a nonnegative integer). The constructions of Lecture 18 will show most of the following properties of the Brouwer topological degree.

i) The *degree* $deg(F; \Omega, \mathbf{p})$ is defined for any continuous mapping F from $\overline{\Omega}$ into R^N which does not take the value $\mathbf{p}$ on $\partial\Omega$, and it depends only upon the restriction of F to $\partial\Omega$.

ii) It is an integer and if $deg(F; \Omega, \mathbf{p}) \neq 0$ then there exists $\mathbf{x} \in \Omega$ with $\mathbf{F}(\mathbf{x}) = \mathbf{p}$.

iii) If the degree is defined for Ω_1 and for Ω_2, and $\Omega_1 \cup \Omega_2 \subset \Omega \subset \overline{\Omega_1 \cup \Omega_2}$, then $deg(F; \Omega, \mathbf{p}) = deg(F; \Omega_1, \mathbf{p}) + deg(F; \Omega_2, \mathbf{p}) - deg(F; \Omega_1 \cap \Omega_2, \mathbf{p})$.

---

[8] Marie Ennemond Camille JORDAN, French mathematician, 1833–1922. He held a chair (Mathématiques) at Collège de France, Paris, France.

iv) The degree is invariant by *homotopy*, i.e. $deg(F; \Omega, \mathbf{p}) = deg(G; \Omega, \mathbf{p})$ if there exists a homotopy between F and G, i.e. there exists a continuous mapping H from $\overline{\Omega} \times [0,1]$ into R^N, such that H does not take the value $\mathbf{p}$ on $\partial\Omega \times [0,1]$, and such that $H(\cdot, 0) = F$ and $H(\cdot, 1) = G$ on $\overline{\Omega}$.

v) In the case where F is continuous from $\overline{\Omega}$ into R^N, of class C^1 in Ω, and only takes the value $\mathbf{p}$ at a finite number of points $\mathbf{a}_j, j = 1, \ldots, r$, of Ω, and if the Jacobian determinant of F is nonzero at each of these points, then $deg(F; \Omega, \mathbf{p}) = \sum_{j=1}^{r} sign(det(\nabla F)(\mathbf{a}_j))$. For instance, if $\mathbf{F}(\mathbf{x}) = \mathbf{x}$ for $\mathbf{x} \in \partial\Omega$ then $deg(F; \Omega, \mathbf{p}) = 1$ if $\mathbf{p} \in \Omega$, 0 if $\mathbf{p} \notin \overline{\Omega}$ (and is not defined if $\mathbf{p} \in \partial\Omega$); if $\mathbf{F}(\mathbf{x}) = -\mathbf{x}$ for $\mathbf{x} \in \partial\Omega$ then $deg(F; \Omega, \mathbf{p}) = (-1)^N$ if $-\mathbf{p} \in \Omega$, 0 if $-\mathbf{p} \notin \overline{\Omega}$ (and is not defined if $-\mathbf{p} \in \partial\Omega$).

Lemma 17.1. *There exists no nonzero continuous vector field from S^2 into R^3 which is everywhere tangent; more generally every nonzero continuous vector field from S^{2N} into R^{2N+1} is normal at (at least) one point of S^{2N}.*

Proof: Suppose that F is a nonzero continuous vector field from S^{2N} into R^{2N+1} which is nowhere normal; then the homotopy H defined by $\mathbf{H}(\mathbf{x}, t) = (1 - t)\mathbf{F}(\mathbf{x}) + t\mathbf{x}$ for $\mathbf{x} \in \overline{\Omega}, t \in [0,1]$ does not take the value $\mathbf{0}$ on S^{2N} so $deg(F; B(0,1), \mathbf{0}) = deg(id; B(0,1), \mathbf{0}) = 1$, and similarly the homotopy K defined by $\mathbf{K}(\mathbf{x}, t) = (1 - t)\mathbf{F}(\mathbf{x}) - t\mathbf{x}$ for $\mathbf{x} \in \overline{\Omega}, t \in [0,1]$ does not take the value $\mathbf{0}$ on S^{2N} so $deg(F; B(0,1), \mathbf{0}) = deg(-id; B(0,1), \mathbf{0}) = -1$, a contradiction.■

Lemma 17.2. *There does not exist a continuous retraction from a bounded open set $\Omega \subset R^N$ onto its boundary $\partial\Omega$, i.e. a continuous mapping F from $\overline{\Omega}$ into $\partial\Omega$ such that $\mathbf{F}(\mathbf{x}) = \mathbf{x}$ on $\partial\Omega$.*

Proof: For $\mathbf{p} \in \Omega$ one has $deg(F; \Omega, \mathbf{p}) = deg(id; \Omega, \mathbf{p}) = 1$ and therefore there exists $\mathbf{x} \in \Omega$ with $\mathbf{F}(\mathbf{x}) = \mathbf{p}$, contradicting the fact that the range of F is inside $\partial\Omega$.■

The result does not extend in infinite dimensions to the closed unit ball of the Hilbert space ℓ^2: one defines Φ by $\Phi(\mathbf{x}) = (\sqrt{1 - |\mathbf{x}|^2}, x_1, \ldots)$ for $\mathbf{x} = (x_1, \ldots, x_n, \ldots)$; then Φ is Lipschitz continuous, maps the closed unit ball into its boundary and has no fixed point; one can deduce that there exists a continuous retraction of the unit ball onto its boundary.

Theorem 17.3. *(Brouwer fixed point theorem) Every continuous mapping Φ from the closed unit ball of R^N into itself has at least one fixed point.*

Proof: Let $\Omega = B(0,1)$, and assume that Φ has no fixed point in $\overline{\Omega}$; then for every $\mathbf{x} \in \overline{\Omega}$ the line joining $\mathbf{x}$ to $\Phi(x)$ is well defined and intersects $\partial\Omega$ in two points, and if one takes $F(\mathbf{x})$ to be that point on the side of $\mathbf{x}$ one sees that F is a continuous retraction from $\overline{\Omega}$ onto its boundary. Analytically,

$F(\mathbf{x}) = (1-t)\mathbf{x} + t\,\Phi(\mathbf{x})$ with $t \leq 0$ and $|(1-t)\mathbf{x} + t\,\Phi(\mathbf{x})| = 1$, i.e. $t^2|\mathbf{x} - \Phi(\mathbf{x})|^2 + 2t(\mathbf{x} - \Phi(\mathbf{x}).\mathbf{x}) + |\mathbf{x}|^2 - 1 = 0.$ ∎

The theorem extends to any nonempty compact convex set $C \subset R^N$: one first restricts attention to the affine subspace generated by C, so that C has a nonempty interior in that subspace, and one notices that a compact convex set with nonempty interior in R^M is homeomorphic to the closed unit ball in R^M (taking 0 inside C, and using the Minkowski[9] functional p_C, the mapping $\mathbf{c} \mapsto \frac{p_C(\mathbf{c})}{\|\mathbf{c}\|}\mathbf{c}$ is a homeomorphism of C onto the closed unit ball).

[Taught on Friday February 19, 1999.]

---

[9] Hermann MINKOWSKI, German mathematician, 1864–1909. He worked in Göttingen, Germany.

18

Brouwer's topological degree

I had read about the topological degree in some lecture notes by J. T. SCHWARTZ[1] of the COURANT Institute, *nonlinear functional analysis*; I did not find these notes very clear, and in 1974 I simplified the exposition of the essential results by the following approach, based on the study of functionals $J_\varphi(u) = \int_\Omega \varphi(\mathbf{u}) det(\nabla u)\, dx$. [In the fall of 1975 I learnt from John M. BALL[2] about null Lagrangians and the property that Jacobian determinants are sequentially weakly continuous and I immediately linked this kind of robustness to that encountered in the study of the topological degree.]

Although the topological degree will be defined for some continuous mappings, the integrals that we start with assume that the mappings are of class C^1, and even have two derivatives in some proofs; a density argument is then necessary for extending the results obtained to continuous mappings.

Let Ω be a bounded regular open set of R^N, i.e. whose boundary is given locally by a Lipschitz function so that the exterior normal $\boldsymbol{\nu}$ to $\partial\Omega$ is defined almost everywhere on $\partial\Omega$ and the formula of integration by parts is valid (and uses the measure $d\sigma$ on $\partial\Omega$). All our mappings are assumed to be continuously extended to $\partial\Omega$, so that they will be defined on $\overline{\Omega}$.

Let u be a C^1 mapping from $\overline{\Omega}$ into R^N and φ a continuous scalar function on R^N; we define the functional J_φ by the formula

$$J_\varphi(u) = \int_\Omega \varphi(\mathbf{u}) det(\nabla u)\, d\mathbf{x}, \qquad (18.1)$$

where $\nabla u(\mathbf{x})$ is the Jacobian matrix at the point $\mathbf{x}$, whose entries are the partial derivatives $\frac{\partial u_i}{\partial x_j}$. Remark that the definition makes sense for mappings u in the Sobolev space $W^{1,N}(\Omega; R^N)$ and φ bounded; Louis NIRENBERG and

---

[1] Jacob Theodore SCHWARTZ, American mathematician. He works at New York University (COURANT Institute of Mathematical Sciences), New York, NY.

[2] John McLeod BALL, British mathematician, born in 1948. He holds the Sedleian chair at Oxford, England, UK.

Haïm BREZIS have recently extended the topological degree to mappings in VMO (Vanishing Mean Oscillation, the closure of C^∞ mappings in BMO). The crucial property of the functional J_φ is the following:

Lemma 18.1. *Assume that u is C^2 from $\overline{\Omega}$ into R^N and that φ is C^1 from R^N into R; let v be a C^1 mapping from $\overline{\Omega}$ into R^N, then*

$$\frac{d\left(J_\varphi(u + \varepsilon\, v)\right)}{d\varepsilon}\bigg|_{\varepsilon=0} = \int_{\partial\Omega} \varphi(\mathbf{u})\left(\sum_{k=1}^{N} \psi_k \nu_k\right) d\sigma, \qquad (18.2)$$

where ψ_k is defined by

$$\psi_k = \det\left(\frac{\partial u}{\partial x_1}, \ldots, \frac{\partial u}{\partial x_{k-1}}, \mathbf{v}, \frac{\partial u}{\partial x_{k+1}}, \ldots, \frac{\partial u}{\partial x_N}\right), \qquad (18.3)$$

i.e. $\psi_k(\mathbf{x})$ is obtained from the Jacobian matrix $\nabla u(\mathbf{x})$ by replacing the kth column by the vector $\mathbf{v}(\mathbf{x})$.

A consequence is that if u and w are C^1 mappings from $\overline{\Omega}$ into R^N which are equal on the boundary $\partial\Omega$ and if φ is continuous, then $J_\varphi(u) = J_\varphi(w)$. Indeed, by an argument of density, it is enough to prove the corollary when u and w are of class C^2 and φ is of class C^1. Lemma 18.1 is used for computing the derivative of $J_\varphi((1 - \theta)u + \theta\, w)$ with respect to θ and it says that the derivative is equal to an integral on $\partial\Omega$, and this integral is 0 as the mappings ψ_k vanish on $\partial\Omega$ because v is $w - u$, which is assumed to be 0 on the boundary.

More generally one can change the values of u on the boundary without changing the value of $J_\varphi(u)$ if one avoids the support of φ and this gives the following property of invariance by homotopy.

Lemma 18.2. *Assume that u and w are C^1 mappings from $\overline{\Omega}$ into R^N and $\varphi \in C_c(R^N)$. We assume that u and w can be joined by a homotopy having the property that on the boundary $\partial\Omega$ it avoids the support of φ, then $J_\varphi(u) = J_\varphi(w)$.*

Proof: The hypothesis means that there exists a continuous mapping F defined on $\overline{\Omega} \times [0, 1]$ with values in R^N such that $F(\cdot, 0) = u$, $F(\cdot, 1) = w$ on $\overline{\Omega}$ and $F(\mathbf{x}, \theta) \notin support(\varphi)$ for $(\mathbf{x}, \theta) \in \partial\Omega \times [0, 1]$), but one can regularize u, w, φ and F and still satisfy the same conditions; then one considers $G(\theta) = J_\varphi\left(F(\cdot, \theta)\right)$ and Lemma 18.1 applies with u replaced by $F(\cdot, \theta)$ and v by $\frac{\partial F}{\partial\theta}$: it says that $G'(\theta)$ is equal to an integral on the boundary and the integrand is 0 because it contains a term $\varphi\left(F(\cdot, \theta)\right)$ which is 0 on the boundary, and therefore $G(0) = G(1)$, which is our assertion.∎

Lemma 18.3. *$J_\varphi(u)$ can be defined when u is a continuous mapping from $\overline{\Omega}$ into R^N satisfying the condition $u(\mathbf{x}) \notin support(\varphi)$ when $\mathbf{x} \in \partial\Omega$.*

Proof: One can define $J_\varphi(u)$ by taking any sequence v_n of C^1 mappings from $\overline{\Omega}$ into R^N which converges uniformly to u, because for n large $J_\varphi(v_n)$ is constant and $J_\varphi(u)$ is defined as this limiting value. Indeed, let $\varepsilon > 0$ be small enough so that for $\mathbf{x} \in \partial\Omega$ the distance of $\mathbf{u}(\mathbf{x})$ to the support of φ is at least 2ε; if two mappings v_n and v_m of class C^1 are in the ball of center u and radius ε in the C^0 distance, then they can be joined by the homotopy $(1-\theta)v_n + \theta v_m$ and then $J_\varphi(v_n) = J_\varphi(v_m)$, so $J_\varphi(v)$ is constant in a ball around u.∎

With this extension of the definition to some continuous mappings, we can see that the preceding results are true for continuous mappings.

Lemma 18.4. *If u is a continuous mapping from $\overline{\Omega}$ into R^N such that $J_\varphi(u) \neq 0$ (the condition $\mathbf{u}(\mathbf{x}) \notin support(\varphi)$ for $\mathbf{x} \in \partial\Omega$ being satisfied in order to define $J_\varphi(u)$), then there exists $\mathbf{x} \in \Omega$ such that $\mathbf{u}(\mathbf{x}) \in support(\varphi)$.*

Proof: As $J_\varphi(v_n) \neq 0$ for large n, $\varphi(v_n)$ cannot vanish identically and so there exists $\mathbf{x_n} \in \Omega$ such that $\mathbf{v_n}(\mathbf{x_n}) \in support(\varphi)$; every limit point $\mathbf{x} \in \overline{\Omega}$ of the sequence $\mathbf{x_n}$ is then such that $\mathbf{u}(\mathbf{x}) \in support(\varphi)$ and $\mathbf{x} \notin \partial\Omega$ by hypothesis.∎

Proof of Lemma 18.1: The derivative in ε that we are considering is

$$\frac{d(J_\varphi(u + \varepsilon\, v))}{d\varepsilon}\Big|_{\varepsilon=0} = \int_\Omega \Big(\sum_{i=1}^N \frac{\partial(\varphi(u))}{\partial u_i} v_i \det(\nabla u) + \varphi(\mathbf{u}) \sum_{k=1}^N H_k \Big)\, d\mathbf{x}, \quad (18.4)$$

where the functions H_k are

$$H_k = \det\Big(\frac{\partial u}{\partial x_1}, \ldots, \frac{\partial u}{\partial x_{k-1}}, \frac{\partial v}{\partial x_k}, \frac{\partial u}{\partial x_{k+1}}, \ldots, \frac{\partial u}{\partial x_N} \Big), \quad (18.5)$$

expressing the multilinearity of the determinant. Lemma 18.1 will be proven by integration by parts if we show that

$$\sum_{i=1}^N \frac{\partial(\varphi(u))}{\partial u_i} v_i \det(\nabla u) + \varphi(\mathbf{u}) \sum_{k=1}^N H_k = \sum_{k=1}^N \frac{\partial(\varphi(u)\psi_k)}{\partial x_k}, \quad (18.6)$$

and this follows from the identities

$$\sum_{k=1}^N \frac{\partial \psi_k}{\partial x_k} = \sum_{k=1}^N H_k, \text{ and } \sum_{i=1}^N \frac{\partial u_i}{\partial x_k} \psi_k = v_i \det(\nabla u). \quad (18.7)$$

The first identity requires u to be of class C^2; once again the multilinearity of the determinant is used and we must show that the sum of the terms containing second derivatives of u is 0. Here it is the antisymmetry of the determinant that is needed because there are two terms showing a given second derivative $\frac{\partial^2 u}{\partial x_i \partial x_k}$: one has it in column i with $\mathbf{v}$ in column k and the other has it in column k with $\mathbf{v}$ in column i, all the other columns being similar. The second identity is linear in $\mathbf{v}$, so we check it in the case where only one component of $\mathbf{v}$, say v_1, is 1 and the others are 0; then the left-hand side consists in developing with respect to the first row a determinant obtained from ∇u by replacing the first row by $grad\, u_i$ so it gives $det(\nabla u)$ if $i = 1$ and, again from antisymmetry, it gives 0 if $i \neq 1$ because two rows are identical.■

The usual definition of the topological degree consists in computing an algebraic number of solutions of $\mathbf{u}(\mathbf{x}) = \mathbf{p}$ for a point $\mathbf{p} \in R^N$ and it is obtained by letting the function φ approach the Dirac mass at the point $\mathbf{p}$. We are led to the following definition.

Definition 18.5. *If u is a continuous mapping from $\overline{\Omega}$ into R^N satisfying the condition $\mathbf{u}(\mathbf{x}) \neq \mathbf{p}$ when $\mathbf{x} \in \partial\Omega$ then one can define $deg(u; \Omega, \mathbf{p})$ as the limit of the values $J_{\varphi_n}(u)$ for a sequence of functions φ_n whose supports converge to the point $\mathbf{p}$ and whose integrals converge to 1.*

Obviously one has $\mathbf{u}(\mathbf{x}) \notin support(\varphi_n)$ for $\mathbf{x} \in \partial\Omega$ for large n so that $J_{\varphi_n}(u)$ has a meaning by Lemma 18.3, but one difficulty is to show that the limit exists; the proof actually gives an important property, namely that the degree $deg(u; \Omega, \mathbf{p})$ is always an integer, for the values $\mathbf{p}$ for which this degree is defined, i.e. $\mathbf{p} \notin u(\partial\Omega)$. The first step is to notice that there is a discrete formula for computing the degree in the case where u is of class C^1 under a slight restriction.

Lemma 18.6. *Let u be of class C^1 from $\overline{\Omega}$ into R^N such that $\nabla u(\mathbf{z})$ is invertible at every point $\mathbf{z}$ solution of $\mathbf{u}(\mathbf{z}) = \mathbf{p}$, none of these solutions being on the boundary, so that there is only a finite number of them; then $deg(u; \Omega, \mathbf{p})$ is an integer, the sum of the signs of the Jacobian ∇u at all these points:*

$$deg(u; \Omega, \mathbf{p}) = \sum_{\mathbf{z}_\alpha | u(\mathbf{z}_\alpha)=p} sign\big(det(\nabla u)(\mathbf{z}_\alpha)\big). \qquad (18.8)$$

Proof: For n large enough, φ_n is 0 except in small disjoint neighborhoods of the $\mathbf{z}_\alpha$ solutions of $\mathbf{u}(\mathbf{z}_\alpha) = \mathbf{p}$; around each $\mathbf{z}_\alpha$ one can use a change of variable in the integral by taking $\mathbf{y} = \mathbf{u}(\mathbf{x})$ as the new variable: one then obtains a contribution $sign\big(det(\nabla u(\mathbf{z}_\alpha))\big) \int_{\mathbf{y} \in R^N} \varphi_n(\mathbf{y})\, d\mathbf{y}$ and this gives our formula in the limit $n \to \infty$.■

The second step is to notice that every C^1 mapping from Ω into R^N (and thus every continuous mapping from Ω into R^N) can be approximated by such special mappings u; this is done by adding a small constant vector to u and using a particular case of *Sard's*[3] *lemma*, which states that the set of critical

---

[3] Arthur SARD, American mathematician, 1909–1980. He worked at Queens College, New York, NY.

values $\mathbf{p}$, such that $\nabla u(\mathbf{x})$ is not invertible at some solution of $\mathbf{u}(\mathbf{x}) = \mathbf{p}$, has measure 0 and so has an empty interior.

The third step is to notice that one can extend all the properties of the functionals J_φ to the topological degree $deg(u; \Omega, \mathbf{p})$ and in particular the invariance by homotopy. Another easy consequence is that the degree is continuous in $\mathbf{p}$ and, because it is an integer, it is locally constant in each connected component of the complement of $u(\partial\Omega)$ (which one needs to avoid in order to give a meaning to the definition).

It is useful to notice that in practical situations one computes the degree by using a homotopy to a simple C^1 mapping for which one can obtain the degree explicitly using the formula (18.8), and so Sard's lemma is only a technical tool used to show that the degree is defined for every continuous mapping. The proof of Sard's lemma in the case of spaces of the same dimension, which is the one that interests us here, is relatively easy: one covers the initial open set with small cubes and notices that around one critical point the image of the corresponding cube is inside a flat cylinder of much smaller volume; using the uniform continuity of derivatives one concludes that the set of critical values is covered by sets of arbitrarily small volume, and so has measure 0.

[Taught on Monday February 22, 1999.]

19

Time-dependent solutions I

Let us look now at the time-dependent equation, first without nonlinearity for the Stokes equation, then with the nonlinearity for the Navier–Stokes equation.

I switch now to more traditional notations and define

$$V = \{u \in H_0^1(\Omega; R^N) \mid div\, u = 0 \text{ in } \Omega\}, \tag{19.1}$$

previously denoted by W, and

$$H = \{u \in L^2(\Omega; R^N) \mid div\, u = 0 \text{ in } \Omega, u.\nu = 0 \text{ on } \partial\Omega\}, \tag{19.2}$$

for which we shall have to show that $u.\nu$ makes sense on the boundary; we shall also have to prove that V is dense in H. Before doing so, we start with some abstract results on evolution equations, where the spaces denoted by V or H do not necessarily mean those above (which are adapted to the treatment of the Stokes equation).

There is an abstract theory for linear evolution equations, the theory of *semi-groups*, which was developed independently in the 1940s by Kôsaku YOSIDA[1] in Japan, and by HILLE[2,3] and then Ralph PHILIPPS[4,5] in the United States, but the theory has proven difficult to generalize to nonlinear equations, apart from situations where the maximum principle plays a role, which is usually not the case for equations of continuum mechanics. One advantage of the theory is that it puts into the same framework many linear

---

[1] Kôsaku YOSIDA, Japanese mathematician, 1909–1990. He worked in Tokyo, Japan.

[2] Einar HILLE, Swedish-born mathematician, 1894–1980. He worked at YALE University, New Haven, CT.

[3] Elihu YALE, English merchant, 1649–1721.

[4] Ralph Saul PHILLIPS, American mathematician, 1913–1998. He worked at STANFORD University, Stanford, CA.

[5] Leland STANFORD, American businessman, 1824–1893.

evolution equations with coefficients independent of t, but that is also a defect as it does not take into account the particular properties that the equations may have: the transport equation $\frac{\partial u}{\partial t} + \sum_{j=1}^{N} a_j \frac{\partial u}{\partial x_j} = 0$, diffusion equations like the heat equation $\frac{\partial u}{\partial t} - \sum_{i,j=1}^{N} \frac{\partial}{\partial x_i}\left(a_{ij} \frac{\partial u}{\partial x_j}\right) = 0$, Schrödinger's equation $i\frac{\partial u}{\partial t} - \Delta u + V u = 0$, the wave equation $\varrho\frac{\partial^2 u}{\partial t^2} - \sum_{i,j=1}^{N} \frac{\partial}{\partial x_i}\left(a_{ij} \frac{\partial u}{\partial x_j}\right) = 0$, the systems of linearized elasticity, Maxwell's equation, or the Stokes equation, can all be considered in such an abstract framework. The framework uses one Banach space, which of course changes from one equation to another, and it is sometimes an important restriction, because a good understanding of some equations from continuum mechanics often requires the use of more than one functional space: for Stokes's equation, the bound on the *kinetic energy* corresponds to a bound in $L^\infty(0, T; H)$, while the bound on the *energy dissipated by viscosity* corresponds to a bound in $L^2(0, T; V)$. Despite these shortcomings, I quickly sketch the main ideas of the semi-group approach.

For an abstract evolution equation $\frac{du}{dt} + A u = 0$, where A is a partial differential operator, one cannot define $e^{-t A}$ by the usual series $\sum_{k=0}^{\infty} \frac{(-t)^k}{k!} A^k$, as in the case where $A \in \mathcal{L}(E, E)$ for a Banach space E, and then write the solution as $u(t) = e^{-t A} u(0)$. Nevertheless, if one finds a way to define the solution in a unique way, one may expect that the mapping $u(0) \mapsto u(t)$ defines an operator $S(t) \in \mathcal{L}(E, E)$ for $t \geq 0$, satisfying $S(0) = I$ and $S(t_1) S(t_2) = S(t_1 + t_2)$ (the semi-group property), and some sort of continuity in t for $S(t)e$ for each $e \in E$, for example $S(t)e \to e$ in E strong as $t \to 0$. Given such a (strongly continuous) semi-group, the *uniform boundedness principle* implies that $||S(t)||$ is bounded for $t \in [0, 1]$, and then that $||S(t)|| \leq M e^{\omega t}$ for $t \geq 0$; putting $u(t) = v(t) e^{\omega t}$ creates a bounded semi-group $S_1(t) = S(t) e^{-\omega t}$ for v, satisfying $||S_1(t)|| \leq M$ for all $t \geq 0$; in the equivalent norm $||e||_1 = \sup_{t \geq 0} ||S_1(t)e||$, S_1 becomes a semi-group of *contractions*.

The *domain* $D(A)$ of the *infinitesimal generator* A of a (strongly continuous) semi-group S is defined as the subspace of elements $e \in E$ for which $S(t)e$ has a derivative at $t = 0$, denoted by $-A e$; one deduces that if $e \in D(A)$ then $S(t)e \in D(A)$ and its derivative is $-A S(t)e$, so that $S(t)$ does play the role of $e^{-t A}$. One then shows that $D(A)$ is *dense* in E, and that A is *closed*.[6] If $S(t)$ is a semi-group of contraction, one then shows that $I + \lambda A$ is invertible for $\lambda \geq 0$ with $||(I + \lambda A)^{-1}|| \leq 1$.

Conversely, if a closed operator A with dense domain is such that $I + \lambda A$ is invertible for $\lambda \geq 0$ with $||(I + \lambda A)^{-1}|| \leq 1$, then one can construct a semi-group S of contractions, of which A is the infinitesimal generator. Without going into the details (what I am sketching is a simplified view of the *Hille–Yosida theorem*), the idea is to consider the implicit approximation scheme $\frac{u_{n+1} - u_n}{\Delta t} + A u_{n+1} = 0$, where u_n serves as an approximation of $u(n \Delta t)$, and as $u_{n+1} = (I + \Delta t A)^{-1} u_n$, the way to use the bounds $||(I + \lambda A)^{-1}|| \leq 1$

---

[6] It means that if $x_n \to x_\infty$ in E with $x_n \in D(A)$ for all n and if $A x_n \to y_\infty$ in E, then one has $x_\infty \in D(A)$ and $A x_\infty = y_\infty$.

for $\lambda \geq 0$ appears easily (the explicit scheme $\frac{u_{n+1}-u_n}{\Delta t} + A u_n = 0$ requires $u_n \in D(A)$, and therefore one needs $u(0) \in D(A^k)$ for all k just for defining all the u_n, so this scheme is not of great use).

I shall present now a different framework, where two Hilbert spaces V and H are used (or three if one counts V', as H' is identified with its dual); this framework is adapted to solving diffusion equations, or Stokes's equation, for example (in semi-group theory it is related to *analytic semi-groups*, which can be extended for t in a sector of the complex plane). I have learnt many of the results that I present from Jacques-Louis LIONS, and I have only improved small technical details.

Let V and H be two (real) Hilbert spaces, with norms $|| \cdot ||$ for V and $| \cdot |$ for H, V being continuously embedded in H and being dense in H; H is identified with its dual H', which is continuously embedded in V' and dense in V' (in some cases the identification of H with its dual H' may create a few problems, as will be explained in Lecture 21); the norm in V' is denoted by $|| \cdot ||_*$. Let $A \in \mathcal{L}(V, V')$ be such that there exist $\alpha > 0$ and $\beta \in R$ for which

$$\langle A u, u \rangle \geq \alpha ||u||^2 - \beta |u|^2 \quad \text{for all } u \in V \tag{19.3}$$

(for simplification, I assume that A is independent of t; in practical situations one may have a bilinear continuous form $a(t, u, v)$ measurable in t).

Lemma 19.1. *Given* $u_0 \in H$, $f_1 \in L^1(0, T; H)$ *and* $f_2 \in L^2(0, T; V')$, *there exists a unique* $u \in C([0, T]; H) \cap L^2(0, T; V)$ *with* $\frac{du}{dt} \in L^1(0, T; H) + L^2(0, T; V')$, *solution of*

$$\frac{du}{dt} + A u = f_1 + f_2 \text{ in } (0, T); \quad u(0) = u_0, \tag{19.4}$$

which in variational form means

$$\int_0^T \left(-\frac{d\varphi}{dt}(u, v) + \varphi \langle A u, v \rangle \right) dt = \varphi(0)(u_0, v) + \int_0^T \varphi [(f_1, v) + \langle f_2, v \rangle] dt$$
for all $v \in V$ *and all* $\varphi \in C^\infty([0, T])$ *satisfying* $\varphi(T) = 0$. \tag{19.5}

Jacques-Louis LIONS always considered $f \in L^2(0, T; V')$, i.e. the case $f_1 = 0$, which gives $\frac{du}{dt} \in L^2(0, T; V')$; because of the natural bounds $u \in L^\infty(0, T; H) \cap L^2(0, T; V)$, I find it natural to take $f \in L^1(0, T; H) + L^2(0, T; V')$.

I shall assume that V *is separable* (and then H is separable as V is dense in H); this is not a restriction for applications, and it avoids some technical difficulties about measurability of functions with values in V, H or V'. Let $e_1, \ldots$ be any Faedo–Ritz–Galerkin basis of V, and let V_m be the subspace generated by $e_1, \ldots, e_m$. One looks for a function u_m from $[0, T]$ into V_m, i.e. $u_m(t) = \sum_{i=1}^m \xi_{mi}(t) e_i$, and the coefficients ξ_{mi} will belong to $W^{1,1}(0, T)$, which is continuously embedded in $C([0, T])$; one asks u_m to satisfy

$$\left(\frac{du_m}{dt}, e_k\right) + \langle A\,u_m, e_k\rangle = (f_1, e_k) + \langle f_2, e_k\rangle \text{ a.e. in } (0, T) \tag{19.6}$$
and $(u_m(0), e_k) = (u_0, e_k)$ for $k = 1, \dots, m$.

This is an ordinary linear differential equation in R^m, of the form $\xi' + A_m\xi = \eta_m$ in $(0, T)$ and $\xi(0) = \xi_{0m}$, with $\xi_{0m} \in R^m$ and $\eta_m \in L^1(0, T; R^m)$; it has a unique solution in $W^{1,1}(0, T; R^m)$, which is given explicitly by $\xi(t) = e^{-t\,A_m}\xi_{0m} + \int_0^t e^{-(t-s)\,A_m}\eta_m(s)\,ds$ for $t \in [0, T]$; one may prefer to deal with classical C^1 solutions in V_m, and that consists in choosing $u_{0m} \in V_m$, $f_{1m}, f_{2m} \in C([0, T]; V_m)$ approaching in a strong or weak way u_0 in H, f_1 in $L^1(0, T; H)$ and f_2 in $L^2(0, T; V')$. Because the equation is linear, we immediately know existence and uniqueness on the whole interval $[0, T]$, but when we shall deal with a nonlinear equation like the Navier–Stokes equation, we shall have to start with a local existence result and then show that the solution exists on $[0, T]$.

We need now precise bounds (independent of m) in order to take the limit $m \to \infty$, and we shall need some technical results.

Lemma 19.2. *i)* $W^{1,1}(0, T) \subset C([0, T])$,

ii) $u, v \in W^{1,1}(0, T)$ *imply* $u\,v \in W^{1,1}(0, T)$ *and* $(u\,v)' = u\,v' + u'v$ *a.e. in* $(0, T)$,

iii) Gronwall's[7] inequality: if $\varphi \in L^\infty(0, T)$ *satisfies* $\varphi(t) \geq 0$ *a.e. on* $(0, T)$ *and* $\varphi(t) \leq \psi(t) = A + \int_0^t (\lambda_1\varphi + \lambda_2)\,ds$ *a.e. in* $(0, T)$, *where* $\lambda_1, \lambda_2 \in L^1(0, T)$, *then* $\psi(t) \leq \left(A + \int_0^t |\lambda_2(s)|\,ds\right)exp\left(\int_0^t |\lambda_1(s)|\,ds\right)$ *for* $t \in [0, T]$.

Proof: For $f \in L^1(0, T)$, let $f_n \in C([0, T])$ converge to f in $L^1(0, T)$, then $u_n \in C^1([0, T])$ defined by $u_n(t) = \int_0^t f_n(s)\,ds$ converges uniformly to u defined by $u(t) = \int_0^t f(s)\,ds$, and as u'_n converges in the sense of distributions to u', one has $u' = f$ a.e. in $(0, T)$; if $v \in W^{1,1}(0, T)$ has $v' = f$, then $v - u$ has derivative 0 and is therefore a constant, showing that v is continuous and that the constant is $v(0)$.

The proof of i) has shown that $C^1([0, T])$ is dense in $W^{1,1}(0, T)$. If $u_n \in C^1([0, T])$ with $u'_n \to u'$ in $L^1(0, T)$, and $v_n \in C^1([0, T])$ with $v'_n \to v'$ in $L^1(0, T)$, then $u_n v_n$ converges uniformly to $u\,v$ and $(u_n v_n)' = u'_n v_n + u_n v'_n$ converges in $L^1(0, T)$ to $u'v + u\,v'$, and to $(u\,v)'$ in the sense of distributions, and therefore $(u\,v)' = u'v + u\,v'$.

As $\varphi \geq 0$, the inequality stays true if one replaces λ_1 by its absolute value, and one may then assume that $\lambda_1 \geq 0$ a.e. on $(0, T)$. Then $\psi' = \lambda_1\varphi + \lambda_2 \leq \lambda_1\psi + \lambda_2$ a.e. on $(0, T)$. As for the proof of i) and ii), if one defines E by $E(t) = exp\left(-\int_0^t \lambda_1(s)\,ds\right)$ then $E \in W^{1,1}(0, T)$ and $E' = -E\,\lambda_1$. One deduces that $(E\,\psi)' \leq E\,\lambda_2$, so that $\psi(t) \leq A\,exp\left(\int_0^t \lambda_1(s)\,ds\right) + \int_0^t \lambda_2(s)exp\left(\int_s^t \lambda_1(\sigma)\,d\sigma\right)ds$, from which the bound follows. ∎

---

[7] Thomas Hakon GRONWALL, Swedish-born mathematician, 1877–1932. He worked at Princeton University, Princeton, NJ.

To obtain estimates on u_m, one replaces e_k by u_m, which is a linear combination of the e_k, and one obtains

$$\frac{1}{2}\frac{d|u_m|^2}{dt} + \alpha\,||u_m||^2 - \beta\,|u_m|^2 \leq |f_{1m}|\,|u_m| + ||f_{2m}||_*||u_m|| \quad \text{a.e. in } (0,T).$$
(19.7)

Using the inequalities $|f_{1m}|\,|u_m| \leq \frac{1}{2}|f_{1m}| + \frac{1}{2}|f_{1m}|\,|u_m|^2$ and $||f_{2m}||_*||u_m|| \leq \frac{\alpha}{2}||u_m||^2 + \frac{1}{2\alpha}||f_{2m}||_*^2$, then gives

$$\frac{d|u_m|^2}{dt} + \alpha\,||u_m||^2 \leq (2\beta + |f_{1m}|)|u_m|^2 + \frac{1}{\alpha}||f_{2m}||_*^2 \quad \text{a.e. in } (0,T), \quad (19.8)$$

and, forgetting for a while the term $\alpha\,||u_m||^2$, Gronwall's inequality applies with $\varphi = |u_m|^2$ after integrating in t, and it gives the bound

$$|u_m(t)|^2 \leq \left(|u_{0m}|^2 + \frac{1}{\alpha}\int_0^t ||f_{2m}(s)||_*^2\,ds\right)exp\left(\int_0^t (2\beta + |f_{1m}(s)|)\,ds\right), t \in [0,T],$$
(19.9)

and then taking into account the term in $\alpha\,||u_m||^2$ gives

$$\alpha\int_0^T ||u_m(t)||^2\,dt \leq \left(|u_{0m}|^2 + \frac{1}{\alpha}\int_0^T ||f_{2m}(t)||_*^2\,dt\right)e^{\int_0^T (2\beta + |f_{1m}(t)|)\,dt} - |u_{0m}|^2.$$
(19.10)

These bounds are good enough for our purpose, but show a strange dependence with respect to the norm of f_{1m}; one way to avoid it is to use linearity, i.e. to consider first the case $f_{1m} = 0$ for which the above bound is acceptable and then the case where $f_{2m} = 0$, where from the bound

$$\frac{1}{2}\frac{d|u_m|^2}{dt} + \alpha\,||u_m||^2 - \beta\,|u_m|^2 \leq |f_{1m}|\,|u_m| \quad \text{a.e. in } (0,T), \quad (19.11)$$

one forgets the term $\alpha\,||u_m||^2$ and one deduces

$$\frac{d|u_m|}{dt} \leq \beta\,|u_m| + |f_{1m}| \quad \text{a.e. in } (0,T), \quad (19.12)$$

giving

$$|u_m(t)| \leq |u_{0m}|e^{\beta t} + \int_0^t |f_{1m}(s)|e^{\beta(t-s)}\,ds, \quad (19.13)$$

and giving the expected affine dependence in the norm of f_{1m}. In our finite-dimensional situation one does have $|u_m| \in W^{1,1}(0,T)$ and $\frac{d|u_m|^2}{dt} = 2|u_m|\frac{d|u_m|}{dt}$, but the argument would not work in infinite dimensions, where it is better to consider $\sqrt{\varepsilon + |u_m|^2}$ for $\varepsilon > 0$, and therefore

$$\frac{d\sqrt{\varepsilon + |u_m|^2}}{dt} = \frac{1}{2\sqrt{\varepsilon + |u_m|^2}}\frac{d|u_m|^2}{dt} \leq (\beta\,|u_m| + |f_{1m}|)\frac{|u_m|}{\sqrt{\varepsilon + |u_m|^2}}$$
$$\leq \beta\sqrt{\varepsilon + |u_m|^2} + |f_{1m}| \quad \text{a.e. in } (0,T), \quad (19.14)$$

and in the bound obtained for $\sqrt{\varepsilon + |u_m|^2}$ one lets ε tend to 0, or one gets the inequality for $\frac{d|u_m|}{dt}$, which shows that $|u_m| \in BV(0, T)$.

This way of getting bounds is not possible if one is dealing with a nonlinear equation; another way to deal with the bounds is to use a form of *Young's inequality*, here $||f_{2m}||_* ||u_m|| \leq \frac{\alpha}{2} ||u_m||^2 + \frac{1}{2\alpha} ||f_{2m}||_*^2$, which gives

$$\frac{d|u_m|^2}{dt} + \alpha ||u_m||^2 \leq 2\beta |u_m|^2 + 2|f_{1m}||u_m| + \frac{1}{\alpha} ||f_{2m}||_*^2 \text{ a.e. in } (0, T), \quad (19.15)$$

which, forgetting again the term $\alpha ||u_m||^2$ for a while, gives

$$|u_m(t)|^2 \leq |u_{0m}|^2 + \frac{1}{\alpha} \int_0^t ||f_{2m}(s)||_*^2 \, ds + \int_0^t (2\beta |u_m(s)|^2 + 2|f_{1m}||u_m(s)|) \, ds, \quad (19.16)$$

and then to use a variant of Gronwall inequality

$$\varphi(t) \leq \psi(t) = A + \int_0^t (2\mu_1 \varphi + 2\mu_2 \sqrt{\varphi}) \, ds \quad (19.17)$$

which gives

$$\psi' \leq 2|\mu_1|\psi + 2|\mu_2|\sqrt{\psi}, \quad (19.18)$$

from which one gets

$$\sqrt{\psi(t)} \leq \left(A + \int_0^t |\mu_2(s)| \, ds \right) exp \left(\int_0^t |\mu_1(s)| \, ds \right), \quad (19.19)$$

and therefore as $A = |u_{0m}|^2 + \frac{1}{\alpha} \int_0^\tau ||f_{2m}(t)||_*^2 \, dt$ as long as $t \leq \tau$, one deduces

$$|u_m(t)| \leq \left(\sqrt{|u_{0m}|^2 + \frac{1}{\alpha} \int_0^t ||f_{2m}(s)||_*^2 \, ds} + \int_0^t |f_{1m}(s)|) \, ds \right) e^{\beta t} \text{ on } [0, T]. \quad (19.20)$$

[Taught on Wednesday February 24, 1999.]

20

Time-dependent solutions II

For u_n defined by (19.6) one has obtained a uniform bound in $C([0,T];H) \cap L^2(0,T;V)$ by taking u_{0n} bounded in H, f_{1n} bounded in $L^1(0,T;H)$ and f_{2n} bounded in $L^2(0,T;V')$, according to (19.9) and (19.10); one can extract a subsequence u_p converging to u_∞ in $L^\infty(0,T;H)$ weak $\star$ and in $L^2(0,T;V)$ weak, i.e.

$$\begin{array}{l} \int_0^T (u_p, v)\, dt \to \int_0^T (u_\infty, v)\, dt \text{ for all } v \in L^1(0,T;H), \\ \int_0^T \langle u_p, v \rangle\, dt \to \int_0^T \langle u_\infty, v \rangle\, dt \text{ for all } v \in L^2(0,T;V). \end{array} \tag{20.1}$$

If u_{0n} converges weakly to u_0 in H, f_{1n} converges weakly to f_1 in $L^1(0,T;H)$ and f_{2n} converges weakly to f_2 in $L^2(0,T;V')$, then one can take the limit in (19.6) as $p \to \infty$. For $\varphi \in C^\infty([0,T])$ satisfying $\varphi(T) = 0$, one rewrites the term $\int_0^T \left(\frac{du_p}{dt}, e_k\right)\varphi\, dt = -\int_0^T (u_p, e_k)\frac{d\varphi}{dt}\, dt - (u_{0p}, e_k)\varphi(0)$, which enables us to obtain the limit equation

$$\begin{array}{l} \text{for all } k, \; -\int_0^T (u_\infty, e_k)\frac{d\varphi}{dt}\, dt + \int_0^T \langle A\, u_\infty, e_k \rangle \varphi\, dt = \\ (u_0, e_k)\varphi(0) + \int_0^T (f_1, e_k)\varphi\, dt + \int_0^T \langle f_2, e_k \rangle \varphi\, dt \\ \text{for all test functions } \varphi \in C^\infty([0,T]) \text{ such that } \varphi(T) = 0. \end{array} \tag{20.2}$$

Using linearity one can replace e_k by any linear combination of the elements of the basis, and then, using an argument of density, by any element $v \in V$. Putting $g_2 = f_2 - A\, u_\infty \in L^2(0,T;V')$, one has

$$-\int_0^T (u_\infty, v)\frac{d\varphi}{dt}\, dt = (u_0, v)\varphi(0) + \int_0^T (f_1, v)\varphi\, dt + \int_0^T \langle g_2, v \rangle \varphi\, dt \text{ for all } v \in V, \tag{20.3}$$

for all $\varphi \in C^\infty([0,T])$ such that $\varphi(T) = 0$. This means that $(u_\infty, v) \in W^{1,1}(0,T)$, that its value at 0 is (u_0, v) and that its derivative is $(f_1, v) + \langle g_2, v \rangle$, but if one defines $u_* \in C([0,T];V')$ by $u_*(t) = u_0 + \int_0^T \big(f_1(s) + g_2(s)\big)\, ds$, then $\langle u_*, v \rangle$ has the same properties as (u_∞, v) and therefore they are equal. This shows that $u_\infty \in W^{1,1}(0,T;V')$ with $\frac{du_\infty}{dt} = f_1 + g_2$, i.e. u_∞ solves the equation $\frac{du_\infty}{dt} + A\,u_\infty = f_1 + f_2$ in $(0,T)$ and $u(0) = u_0$. As we shall show that this equation has a unique solution, all the sequence does converge weakly to u_∞.

Uniqueness, which could have been proven before proving existence, follows from the formula

$$\left\langle \frac{du}{dt}, u \right\rangle = \frac{1}{2} \frac{d|u|^2}{dt} \quad \text{in } (0,T), \tag{20.4}$$

as $\frac{du}{dt} + A\,u = 0$ then implies $\frac{1}{2}\frac{d|u|^2}{dt} + \alpha\,||u||^2 - \beta\,|u|^2 \le 0$, and therefore $|u(t)| \le |u(0)|e^{\beta t}$, proving that $u = 0$ if $u(0) = 0$. The formula is valid if $u \in W_1(0,T)$, where

$$W_1(0,T) = \left\{ u \in L^2(0,T;V) \mid \frac{du}{dt} \in L^1(0,T;H) + L^2(0,T;V') \right\}, \tag{20.5}$$

or in the smaller space used by Jacques-Louis LIONS,

$$W(0,T) = \left\{ u \in L^2(0,T;V) \mid \frac{du}{dt} \in L^2(0,T;V') \right\}. \tag{20.6}$$

The formula to be proven is true pointwise if $u \in C^1([0,T];V)$, and in weak formulation it can be written as $\int_0^T \langle \frac{du}{dt}, u \rangle \varphi\,dt = -\frac{1}{2}\int_0^T |u|^2 \frac{d\varphi}{dt}\,dt$ for all $\varphi \in C_c^\infty(0,T)$.

We shall first prove that $C^\infty([0,T];V)$ is dense in $W_1(0,T)$; then we shall use the density in order to prove that $W_1(0,T) \subset C([0,T];H)$; then we shall deduce that the formula is true in $W_1(0,T)$, because both sides of the variational formulation are continuous bilinear forms on $W_1(0,T)$ (the formula implies that $|u|^2 \in W^{1,1}(0,T)$ for $u \in W_1(0,T)$).

1) One notices that the space $W_1(0,T)$ is local, i.e. if $\psi \in C^\infty([0,T])$ and $u \in W_1(0,T)$ then $\psi\,u \in W_1(0,T)$, because u being in $L^2(0,T;V)$ is automatically in $L^1(0,T;H)$ or in $L^2(0,T;V')$. Choosing $\theta \in C^\infty([0,T])$, equal to 1 on $[0,T/3]$ and 0 on $[2T/3,T]$, one can consider $\theta\,u$ as being 0 on $[T,\infty)$ and $(1-\theta)u$ as being 0 on $(-\infty,0]$. One then regularizes $\theta\,u$ by convolution with a regularizing sequence with support on $(-\infty,0)$, and similarly one regularizes $(1-\theta)u$ by convolution with a regularizing sequence with support on $(0,\infty)$, and the usual properties of regularization show that $C^\infty([0,T],V)$ is dense in $W_1(0,T)$.

2) In order to prove that $W_1(0,T)$ is continuously embedded in $C([0,T];H)$, one only needs to show that there exists C such that $||u||_{C([0,T];H)} \le C\,||u||_{L^2(0,T;V)} + C||\frac{du}{dt}||_{L^1(0,T;H)+L^2(0,T;V')}$ for all functions $u \in C^\infty([0,T];V)$. The norm in $L^1(0,T;H) + L^2(0,T;V')$ is the infimum of $||h_1||_{L^1(0,T;H)} + ||h_2||_{L^2(0,T;V')}$ over all the decompositions $\frac{du}{dt} = h_1 + h_2$ with $h_1 \in L^1(0,T;H)$ and $h_2 \in L^2(0,T;V')$. By reasoning on $\theta\,u$ and then $(1-\theta)u$, one may assume that u is 0 at one end of the interval; for example, assuming that $u(0) = 0$

one has $|u(t)|^2 = 2\int_0^t \left(\frac{du}{dt}, u\right) ds = 2\int_0^t \langle \frac{du}{dt}, u \rangle \, ds = 2\int_0^t \left(\langle h_1, u \rangle + \langle h_2, u \rangle\right) ds$,
from which one deduces

$$\|u\|^2_{C([0,T];H)} \leq 2\|h_1\|_{L^1(0,T;H)}\|u\|_{C([0,T];H)} + 2\|h_2\|_{L^2(0,T;V')}\|u\|_{L^2(0,T;V)}$$
$$\leq 2\|h_1\|^2_{L^1(0,T;H)} + \tfrac{1}{2}\|u\|^2_{C([0,T];H)} + 2\|h_2\|_{L^2(0,T;V')}\|u\|_{L^2(0,T;V)},$$

$$(20.7)$$

and therefore

$$\|u\|^2_{C([0,T];H)} \leq 4\|h_1\|^2_{L^1(0,T;H)} + 4\|h_2\|_{L^2(0,T;V')}\|u\|_{L^2(0,T;V)}, \qquad (20.8)$$

and by taking the infimum on the decompositions $\frac{du}{dt} = h_1 + h_2$ with $h_1 \in L^1(0,T;H)$ and $h_2 \in L^2(0,T;V')$, it proves the continuous embedding of $W_1(0,T)$ into $C([0,T];H)$.

3) All the terms of the weak formulation are then seen to be continuous on $W_1(0,T)$, and the formula is true by density.

It must be noticed that one does not have in general $\frac{du_n}{dt}$ bounded in $L^1(0,T;H) + L^2(0,T;V')$. In order to obtain bounds for $\frac{du_n}{dt}$, one can either make time regularity hypotheses on f_1 and f_2 and a regularity hypothesis on u_0 (which corresponds to regularity in space variables in applications to partial differential equations), or use a special Faedo–Ritz–Galerkin basis.

The first idea consists in noticing that formally $u' = \frac{du}{dt}$ satisfies $\frac{du'}{dt} + A u' = f_1' + f_2'$ and $u'(0) = f_1(0) + f_2(0) - A u_0$, which suggests that if $f_1 \in W^{1,1}(0,T;H)$, $f_2 \in H^1(0,T;V')$ and $u_0 \in V$ with $A u_0 - f_2(0) \in H$, then $u' \in W_1(0,T)$ and one can expect a bound on $\frac{du_n}{dt}$. Indeed, one can easily choose $f_{1n} \in C^\infty([0,T];H)$ and $f_{2n} \in C^\infty(0,T;V')$ converging respectively to f_1 and f_2, so that $u_n \in C^\infty(0,T;V_n)$, but one must be a little careful for the bound on $u_n'(0)$; Jacques-Louis LIONS taught the trick of taking u_0 as the first element of the basis (if it is not 0), so that one can take $u_{0n} = u_0$ and then one asks that $f_{2n}(0) - f_2(0)$ converges weakly to 0 in H.

It is useful to notice that if $f_1 \in W^{1,1}(0,T;H)$, $f_2 \in H^1(0,T;V')$, but $u_0 \in H$ only, then one does not obtain $u' \in W_1(0,T)$ by lack of the needed regularity on u_0, but one has $t\,u' \in W_1(0,T)$, as $v = t\,u'$ satisfies $\frac{dv}{dt} + A v = t\frac{df_1}{dt} + t\frac{df_2}{dt} + u'$ and $v(0) = 0$. This is a form of regularization effect for the solutions of the equation, already apparent from the fact that one does not need $u_0 \in V$ in order to have the solution taking its values in V. To obtain the corresponding estimate for $t\,u_n'$, it is better to use $w = t\,u' - u$, which satisfies $\frac{dw}{dt} + A w = t\frac{df_1}{dt} + t\frac{df_2}{dt} - A u$ and $v(0) = 0$, and the corresponding bounds for $t\frac{du_n}{dt} - u_n$ are obtained easily.

The choice of a special Faedo–Ritz–Galerkin basis is a different trick, and we shall use it for the Navier–Stokes equation (at least in 3 dimensions, as the case of 2 dimensions may be handled more easily), but it requires the symmetry of A, or simply $A^T - A \in \mathcal{L}(V,H)$, and the compact injection of V into H. One assumes that $A = A_0 + B$ with A_0 symmetric V-elliptic and $B \in \mathcal{L}(V,H)$. As A_0 is an isomorphism from V onto V', its inverse A_0^{-1} maps V' and therefore H into V, so A_0^{-1} is a compact operator on H, and as it is symmetric, (F.) Riesz theory asserts that H has an orthonormal basis made of eigenvectors of A_0^{-1}, $e_n, n \geq 1$, with real positive eigenvalues μ_n converging to 0. Therefore $A_0 e_n = \lambda_n e_n$ with $\lambda_n = \frac{1}{\mu_n}$ tending to $+\infty$, and if one replaces the norm $||u||$ on V by the equivalent norm $\sqrt{\langle A_0 u, u \rangle}$, then the basis is also orthogonal in V, and therefore it is also orthogonal in V'. The estimate for $\frac{du_n}{dt}$ comes easily once one has observed that for a finite linear combination $v = \sum_i v_i e_i$, one has $||v||^2 = \sum_i \lambda_i |v_i|^2$, $|v|^2 = \sum_i |v_i|^2$, and $||v||_*^2 = \sum_i \frac{1}{\lambda_i} |v_i|^2$.

It is not necessary to take this special basis of eigenvectors in order to deduce estimates on $\frac{du_n}{dt}$, and a more general condition is obtained in the following way. Let P_n be the orthogonal projection of H onto the subspace V_n (which is closed, as all finite-dimensional subspaces), where orthogonality is understood in the scalar product of H, so that P_n is a contraction if one uses the norm of H for V_n; let C_n be the norm of P_n considered as a mapping from V onto V_n equipped with the norm of V; then the basis is special enough in order to obtain a bound for $\frac{du_n}{dt}$ if C_n is bounded (the choice of eigenvectors of A_0 gives $C_n = 1$ if one uses $\sqrt{\langle A_0 \cdot, \cdot \rangle}$ for the norm on V). Indeed, if $k \in V'$ and $k_n \in V_n$ is defined by $(k_n, v) = \langle k, v \rangle$ for all $v \in V_n$, then for $w \in V$, one has $\langle k_n, w \rangle = (k_n, w) = (k_n, P_n w) = \langle k, P_n w \rangle = \langle P_n^T k, w \rangle$, so that $||k_n||_* \leq C_n ||k||_*$.

There are interesting results of uniqueness which are not based on the ellipticity of A but on its symmetry, and they may be used as well for A or $-A$ with data at 0, or for A with either initial data at 0 or final data at T (one then talks of *backward uniqueness*). One approach, due to Shmuel AGMON[1] and Louis NIRENBERG consists in proving that if u is a nonvanishing solution of $u' + A u = 0$, then $|u|$ is log-convex, i.e. $t \mapsto \log|u(t)|$ is convex. Indeed, the derivative of $\log|u|$ is $\frac{(u', u)}{|u|^2}$, whose derivative is $\frac{(u'', u) + |u'|^2}{|u|^2} - 2\frac{(u', u)^2}{|u|^4}$, and as $(u'', u) = (A^2 u, u) = |A u|^2 = |u'|^2$, one concludes by using the Cauchy–Schwarz inequality. In the early 1970s I extended this method with Claude BARDOS,[2] and we could apply it to the Navier–Stokes equation in 2 dimensions. I have noticed that the log-convexity property is true if A is normal (i.e. A commutes with A^T, or $|A^T v| = |A v|$ for all v), and it is equivalent in finite dimensions, while Shmuel FRIEDLAND[3] has noticed that a sufficient condition is $|A^T v| \leq |A v|$ for all v, which may happen without equality in infinite dimensions. There is a second approach by Jacques-Louis

---

[1] Shmuel AGMON, Israeli mathematician, born in 1922. He worked at The Hebrew University, Jerusalem, Israel.

[2] Claude W. BARDOS, French mathematician, born in 1940. He works at Université Paris 7 (Denis DIDEROT), Paris, France.

[3] Shmuel FRIEDLAND, Uzbekistan-born mathematician, born in 1944. He works at University of Illinois, Chicago, IL.

LIONS and Bernard MALGRANGE,[4] also valid for symmetric A, and based on *Carleman*[5] *estimates.*

I have also introduced another approach, still valid for symmetric A, which is useful for improving the localization of the trajectory which results from the log-convexity property: if the solution exists on $[0, T]$, then for $0 \leq \tau_1 < \tau_2 \leq T$, the trajectory for $t \in (\tau_1, \tau_2)$ lies inside the closed ball with diameter the segment $[u(\tau_1), u(\tau_2)]$; this could certainly be more useful if one knew how to prove similar results for nonlinear equations.

[Taught on Friday February 26, 1999.]

---

[4] Bernard MALGRANGE, French mathematician, born in 1928. He worked at Université de Grenoble I (Joseph FOURIER), Saint-Martin-d'Hères, France.

[5] Tage Gillis Torsten CARLEMAN, Swedish mathematician, 1892–1949. He worked in Lund, Sweden.

21

Time-dependent solutions III

To apply the preceding abstract framework to Stokes's equation, there are questions about the functional spaces. We are already familiar with V, defined in (19.1) (which was denoted by W, defined in (12.6) in the stationary case), but we have to identify its closure in $L^2(\Omega; R^N)$, which is the space H of the abstract theory. As we shall see in Lemma 23.2, $H = \{u \in L^2(\Omega; R^N) \mid div\, u = 0,$ and $u.\nu = 0$ on $\partial\Omega\}$, and we shall have to explain the meaning of $u.\nu$ on the boundary, the *normal trace* of u (physically, $\mathbf{u}.\nu = 0$ means that the flow is tangent, the so-called *slip condition*, while the condition $\mathbf{u} = \mathbf{0}$ is called the *no-slip condition*).

There is, however, another question, which is about the "pressure": have we really solved Stokes's equation

$$\frac{\partial u_i}{\partial t} - \nu\, \Delta u_i + \frac{\partial p}{\partial x_i} = f_i, i = 1, \dots, N, \text{ in } \Omega$$
$$div\, u \text{ in } \Omega \tag{21.1}$$
$$u(\cdot, 0) = u_0 \text{ in } \Omega?$$

Certainly, if $f \in L^2(0, T; H^{-1}(\Omega; R^N))$ and if the solution satisfies $u \in L^2(0, T; V) \cap C^0([0, T]; H)$, $\frac{\partial u}{\partial t} \in L^2(0, T; H^{-1}(\Omega; R^N))$, $p \in L^2(0, T; L^2(\Omega))$ $(\simeq L^2((0, T) \times \Omega))$, then using a test function $v \in V$, one deduces that

$$\frac{d(u, v)}{dt} + \nu\, a(u, v) = \langle f, v \rangle \text{ in } (0, T), \tag{21.2}$$

and with the data $u_0 \in H$, one has a solution of the abstract problem, solution which we know to be unique. The question is: can we deduce from the abstract formulation that $\frac{\partial u}{\partial t} \in L^2(0, T; H^{-1}(\Omega; R^N))$ and $p \in L^2(0, T; L^2(\Omega)) = L^2((0, T) \times \Omega)$? Of course, as one may add to p an arbitrary function of t without changing the equation (which is quite nonphysical, but is the price to pay for the unrealistic hypothesis of incompressibility), one must normalize p by asking, for example, that $\int_\Omega p(x, t)\, d\mathbf{x} = 0$ in $(0, T)$ (if Ω is bounded).

If $f_1 = 0$, the abstract formulation has given $\frac{du}{dt} = g \in L^2(0, T; V')$, and the problem comes from the fact that V' is not a space of distributions in

Ω, as $C_c^\infty(\Omega; R^N)$ is certainly not dense in V, as it is not even included in V because of the constraint $div\, u = 0$. I have mentioned that for a bounded open set Ω with Lipschitz boundary the elements of $H^{-1}(\Omega; R^N)$ orthogonal to V have the form $grad\, q$ with $q \in L^2(\Omega)$ (I have deduced it in the case where $meas(\Omega) < \infty$ from $X(\Omega) = L^2(\Omega)$, but I have not proven that last assertion). For any $L \in V'$, one can define $w_L \in V$ as the unique solution of $a(w_L, v) = L(v)$ for every $v \in V$, and even without the interpretation of this equation using the gradient of a pressure, one sees that $-\Delta w_L \in H^{-1}(\Omega; R^N)$, that it defines the same linear form than L on V, and that $||w_L|| = ||L||_*$. The element $g \in L^2(0, T; V')$ can be transformed in this way into a $w_g \in L^2(0, T : V)$ and for every $v \in V$ and every $\varphi \in C_c^\infty(0, T)$, one has

$$-\int_0^T (u(t), v)\frac{d\varphi}{dt}\, dt = \int_0^T \left\langle \frac{du(t)}{dt}, v\right\rangle \varphi\, dt = \int_0^T a(w_g, v)\varphi\, dt. \qquad (21.3)$$

One then defines $W_g \in H^1(0, T; V)$ by

$$W_g(t) = \int_0^t w_g(s)\, ds, \qquad (21.4)$$

and as $\int_0^T a(w_g, v)\varphi\, dt = -\int_0^T a(W_g, v)\frac{d\varphi}{dt}\, dt$ for $\varphi \in C_c^\infty(0, T)$, one deduces that $(u(t), v) - a(W_g, v)$ is a constant, and therefore taking $t = 0$ one has

$$(u(t), v) = (u_0, v) + a(W_g, v) \text{ in } (0, T), \text{ for every } v \in V. \qquad (21.5)$$

As $u - u_0 + \Delta W_g \in C^0(0, T; H^{-1}(\Omega; R^N))$, it shows that

$$u - u_0 + \Delta W_g = grad\, q \text{ for some } q \in C^0(0, T; L^2(\Omega)). \qquad (21.6)$$

Taking the derivative in t (in the sense of distributions), one finds that

$$\frac{\partial u}{\partial t} + \Delta w_g = grad\, p, \text{ but } p = \frac{\partial q}{\partial t}. \qquad (21.7)$$

In order to avoid having the "pressure" in a space of distributions, one can use a regularity theorem.

Lemma 21.1. *Assume that* $A^T = A$. *If* $u_0 \in V$ *and* $f \in L^2(0, T; H)$, *then* $\frac{\partial u}{\partial t}, A\, u \in L^2(0, T; H)$ *and* $u \in C^0([0, T]; V)$. *If* $u_0 \in H$ *and* $\sqrt{t} f \in L^2(0, T; H)$, *then* $\sqrt{t}\frac{\partial u}{\partial t}, \sqrt{t}\, A\, u \in L^2(0, T; H)$ *and* $\sqrt{t}\, u \in C^0([0, T]; V)$.

Proof: Formally, one can multiply the equation by $\frac{\partial u}{\partial t}$ or by $A\, u$, and one gets either $|u'|^2 + \frac{1}{2}a(u, u)' = (f, u')$ or $\frac{1}{2}a(u, u)' + |A\, u|^2 = (f, A\, u)$, and each implies $a(u, u)' \leq \frac{1}{2}|f|^2$, giving the bound of u in $L^\infty(0, T; V)$; then one gets either a bound for u' or a bound for $A\, u$ in $L^2(0, T; H)$, the other bound being given by the equation. Multiplying by u' can be done in the Faedo–Ritz–Galerkin approach, but not multiplying by $A\, u$ unless one uses a special basis; the same estimates are obtained in the finite-dimensional case, and the limit satisfies these bounds, but a little work is necessary in order to improve $u \in L^\infty(0,T;V)$ into $u \in C^0([0,T];V)$. For example, if $f \in H^1(0,T;H)$ and $u_0 \in D(A)$, then the hypotheses for time regularity are satisfied and $u \in H^1(0,T;V) \subset C^0([0,T];V)$, and one concludes by a density argument.

The regularizing effect in the case where one only has $u_0 \in H$, is obtained by multiplying by $t\,u'$ or $t\,A\,u$, the first one being more adapted to the Faedo–Ritz–Galerkin approach.■

In the case where $A^T \neq A$, one has the same result by replacing the hypothesis $u_0 \in V$ by $u_0 \in [D(A),H]_{1/2} = (D(A),H)_{1/2,2}$; however, Jacques-Louis LIONS has shown that if $D(A^T) = D(A)$, then the interpolation space mentioned is actually V.

For the application to Stokes's equation (or to the Navier–Stokes equation), one must be careful about the strange consequences of having identified H and its dual, as *this identification is not compatible with the usual basic identification of $L^2(\Omega)$ with its dual*. The hypothesis $f \in L^2(0,T;H)$ actually means $f \in L^2(0,T;H')$, so that $f \in L^2\big(0,T;L^2(\Omega;R^N)\big) = L^2\big((0,T) \times \Omega;R^N\big)$ is actually possible, without imposing $div\,f = 0$, which is one condition for taking values in H. One way to think about this question is to remember that gradients have no effect on V or H (if $p \in H^1(\Omega)$) and therefore any element of the form $h + grad\,p$ with $h \in H$ and $p \in H^1(\Omega)$ belongs to H'; when we shall study the space H, we shall actually prove in Lemma 23.1 that the orthogonal of H in $L^2(\Omega;R^N)$ is $\{grad\,p \mid p \in H^1(\Omega)\}$. When we interpret $A\,u \in L^2(0,T;H)$, it means $L^2(0,T;H')$, because $A\,u$ is only defined through the bilinear form $a(u,v)$ with $v \in V$ (although with a constant viscosity, the divergence of Δu is 0, the normal trace is not 0 in general). However, when we interpret $u' \in L^2(0,T;H)$, it does mean H and not H', as u takes values in $V \subset H$, and u' is a limit of $\frac{u(\cdot+\tau)-u(\cdot)}{\tau}$, which takes values in V.

If $f \in L^2\big(0,T;L^2(\Omega;R^N)\big) = L^2\big((0,T)\times\Omega;R^N\big)$ and $u_0 \in V$, then one has $u' \in L^2\big(0,T;H\big) \subset L^2\big(0,T;L^2(\Omega;R^N)\big) = L^2\big((0,T) \times \Omega;R^N\big)$, and therefore $S = \frac{\partial u}{\partial t} - \nu\,\Delta u - f \in L^2\big(0,T;H^{-1}(\Omega;R^N)\big)$. As $\int_0^T \langle S(t),v\rangle\varphi\,dt = 0$ for all $v \in V$ and all $\varphi \in C_c^\infty(0,T)$, one deduces that for almost every $t \in (0,T)$, $S(t)$ is orthogonal to V (using the separability of V), and therefore $S(t) = -grad\big(p(t)\big)$ with $p(t) \in L^2(\Omega)$; if one normalizes $p(t)$ by adding a constant so that its integral in Ω is 0 (if Ω is bounded), one has $\|p(t)\|_{L^2(\Omega)} \leq C\,\|S(t)\|_{H^{-1}(\Omega;R^N)}$ and therefore $S = -grad\,p$ with $p \in L^2\big(0,T;L^2(\Omega)\big) = L^2\big((0,T) \times \Omega\big)$.

Another case where the "pressure" may be estimated easily is the case $\Omega = R^N$, where one can use Fourier transform (in $\mathbf{x}$ alone): Stokes's equation becomes

$$\frac{\partial \mathcal{F}u}{\partial t} + 4\nu \pi^2 |\boldsymbol{\xi}|^2 \mathcal{F}u + 2i\pi \, \boldsymbol{\xi} \mathcal{F}p = \mathcal{F}f \text{ in } R^N \times (0,T)$$
$$(\mathcal{F}u.\boldsymbol{\xi}) = 0 \text{ in } R^N \times (0,T) \tag{21.8}$$
$$\mathcal{F}u(\cdot,0) = \mathcal{F}u_0,$$

and taking the scalar product with $\boldsymbol{\xi}$ gives

$$\mathcal{F}p = \frac{1}{2i\pi} \frac{(\mathcal{F}f.\boldsymbol{\xi})}{|\boldsymbol{\xi}|^2} \text{ in } R^N \times (0,T). \tag{21.9}$$

Of course, as Poincaré's inequality does not hold for R^N, one must be careful: if $f \in L^2(0,T; L^2(R^N; R^N)) = L^2((0,T) \times R^N; R^N)$, then one finds a bound for $grad\, p$, but not for p, and therefore one does not find that p takes values in $H^1(R^N)$, but in a different space (which was studied first by Jacques DENY[1] and Jacques-Louis LIONS, and presents a particular difficulty for $N = 2$); however, if

$$f_i = \sum_{j=1}^{N} \frac{\partial g_{ij}}{\partial x_j} \text{ in } R^N \times (0,T),$$
$$\text{with } g_{i,j} \in L^2(0,T; L^2(R^N)) = L^2((0,T) \times R^N), i,j = 1,\ldots,N, \tag{21.10}$$

then

$$p \in L^2(0,T; L^2(R^N)) = L^2((0,T) \times R^N) \text{ with}$$
$$\|p(\cdot,t)\|_{L^2(R^N)} \leq C \sum_{i,j=1}^{N} \|g_{ij}(\cdot,t)\|_{L^2(R^N)} \text{ a.e. } t \in (0,T). \tag{21.11}$$

For the Navier–Stokes equation, the dimension N plays a very important role, more important than for the stationary case. For $N = 2$, we shall be able to prove an existence and uniqueness result. For $N \geq 3$, we shall prove existence of weak solutions defined on $(0,T)$, but the uniqueness of weak solutions is an open question; for smooth data, we shall also prove that strong solutions exist locally, and that they are unique, but it is an open question to show that they can be extended up to T; however, for small smooth data the strong solution exists globally.

It should be noticed, however, that the approach for proving existence goes against physical intuition: there is a transport operator

$$\frac{D}{Dt} = \frac{\partial}{\partial t} + \sum_{j=1}^{N} u_j \frac{\partial}{\partial x_j}, \tag{21.12}$$

and there are various physical quantities transported along the flow, like *mass*, *momentum*, *energy* (or *vorticity*, *helicity*, *thermodynamic entropy*); each component of the velocity satisfies an equation

---

[1] Jacques DENY, French mathematician, born in 1918?. He worked at Université Paris XI (Paris-Sud), Orsay, France, where he was my colleague from 1975 to 1982.

$$\left(\frac{D}{DT} - \nu\Delta\right)u_i = f_i - \frac{\partial p}{\partial x_i} \quad \text{in } \Omega \times (0,T), \tag{21.13}$$

and the operator $\frac{D}{DT} - \nu\Delta$ which is applied to each u_i has good properties, some of the bound using the maximum principle and requiring little smoothness of the coefficients u_j, but as $\frac{\partial p}{\partial x_i}$ is needed and the equations are coupled via $div\, u = 0$, it would be useful to have an equation for p; taking the divergence of the equation gives

$$-\Delta p = -div\, f + \sum_{i,j=1}^{N} \frac{\partial u_i}{\partial x_j}\frac{\partial u_j}{\partial x_i}, \tag{21.14}$$

where one has used $div\, u = 0$ for simplifying the divergence of the nonlinear term. The difficulty comes from the fact that one does not have adequate boundary conditions for p. The nonlinearity appearing in the equation for p is actually a little special, with slightly better bounds than expected.

The usual approach, however, does not work with the operator $\frac{D}{DT} - \nu\Delta$, but cuts the operator $\frac{D}{DT}$ into two parts: sending the nonlinear term to play with f, one considers the Navier–Stokes equation as a perturbation of Stokes's equation, and this is obviously not a good idea (although it works in 2 dimensions), but no one has really found how to do better yet.

We have seen in studying the stationary Navier–Stokes equation that the nonlinear operator B defined by

$$\langle B(u,v), w\rangle = \int_{\Omega} u_j \frac{\partial v_i}{\partial x_j} w_i\, dx, \tag{21.15}$$

is continuous from $V \times V$ into V' for $N \le 4$, and satisfies $\langle B(u,v), w\rangle + \langle B(u,w), v\rangle = 0$ (in particular $\langle B(u,v), v\rangle = 0$, but for the evolution problem we shall need more precise bounds. As $||B(u,v)||_{V'} \le C \sum_{i,j=1}^{N} ||u_j v_i||_{L^4(\Omega)}$, one deduces

$$
\begin{aligned}
||B(u,u)||_* &\le C\,||u||^2 \text{ in dimension } N = 4,\\
||B(u,u)||_* &\le C\,||u||^{3/2}|u|^{1/2} \text{ in dimension } N = 3,\\
||B(u,u)||_* &\le C\,||u||\,|u| \text{ in dimension } N = 2.
\end{aligned}
\tag{21.16}
$$

The first two inequalities follow from the Sobolev embedding theorem, which implies $H_0^1(\Omega) \subset L^4(\Omega)$ in dimension $N = 4$, $H_0^1(\Omega) \subset L^6(\Omega)$ in dimension $N = 3$, the second using also Hölder's inequality $||v||_{L^4} \le ||v||_{L^6}^{3/4}||u||_{L^2}^{1/4}$. The third inequality, which Jacques-Louis LIONS attributed to Olga LADYZHENS-KAYA, uses the same method with which Emilio GAGLIARDO and Louis NIRENBERG independently proved the Sobolev embedding theorem 16.1: $|u|^2 \le F(x_2) = \int_R |u|\left|\frac{\partial u}{\partial x_1}\right| dx_1$ and $|u|^2 \le G(x_1) = \int_R |u|\left|\frac{\partial u}{\partial x_2}\right| dx_2$ give $\int_{R^2} |u|^4\, dx \le \int_{R^2} F(x_2)G(x_1)\, dx_1\, dx_2 = \int_{R^2} |u|\left|\frac{\partial u}{\partial x_1}\right| dx \int_{R^2} |u|\left|\frac{\partial u}{\partial x_1}\right| dx$, implying the desired result by using the Cauchy–Schwarz inequality.

The natural bounds for a solution are $u \in L^\infty(0, T; H)$, which corresponds to the fact that the kinetic energy is bounded, and $u \in L^2(0, T; V)$, which corresponds to the fact that the energy dissipated by viscosity between time 0 and T is bounded. The dependence with N becomes then

$$u \in L^2(0, T; V) \cap L^\infty(0, T; H) \text{ imply}$$
$$B(u, u) \in \begin{cases} L^1(0, T; V') \text{ in dimension } N = 4, \\ L^{4/3}(0, T; V') \text{ in dimension } N = 3, \\ L^2(0, T; V') \text{ in dimension } N = 2, \end{cases} \tag{21.17}$$

and therefore it is only for $N = 2$ that $B(u, u)$ falls into a space which is allowed for the (abstract) Stokes equation; for $N \geq 3$, the nonlinearity is then a much too strong nonlinear operator, and the Navier–Stokes equation is then not a mere perturbation of Stokes's equation.

[Taught on Wednesday March 3, 1999. There were no classes on Monday March 1, which was the spring mid-semester break.]

22

Uniqueness in 2 dimensions

For $\Omega \subset R^N$, we have defined $V = \{u \in H_0^1(\Omega; R^N) \mid div\, u = 0 \text{ in } \Omega\}$ and $H = \{u \in L^2(\Omega; R^N) \mid div\, u = 0 \text{ in } \Omega \text{ and } u.\nu = 0 \text{ on } \partial\Omega\}$, and we shall see in Lecture 23 the meaning of $u.\nu$, where ν is the exterior normal to $\partial\Omega$. If V is dense in H (which we shall show if Ω is bounded with $\partial\Omega$ smooth enough), then one can consider the Navier–Stokes equation (in the incompressible case) as an abstract evolution equation $u' + B(u, u) + A\, u = f$ in $(0, T)$, where $A \in \mathcal{L}(V, V')$ is given by $\langle A\, u, v \rangle = \nu \int_\Omega (\sum_{i,j=1}^N \frac{\partial u_i}{\partial x_j} \frac{\partial v_i}{\partial x_j})\, dx$ for all $u, v \in V$, and B is the bilinear continuous mapping from $V \times V$ into V' (for $N \leq 4$), given by $\langle B(u, v), w \rangle = \int_\Omega (\sum_{i,j=1}^N u_j \frac{\partial v_i}{\partial x_j} w_i)\, dx$ for all $u, v, w \in V$, and it satisfies $\langle B(u, v), w \rangle + \langle B(u, w), v \rangle = 0$, and in particular $\langle B(u, v), v \rangle = 0$. As mentioned before, $u' + B(u, u)$ is $\frac{Du}{Dt}$ where $\frac{D}{Dt} = \frac{\partial}{\partial t} + \sum_{j=1}^N u_j \frac{\partial}{\partial x_j}$ is the operator of derivation along the flow, and it is not a good idea to handle this operator by cutting it into two parts; geometers use a formalism involving *affine connections*, but up to now it has not helped us understand more on that question of transport of various quantities along the flow, and it is therefore not clear yet what the right way to handle this operator could be (the most difficult test for knowing if a framework is right is that it should work for turbulent flows; this is a question of identifying *effective equations*, i.e. a problem of *homogenization*, which I shall describe in Lecture 35). However, this bad way of treating the nonlinearity does not hurt for $N = 2$ and one can prove *uniqueness*, or more precisely the following *continuous dependence* with respect to the data.

Lemma 22.1. *For $N = 2$, Ω being any open set in R^2, if*

$$f_j = \sum_{k=1}^2 \frac{\partial g_{jk}}{\partial x_k}, j = 1, 2, \text{ with } g_{jk} \in L^2(0, T; L^2(\Omega; R^2)) = L^2((0, T) \times \Omega; R^2),$$

$$(22.1)$$

and if

$$u_j' + B(u_j, u_j) + A u_j = f_j, j = 1, 2, \ in \ (0, T)$$
$$u_j(0) = u_{0j} \in H, j = 1, 2,$$

(22.2)

with $u_j \in L^2(0, T; V) \cap C^0([0, T]; H), j = 1, 2,$ then one has

$$||u_2 - u_1||^2_{C^0([0,T];H)} + \nu ||u_2 - u_1||^2_{L^2(0,T;V)} \le C \big(|u_{02} - u_{01}|^2_H +$$
$$\tfrac{1}{\nu} \sum_{k=1}^2 ||g_{2k} - g_{1k}||^2_{L^2(0,T;L^2(\Omega;R^2))} \big) K$$

(22.3)

with $K(u_1, u_2) = exp \big(\tfrac{C}{\nu} \int_0^T \min\{|grad \, u_1|^2_{L^2}, |grad \, u_2|^2_{L^2}\} \, dt \big),$

where C is a universal constant.

Proof: As $u_j \in L^2(0, T; V) \cap C^0([0, T]; H)$ implies $B(u_j, u_j) \in L^2(0, T; V')$ in 2 dimensions, one has $u_j \in W(0, T)$. Subtracting the two equations and multiplying by $u_2 - u_1$ gives

$$\tfrac{1}{2} \tfrac{d}{dt} \big(|u_2 - u_1|^2_H \big) + \langle B(u_2, u_2) - B(u_1, u_1), u_2 - u_1 \rangle + \nu |grad(u_2 - u_1)|^2_{L^2} =$$
$$- \sum_{k=1}^2 \big(g_{2k} - g_{1k}, grad(u_2 - u_1) \big).$$

(22.4)

Because one has $- \sum_{k=1}^2 \big(g_{2k} - g_{1k}, grad(u_2 - u_1) \big) \le \tfrac{\nu}{3} |grad(u_2 - u_1)|^2_{L^2} + \tfrac{3}{4\nu} \sum_{k=1}^2 |g_{2k} - g_{1k}|^2_{L^2}$ and $\langle B(u_2, u_2) - B(u_1, u_1), u_2 - u_1 \rangle = \langle B(u_2, u_2 - u_1) + B(u_2 - u_1, u_1), u_2 - u_1 \rangle = \langle B(u_2 - u_1, u_1), u_2 - u_1 \rangle = B(u_2 - u_1, u_2), u_2 - u_1 \rangle,$ one deduces that

$$|\langle B(u_2, u_2) - B(u_1, u_1), u_2 - u_1 \rangle| \le$$
$$C \min\{|grad \, u_1|_{L^2}, |grad \, u_2|_{L^2}\} |u_2 - u_1|_H \, |grad(u_2 - u_1)|_{L^2},$$

(22.5)

where C is a universal constant, independent of Ω (one does not assume here that Poincaré's inequality holds, which is the reason for the restriction on f_j). Using the bound $|\langle B(u_2, u_2) - B(u_1, u_1), u_2 - u_1 \rangle| \le \tfrac{\nu}{3} |grad(u_2 - u_1)|^2_{L^2} + \tfrac{3C^2}{4\nu} \min\{|grad \, u_1|^2_{L^2}, |grad \, u_2|^2_{L^2}\} |u_2 - u_1|^2_H,$ one obtains

$$\tfrac{1}{2} \tfrac{d}{dt} \big(|u_2 - u_1|^2_H \big) + \tfrac{\nu}{3} |grad(u_2 - u_1)|^2_{L^2} \le \tfrac{3}{4\nu} \sum_{k=1}^2 |g_{2k} - g_{1k}|^2_{L^2} +$$
$$\tfrac{3C^2}{4\nu} \min\{|grad \, u_1|^2_{L^2}, |grad \, u_2|^2_{L^2}\} |u_2 - u_1|^2_H,$$

(22.6)

and one concludes with Gronwall's inequality.∎

The same type of proof applies in dimension $N = 3$ or higher, if the solutions are regular enough.

It would be better if one did not cut the operator $\tfrac{D}{Dt}$ into two pieces, and if one could use the fact that the maximum principle applies to $\tfrac{D}{Dt} - \nu \Delta$, but a difficulty appears because of the "pressure". As a way to handle a similar situation, I want to show a uniqueness result of Michel ARTOLA[1] for an equation

---

[1] Michel ARTOLA, French mathematician, born in 1932. He worked at Université de Bordeaux I, Talence, France.

$$\frac{\partial u}{\partial t} - div(A(\mathbf{x}, u)grad\, u) = f, \qquad (22.7)$$

with Dirichlet conditions (there is an extension to the case $f(\mathbf{x}, u)$ that we worked out together, with the type of proof that I show, which is a little more general than Michel ARTOLA's original proof, which extended a result of Neil TRUDINGER). Of course A is a Carathéodory[2] function; one assumes that $|A(\mathbf{x}, u)| \leq \beta$ and $(A(\mathbf{x}, u)\boldsymbol{\xi}.\boldsymbol{\xi}) \geq \alpha|\boldsymbol{\xi}|^2$ for all $\boldsymbol{\xi} \in R^N$ with $\alpha > 0$, and $|A(\mathbf{x}, u) - A(\mathbf{x}, v)| \leq \omega(|v - u|)$ with ω nondecreasing and satisfying $\int_0^1 \frac{ds}{\omega^2(s)} = +\infty$. Under these conditions, if $f = div\, g$ with $g \in L^2(0, T; L^2(\Omega; R^N)) = L^2((0, T) \times \Omega; R^N)$ and $u_0 \in L^2(\Omega)$ there is a unique solution $u \in C^0([0, T]; L^2(\Omega)) \cap L^2(0, T; H_0^1(\Omega))$ (actually, we also showed that there is a contraction property in $L^1(\Omega)$). If u_1 and u_2 are two such solutions, one subtracts the two equations and one multiplies by $\varphi'(u_2 - u_1)$ where φ is convex, $\varphi'(0) = 0$ and φ' is bounded, which gives

$$\frac{d}{dt}\left(\int_\Omega \varphi(u_2 - u_1)\, d\mathbf{x}\right) + \int_\Omega \varphi''(u_2 - u_1)\big(A(\mathbf{x}, u_2)grad\, u_2 - A(\mathbf{x}, u_1)grad\, u_1, grad(u_2 - u_1)\big)\, d\mathbf{x} = 0. \qquad (22.8)$$

Using $A(\mathbf{x}, u_2)grad\, u_2 - A(\mathbf{x}, u_1)grad\, u_1 = A(\mathbf{x}, u_2)grad(u_2 - u_1) + (A(\mathbf{x}, u_2) - A(\mathbf{x}, u_1))grad\, u_1$, one obtains

$$\frac{d}{dt}\left(\int_\Omega \varphi(u_2 - u_1)\, d\mathbf{x}\right) + \alpha \int_\Omega \varphi''(u_2 - u_1)|grad(u_2 - u_1)|^2\, d\mathbf{x} \leq \int_\Omega \varphi''(u_2 - u_1)\omega(|u_2 - u_1|)|grad\, u_1|\,|grad(u_2 - u_1)|\, d\mathbf{x}, \qquad (22.9)$$

from which one deduces

$$\frac{d}{dt}\left(\int_\Omega \varphi(u_2 - u_1)\, d\mathbf{x}\right) \leq C \int_\Omega \varphi''(u_2 - u_1)\omega^2(|u_2 - u_1|)|grad\, u_1|^2\, d\mathbf{x} \leq C \max_{s \in R}\{\varphi''(s)\omega^2(|s|)\} \int_\Omega |grad\, u_1|^2\, d\mathbf{x}. \qquad (22.10)$$

One then chooses $0 < \varepsilon < \eta$, and $\varphi_{\varepsilon\eta}$ even and defined by $\varphi_{\varepsilon\eta}(0) = \varphi'_{\varepsilon\eta}(0) = 0$ and $\varphi''_{\varepsilon\eta}(s) = \frac{1}{\omega^2(s)}$ for $\varepsilon < s < \eta$ and $\varphi''_{\varepsilon\eta} = 0$ in $(0, \varepsilon)$ and on $(\eta, +\infty)$; this gives after integration

$$\int_\Omega \varphi_{\varepsilon\eta}(|u_2(\mathbf{x}, t) - u_1(\mathbf{x}, t)|)\, d\mathbf{x} \leq C \int_0^t \int_\Omega |grad\, u_1|^2\, d\mathbf{x}\, dt \quad \text{for } t \in [0, T]. \qquad (22.11)$$

Fixing $\eta > 0$, one lets ε tend to 0, and as $\varphi_{\varepsilon\eta} \to +\infty$ on $(\eta, +\infty)$, one deduces that $|u_2(\mathbf{x}, t) - u_1(\mathbf{x}, t)| \leq \eta$ for almost every $\mathbf{x} \in \Omega$, and then letting η tend to 0, one deduces that $u_2 = u_1$.

Our next step is to study the functional space H, and show that there is a notion of normal trace $(u.\nu)$ on the boundary, so that the definition of H makes sense. Then we shall check that V is dense in H.

[Taught on Friday March 5, 1999.]

---

[2] Constantin CARATHÉODORY, Greek-born mathematician, 1873–1950. He worked in Berlin, Germany, Smyrna (then in Greece, now Izmir, Turkey), and München (Munich), Germany.

23

Traces

I think that it was Jacques-Louis LIONS who introduced the space

$$H(div; \Omega) = \{u \in L^2(\Omega; R^N) \mid div\, u \in L^2(\Omega)\}, \qquad (23.1)$$

and proved that one can give a meaning to $u.\nu$ on the boundary $\partial\Omega$ if Ω has a Lipschitz boundary.

1) One notices that $H(div; \Omega)$ is a local space, i.e. $\theta\, u \in H(div, \Omega)$ for all $u \in H(div; \Omega)$ and $\theta \in C^\infty(R^N)$ as $div(\theta\, u) = \theta\, div\, u + (grad\, \theta.\mathbf{u})$ (notice that we plan to use the results for the case $div\, u = 0$, but that property is lost by multiplication by smooth functions).

2) One shows that $C^\infty(\overline{\Omega}; R^N)$ is dense in $H(div, \Omega)$ if the boundary is smooth enough: after using a partition of unity for localizing the problem, one regularizes each $\theta_i\, u$ by convolution with a suitable regularizing sequence adapted to the support of θ_i, and this is possible if Ω is an open set with compact boundary and if near each point of the boundary Ω is only on one side of the boundary and the boundary has an equation $x_N = F(\mathbf{x'})$ in some basis with F continuous (one can have unbounded boundaries if one asks for a global equation of the corresponding piece with F uniformly continuous).

3) Assuming that Ω has a Lipschitz (compact) boundary, so that one can define the normal to the boundary and traces on the boundary for functions in $H^1(\Omega)$, one has the formula

$$\int_\Omega \left((\mathbf{u}.grad\, \varphi) + \varphi\, div\, u\right) d\mathbf{x} = \int_{\partial\Omega} (\mathbf{u}.\boldsymbol{\nu})\varphi\, d\sigma, \qquad (23.2)$$

for $u \in C^\infty(\overline{\Omega}; R^N)$ and $\varphi \in H^1(\Omega)$. The left side of the equation is a bilinear continuous form on $H(div; \Omega) \times H^1(\Omega)$ and therefore the right side is also continuous for that topology, but the right side is 0 for $\varphi \in H_0^1(\Omega)$ and therefore it is actually defined on the quotient $H^1(\Omega)/H_0^1(\Omega)$; here a natural choice is to use $T(\Omega)$, the space of traces of functions of $H^1(\Omega)$, equipped with the quotient norm

$$||v||_{T(\Omega)} = \inf\{||w||_{H^1(\Omega)} \mid trace(w) = v\}, \tag{23.3}$$

and then the right side is continuous for the norm of $H(div; \Omega) \times T(\Omega)$, and therefore by density $u.\nu$ is defined on $H(div; \Omega)$ as a linear continuous form on $T(\Omega)$.

If Ω has a compact Lipschitz boundary, then $T(\Omega) = H^{1/2}(\partial\Omega)$, and the proof (and definition of the space) is easily derived from the property for R^N_+, which I review below using the Fourier transform. The interest of the preceding result is that it applies even if the boundary is not so smooth and the trace space $T(\Omega)$ has not been characterized.

It is important to notice that one cannot define each of the terms $u_j\nu_j$ on the boundary, but only their sum. There is a framework using *differential forms* which is also useful to know (two years ago, I asked Jacques-Louis LIONS if he was aware of this aspect when he worked on the preceding question, and he said that he was not). For smooth functions, one considers the $(N-1)$-form $\omega = \sum_{i=1}^{N}(-1)^{i-1}u_i\,dx_1 \wedge \ldots \wedge dx_{i-1} \wedge dx_{i+1} \wedge \ldots \wedge dx_N$ (also written $\sum_{i=1}^{N}(-1)^{i-1}u_i\,\widehat{dx_i}$), whose exterior derivative is $d\omega = (div\,u)\,d\mathbf{x}$. In the case of smooth coefficients, one can restrict a p-form to a smooth manifold, as it is a p-linear alternating form and therefore it needs p vectors to act upon and its restriction on the manifold uses only vectors from the tangent space to the manifold; if one restricts the $(N-1)$-form ω to the (smooth) boundary one obtains a $(N-1)$-form which has only one coefficient (as the dimension of the boundary is $N-1$) and that coefficient is $\mathbf{u}.\boldsymbol{\nu}$; it is natural, but not straightforward, that one can relax the hypotheses of regularity and still be able to define the intrinsic quantity $u.\nu$.

There is another space which is important in applications (to electromagnetism, but also to fluids once one considers the vorticity), i.e.

$$H(curl; \Omega) = \{u \in L^2(\Omega; R^3) \mid curl\,u \in L^2(\Omega; R^3)\}, \tag{23.4}$$

and here one should consider the 1-form $\omega = \sum_{i=1}^{3} u_i\,dx_i$, for which one has $d\omega = \sum_{i=1}^{3}(curl\,u)_i\,\widehat{dx_i}$, and $div(curl\,u) = 0$ expresses the fact that $d\,d = 0$; in the smooth case the exterior derivative commutes with the restriction and therefore the restriction is a 1-form on the boundary which has its exterior derivative well defined. It is the tangential component of u which is well defined on $H(curl; \Omega)$, with a differential restriction corresponding to writing the exterior derivative; in the mid 1980s, in answer to a question of Michel CESSENAT,[1] I characterized the traces in the case where Ω is a half space, and he discovered later that it had already been done in the smooth case by

---

[1] Michel CESSENAT, French mathematician. He works at Commissariat à l'Énergie Atomique, France.

PAQUET;[2] a few years ago I extended the result to the case with Lipschitz boundary.[3]

For $u \in H^1(R_+^N)$ its trace on $x_N = 0$ has been shown to belong to $L^2(R^{N-1})$ (after extending u to a function in $H^1(R^N)$); we want to show that it actually belongs to $H^{1/2}(R^{N-1})$, and that all elements of $H^{1/2}(R^{N-1})$ can be traces. Of course for $s \geq 0$, the space $H^s(R^m)$ is defined by the Fourier transform as

$$H^s(R^m) = \{u \in L^2(R^m) \mid |\boldsymbol{\xi}|^s |\mathcal{F}u| \in L^2(R^N)\}, \tag{23.5}$$

but for $s < 0$ it is $\{u \in \mathcal{S}'(R^N) \mid (1 + |\boldsymbol{\xi}|^2)^{s/2} \mathcal{F}u \in L^2(R^N)\}$ (because $|\boldsymbol{\xi}|^s$ is not smooth enough to be a multiplier on $\mathcal{S}'(R^N)$).

For $u \in C_c^\infty(R^N)$, let $v \in C_c^\infty(R^{N-1})$ be the restriction of u to $x_N = 0$; one defines the Fourier transforms of u and v

$$\begin{aligned} \mathcal{F}u(\boldsymbol{\xi}', \xi_N) &= \int_{\mathbf{x} \in R^N} u(\mathbf{x}', x_N) e^{-2i\pi(\mathbf{x}' \cdot \boldsymbol{\xi}') - 2i\pi\, x_N \xi_N} \, d\mathbf{x}' \, dx_N \\ \mathcal{F}v(\boldsymbol{\xi}') &= \int_{\mathbf{x}' \in R^{N-1}} u(\mathbf{x}', 0) e^{-2i\pi(\mathbf{x}' \cdot \boldsymbol{\xi}')} \, d\mathbf{x}', \end{aligned} \tag{23.6}$$

and the critical relation is

$$\mathcal{F}v(\boldsymbol{\xi}') = \int_R \mathcal{F}u(\boldsymbol{\xi}', \xi_N) \, d\xi_N. \tag{23.7}$$

Indeed, one defines $w_{\boldsymbol{\xi}'}$ by $w_{\boldsymbol{\xi}'}(x_N) = \int_{\mathbf{x}' \in R^{N-1}} u(\mathbf{x}', x_N) e^{-2i\pi(\mathbf{x}' \cdot \boldsymbol{\xi}')} \, d\mathbf{x}'$, then $w_{\boldsymbol{\xi}'} \in \mathcal{S}(R)$ and therefore $w_{\boldsymbol{\xi}'}(0) = \int_R \mathcal{F}w_{\boldsymbol{\xi}'}(\xi_N) d\xi_N$, because $w_{\boldsymbol{\xi}'} = \overline{\mathcal{F}}\mathcal{F}w$ or $\mathcal{F}1 = \delta_0$ (guessed by good physicists like DIRAC, but only proved by Laurent SCHWARTZ in his extension of the definition of the Fourier transform); after putting back explicitly the dependence in $\boldsymbol{\xi}'$, it is exactly the relation mentioned. Using the Cauchy–Schwarz inequality, one deduces that

$$\begin{aligned} |\mathcal{F}v(\boldsymbol{\xi}')| &\leq \int_R \frac{1}{\sqrt{1 + |\boldsymbol{\xi}'|^2 + |\xi_N|^2}} \sqrt{1 + |\boldsymbol{\xi}'|^2 + |\xi_N|^2} \, |\mathcal{F}u(\boldsymbol{\xi}', \xi_N)| \, d\xi_N \\ &\leq \left(\int_R \frac{d\xi_N}{1 + |\boldsymbol{\xi}'|^2 + |\xi_N|^2} \right)^{1/2} \left(\int_R (1 + |\boldsymbol{\xi}'|^2 + |\xi_N|^2) |\mathcal{F}u(\boldsymbol{\xi}', \xi_N)|^2 \, d\xi_N \right)^{1/2} \\ &= \left(\frac{\pi}{\sqrt{1 + |\boldsymbol{\xi}'|^2}} \right)^{1/2} \left(\int_R (1 + |\boldsymbol{\xi}'|^2 + |\xi_N|^2) |\mathcal{F}u(\boldsymbol{\xi}', \xi_N)|^2 \, d\xi_N \right)^{1/2}, \end{aligned} \tag{23.8}$$

and therefore $(1 + |\boldsymbol{\xi}'|^2)^{1/4} \mathcal{F}v \in L^2(R^{N-1})$.

Conversely, given $v \in H^{1/2}(R^{N-1})$, one must find $u \in H^1(R^N)$ such that $\mathcal{F}v(\boldsymbol{\xi}') = \int_R \mathcal{F}u(\boldsymbol{\xi}', \xi_N) \, d\xi_N$, and one chooses

$$\mathcal{F}u(\boldsymbol{\xi}', \xi_N) = \mathcal{F}v(\boldsymbol{\xi}') \varphi\left(\frac{\xi_N}{\sqrt{1 + |\boldsymbol{\xi}'|^2}} \right) \frac{1}{\sqrt{1 + |\boldsymbol{\xi}'|^2}}, \tag{23.9}$$

---

[2] Luc PAQUET, Belgian-born mathematician. He works at Université de Valenciennes et du Hainaut-Cambrésis, Valenciennes, France.

[3] I taught it in a course where I had discussed spaces useful for electromagnetism, and then wrote it for a conference in honor of Michel ARTOLA; the proceedings have not appeared yet.

where $\varphi \in C_c^\infty(R)$ satisfies $\int_R \varphi(s)\,ds = 1$. It remains to check that $u \in H^1(R^N)$, and this follows from

$$
\begin{aligned}
\int_{R^N}(1+|\boldsymbol{\xi}|^2)|\mathcal{F}u(\boldsymbol{\xi})|^2\,d\boldsymbol{\xi} &= \\
\int_{R^N}(1+|\boldsymbol{\xi}'|^2+|\xi_N|^2)|\mathcal{F}v(\boldsymbol{\xi}')|^2\varphi^2\left(\frac{\xi_N}{\sqrt{1+|\boldsymbol{\xi}'|^2}}\right)\frac{1}{1+|\boldsymbol{\xi}'|^2}\,d\boldsymbol{\xi}'\,d\xi_N & \\
= \int_{R^{N-1}}|\mathcal{F}v(\boldsymbol{\xi}')|^2\left(\int_R \frac{1+|\boldsymbol{\xi}'|^2+|\xi_N|^2}{1+|\boldsymbol{\xi}'|^2}\varphi^2\left(\frac{\xi_N}{\sqrt{1+|\boldsymbol{\xi}'|^2}}\right)d\xi_N\right)d\boldsymbol{\xi}' & \qquad (23.10) \\
= \left(\int_R(1+s^2)\varphi^2(s)\,ds\right)\int_{R^{N-1}}\sqrt{1+|\boldsymbol{\xi}'|^2}|\mathcal{F}v(\boldsymbol{\xi}')|^2\,d\boldsymbol{\xi}'. &
\end{aligned}
$$

One defines $H = \{u \in L^2(\Omega; R^N) \mid div\,u = 0 \text{ in } \Omega, u.\nu = 0 \text{ on } \partial\Omega\}$; as one imposes $div\,u = 0$, one has $u \in H(div; \Omega)$ and therefore $u.\nu$ has a meaning, and more precisely by (23.2), $u.\nu = 0$ means that for all $\varphi \in H^1(\Omega)$ one has $\int_\Omega ((grad\,\varphi.u) + \varphi\,div\,u)\,dx = 0$ for all $\varphi \in H^1(\Omega)$, and $u \in H$ implies then $\int_\Omega (grad\,\varphi.u)\,dx = 0$ for all $\varphi \in H^1(\Omega)$, as $div\,u = 0$. One sees then that H is orthogonal to the subspace of gradients of functions of $H^1(\Omega)$.

Lemma 23.1. *If the injection of $H^1(\Omega)$ into $L^2(\Omega)$ is compact, then the orthogonal of H in $L^2(\Omega; R^N)$ is the (closed) subspace of $grad\,\varphi$ for $\varphi \in H^1(\Omega)$.*

Proof: One can apply the equivalence lemma 13.3 to the case where $E_1 = H^1(\Omega)$, $A = grad$ with $E_2 = L^2(\Omega; R^N)$, and B is the (compact) injection of $H^1(\Omega)$ into $E_3 = L^2(\Omega)$; the equivalence lemma 13.3 asserts that the range of A is closed. Let us check that the orthogonal of $R(A)$ is H, which proves that the orthogonal of H is the closure of $R(A)$, i.e. $R(A)$ itself.

Assume that $u \in L^2(\Omega; R^N)$ is orthogonal to $R(A)$; taking $\varphi \in C_c^\infty(\Omega)$ and noticing that $\langle div\,u, \varphi\rangle = -\langle u, grad\,\varphi\rangle = 0$ shows that $div\,u = 0$ in the sense of distributions. This proves that $u \in H(div; \Omega)$, and using already the information that $div\,u = 0$ in Ω, one deduces that $\int_\Omega (grad\,\varphi.u)\,dx = \langle u.\nu, trace(\varphi)\rangle$ for all $\varphi \in H^1(\Omega)$, and as the left side is 0 by definition of u, the right side is 0 and therefore $u.\nu = 0$ (as a linear continuous form on the space of traces of functions of $H^1(\Omega)$), i.e. $u \in H$.∎

We can now look at the important question of density of V into H, as this is basic to the framework used.

Lemma 23.2. *If $meas(\Omega) < \infty$ and if $L^2(\Omega) = X(\Omega)$ (which is $\{u \in H^{-1}(\Omega) \mid \frac{\partial u}{\partial x_j} \in H^1(\Omega) \text{ for } j = 1, \ldots, N\}$), then V is dense in H.*

Proof: Let $h \in L^2(\Omega; R^N)$ belong to the orthogonal of V in $L^2(\Omega; R^N)$; then h can be considered an element of $H^{-1}(\Omega; R^N)$, orthogonal to V for the duality product, and therefore of the form $grad\,p$ with $p \in L^2(\Omega)$, and as $grad\,p \in L^2(\Omega; R^N)$ it means that $p \in H^1(\Omega)$. Therefore $h = grad\,p$ with $p \in H^1(\Omega)$ and so h is orthogonal to H, which proves that V is dense in H.∎

The fact that $X(\Omega) = L^2(\Omega)$ requires more regularity of the boundary than the simple compactness of $H^1(\Omega)$ into $L^2(\Omega)$: I had noticed in the fall that it is not true in a plane domain of the form $\{(x,y) \mid 0 < x < 1, 0 < y < x^2\}$, and François MURAT had pointed out that it had already been observed for similar domains by Giuseppe GEYMONAT and Gianni GILARDI, by a much more technical proof than mine (already sketched in footnote 15.2).

One may avoid this problem by taking a definition of H which does not mention a trace on the boundary: $H = \{u \in L^2(\Omega; R^N) \mid div\,u = 0$ in Ω, and $\int_\Omega (grad\,\varphi.u)\,dx = 0$ for all $\varphi \in H^1(\Omega)\}$. Then as in Lemma 23.1, if the injection of $H^1(\Omega)$ into $L^2(\Omega)$ is compact, the subspace of gradients of functions of $H^1(\Omega)$ is closed and is the orthogonal of H in $L^2(\Omega; R^N)$. The definition of V involves $H_0^1(\Omega)$, which is defined without any reference to the regularity of the boundary, as the closure of $C_c^\infty(\Omega)$ into $H^1(\Omega)$. If $h \in L^2(\Omega; R^N)$ is orthogonal to V, then without knowing anything on the regularity of the boundary, one has $h = grad\,S$ for a distribution S, if one invokes a theorem of DE RHAM, or for $S \in L_{loc}^2(\Omega)$, if one uses $X(\omega_k) = L^2(\omega_k)$ for an increasing sequence ω_k of connected open sets with smooth boundaries, whose union is Ω (if a distribution S has its gradient in $L^2(\Omega; R^N)$), one also deduces that $S \in H_{loc}^1(\Omega)$ by classical methods of partial differential equations). It remains to look for hypotheses which imply that every $S \in L_{loc}^2(\Omega)$ with $grad\,S \in L^2(\Omega; R^N)$ actually belongs to $H^1(\Omega)$, and that can be proven for a bounded open set which is locally on one side of the boundary, with a local equation $x_N > F(\mathbf{x}')$ with F continuous, by using the techniques already described.

Then, in interpreting the solution of Stokes's equation with $u_0 \in V$ and $f \in L^2(0, T; L^2(\Omega; R^N)) = L^2((0, T) \times \Omega; R^N)$, one finds $\frac{\partial u}{\partial t} \in L^2(0, T; H)$ and $u \in C^0([0, T]; V)$, and therefore there exists $g \in L^2(0, T; L^2(\Omega; R^N)) = L^2((0, T) \times \Omega; R^N)$ such that $a(u(t), v) = (g(t), v)$ for almost every $t \in (0, T)$, and for every $v \in V$. For t outside a set of measure 0, one has then $-\Delta u + grad\,p = g$ and $div\,u = 0$, and one deduces $\Delta p = div\,g$ in Ω and therefore $p \in H_{loc}^1(\Omega)$, and $u \in H_{loc}^2(\Omega; R^N)$. Even if Ω is a bounded open set with Lipschitz boundary, one cannot always deduce that $p \in H^1(\Omega)$, as the $H^2(\Omega)$ regularity for u which is implied is known to be false for some open sets with Lipschitz boundary (in the case $N = 2$, it may be checked in polar coordinates).

There is another type of density which I have not discussed in this course, which is the density of $\mathcal{V} = V \cap C_c^\infty(\Omega; R^N)$ into V. It is true for bounded domains with smooth boundary, but John HEYWOOD[4] has noticed that it is not true for some unbounded domains in R^3. For example, if one considers a thick screen with a smooth hole in it and Ω is the complement, then all functions in $\mathcal{V}$ have zero flux through the hole; however, there are functions in V with nonzero flux through the hole (and that corresponds to mass

---

[4] John Growes HEYWOOD, Canadian mathematician. He works at UBC (University of British Columbia) in Vancouver, British Columbia (Canada).

coming from ∞ on one side and going to ∞ on the other side), and they cannot be approached by elements from $\mathcal{V}$. Olga LADYZHENSKAYA and Seva SOLONNIKOV[5] have extended his analysis to other unbounded open sets in R^N with $N \geq 3$.

The application to existence of weak solutions to the Navier–Stokes equation is then almost straightforward, but we shall have to prove a compactness result. One takes a special basis in the case where $f \in L^2(0, T; H^{-1}(\Omega; R^N))$ and one solves the approximate equation for u_n a combination of $e_1, \ldots, e_n$,

$$\left\langle \frac{du_n}{dt}, e_k \right\rangle + \langle B(u_n, u_n), e_k \rangle + \langle A\, u_n, e_k \rangle = \langle f, e_k \rangle \text{ and} \tag{23.11}$$
$$(u_n(0), e_k) = (u_0, e_k), \text{ for } k = 1, \ldots, n.$$

The solution exists on an interval $(0, T_c)$ with $T_c \leq T$, and one deduces that $T_c = T$ from the bound obtained by taking the combination of e_k corresponding to u_n, i.e.

$$\frac{1}{2} \frac{d|u_n|^2}{dt} + \nu \,|grad\, u_n|^2 = \langle f, u_n \rangle, \tag{23.12}$$

and using Gronwall's inequality, one obtains a bound independent of t, so that $T_c = T$, and independent of n: u_n stays in a bounded set of $C^0([0, T]; H)$ and in a bounded set of $L^2(0, T; V)$ so that one can extract a weakly converging subsequence. In order to pass to the limit in the nonlinear term $B(u_n, u_n)$, we shall use a compactness argument which needs information on the derivative in t of u_n, and this is the reason for the special choice of the basis: $B(u_n, u_n)$ stays bounded in $L^p(0, T; V')$ with $p = 2$ for $N = 2$, $p = 4/3$ for $N = 3$, and $p = 1$ for $N = 4$, and therefore $\frac{du_n}{dt}$ stays bounded in that space too.

The question of regularity mentioned previously has to do with corners (it is a question which has been extensively studied by Pierre GRISVARD[6]). In R^2, if $v = r^\alpha f(\theta)$ then from $dv = \frac{\partial v}{\partial r} dr + \frac{\partial v}{\partial \theta} d\theta = \frac{\partial v}{\partial x} dx + \frac{\partial v}{\partial y} dy$, together with $r\, dr = x\, dx + y\, dy$ and $r^2 \, d\theta = x\, dy - y\, dx$, one finds that $\frac{\partial v}{\partial x} = r^{\alpha-1}(\alpha f(\theta) \cos\theta - f'(\theta) \sin\theta)$ and $\frac{\partial v}{\partial y} = r^{\alpha-1}(\alpha f(\theta) \sin\theta + f'(\theta) \cos\theta)$, and then $\Delta v = r^{\alpha-2}(f''(\theta) + \alpha^2 f(\theta))$. We look for

$$p = r^\alpha g(\theta), \quad u_1 = r^{\alpha+1} f_1(\theta), \quad u_2 = r^{\alpha+1} f_2(\theta), \tag{23.13}$$

satisfying

$$-\Delta u + grad\, p = 0; \quad div\, u = 0, \tag{23.14}$$

and as $\Delta p = 0$, it means $g''(\theta) + \alpha^2 g(\theta) = 0$, i.e.

$$g(\theta) = A \cos\alpha\theta + B \sin\alpha\theta, \tag{23.15}$$

and the equation $-\Delta u + grad\, p = 0$ gives

[5] Vsevolod Alekseevich SOLONNIKOV, Russian-born mathematician, born in 1933. He works at Università di Ferrara, Ferrara, Italy.
[6] Pierre GRISVARD, French mathematician, 1940–1994. He worked at Université de Nice Sophia-Antipolis, Nice, France.

$$f_1''(\theta) + (\alpha + 1)^2 f_1(\theta) =$$
$$\alpha\, g(\theta) \cos\theta - g'(\theta) \sin\theta =$$
$$\alpha\, A(\cos\alpha\theta \cos\theta + \sin\alpha\theta \sin\theta) + \alpha\, B(\sin\alpha\theta \cos\theta - \cos\alpha\theta \sin\theta) =$$
$$\alpha\, A \cos(\alpha - 1)\theta + \alpha\, B \sin(\alpha - 1)\theta$$
$$f_2''(\theta) + (\alpha + 1)^2 f_2(\theta) = \alpha\, g(\theta) \sin\theta + g'(\theta) \cos\theta =$$
$$\alpha\, A(\cos\alpha\theta \sin\theta - \sin\alpha\theta \cos\theta) + \alpha\, B(\sin\alpha\theta \sin\theta + \cos\alpha\theta \cos\theta) =$$
$$-\alpha\, A \sin(\alpha - 1)\theta + \alpha\, B \cos(\alpha - 1)\theta.$$

$$(23.16)$$

If $\alpha \neq 0$, this gives

$$f_1(\theta) = \tfrac{A}{2} \cos(\alpha - 1)\theta + \tfrac{B}{2} \sin(\alpha - 1)\theta + C_1 \cos(\alpha + 1)\theta + D_1 \sin(\alpha + 1)\theta$$
$$f_2(\theta) = -\tfrac{A}{2} \sin(\alpha - 1)\theta + \tfrac{B}{2} \cos(\alpha - 1)\theta + C_2 \cos(\alpha + 1)\theta + D_2 \sin(\alpha + 1)\theta.$$

$$(23.17)$$

The condition $div\, u = 0$ means

$$(\alpha + 1)f_1(\theta) \cos\theta - f_1'(\theta) \sin\theta + (\alpha + 1)f_2(\theta) \sin\theta + f_2'(\theta) \cos\theta = 0. \quad (23.18)$$

The coefficient of $\tfrac{A}{2}$ is $(\alpha + 1)\cos(\alpha - 1)\theta \cos\theta + (\alpha - 1)\sin(\alpha - 1)\theta \sin\theta - (\alpha + 1)\sin(\alpha - 1)\theta \sin\theta - (\alpha - 1)\cos(\alpha - 1)\theta \cos\theta = 2\cos\alpha\theta$, the coefficient of $\tfrac{B}{2}$ is $(\alpha + 1)\sin(\alpha - 1)\theta \cos\theta - (\alpha - 1)\cos(\alpha - 1)\theta \sin\theta + (\alpha + 1)\cos(\alpha - 1)\theta \sin\theta - (\alpha - 1)\sin(\alpha - 1)\theta \cos\theta = 2\sin\alpha\theta$, the coefficient of $(\alpha + 1)C_1$ is $\cos(\alpha + 1)\theta \cos\theta + \sin(\alpha + 1)\theta \sin\theta = \cos\alpha\theta$, the coefficient of $(\alpha + 1)D_1$ is $\sin(\alpha + 1)\theta \cos\theta - \cos(\alpha + 1)\theta \sin\theta = \sin\alpha\theta$, the coefficient of $(\alpha + 1)C_2$ is $\cos(\alpha + 1)\theta \sin\theta - \sin(\alpha + 1)\theta \cos\theta = -\sin\alpha\theta$, and the coefficient of $(\alpha + 1)D_2$ is $\sin(\alpha + 1)\theta \sin\theta + \cos(\alpha + 1)\theta \cos\theta = \cos\alpha\theta$. Therefore the coefficient of $\cos\alpha\theta$ is $A + (\alpha + 1)C_1 + (\alpha + 1)D_2$ and the coefficient of $\sin\alpha\theta$ is $B + (\alpha + 1)D_1 - (\alpha + 1)C_2$ and these two coefficients must be 0. If one works for $0 < \theta < \theta_0$, then f_1 and f_2 must be 0 for $\theta = 0$ and for $\theta = \theta_0$; for $\theta = 0$ one finds $\tfrac{A}{2} + C_1 = 0$ and $\tfrac{B}{2} + C_2 = 0$; this gives $A = -2(\alpha + 1)E, B = -2(\alpha + 1)F, C_1 = (\alpha + 1)E, C_2 = (\alpha + 1)F, D_1 = (\alpha + 3)F, D_2 = (1 - \alpha)E$, and writing that f_1 and f_2 are 0 for $\theta = \theta_0$ gives two equations for the two unknowns E, F and a nontrivial solution requires a determinant equal to 0, which gives a transcendental equation relating θ_0 to α, which I am not courageous enough to check.

[Taught on Monday March 8, 1999.]

24

Using compactness

Almost all compactness results use a variant of Ascoli's[1] theorem, but one also needs to learn a few technical tricks to add to the classical theorems; one trick is about approximation questions, so that one can replace spaces of continuous functions by spaces of integrable functions; another trick, which I show first, permits us to transfer compactness from one space to another.

Lemma 24.1. *(Jacques-Louis LIONS) If E_1, E_2, E_3, are three normed spaces with $E_1 \subset E_2$ with (continuous) compact injection, and $E_2 \subset E_3$ with continuous injection, then for every $\varepsilon > 0$ there exists $C(\varepsilon)$ such that*

$$||u||_{E_2} \leq \varepsilon\, ||u||_{E_1} + C(\varepsilon)||u||_{E_3} \text{ for all } u \in E_1. \tag{24.1}$$

Proof: If it was not true, there would exist $\varepsilon_0 > 0$ such that for every n one could find $u_n \in E_1$ with $||u_n||_{E_2} > \varepsilon_0||u_n||_{E_1} + n\,||u_n||_{E_3}$. By homogeneity, one may normalize u_n in order to have $||u_n||_{E_1} = 1$, and by continuity one obtains $||u_n||_{E_2} \leq C$, and the inequality implies then that $||u_n||_{E_3} \leq C/n$, so that $u_n \to 0$ in E_3. The sequence u_n is bounded in E_1, and therefore belongs to a compact subset of E_2; one can then extract a subsequence u_m which converges (strongly) in E_2 to a limit z, but one finds a contradiction because one must have $||z||_{E_2} \geq \varepsilon_0 > 0$ and $z = 0$ because u_n must converge to z in E_3.■

Lemma 24.2. *Let E_1, E_2, E_3 be three Banach spaces with $E_1 \subset E_2$ with continuous and compact injection, and $E_2 \subset E_3$ with continuous injection, and let $p \in [1, \infty]$. If a sequence is bounded in $L^p(0,T;E_1)$ and belongs to a compact of $L^p(0,T;E_3)$, then it belongs to a compact of $L^p(0,T;E_2)$.*

---

[1] Giulio ASCOLI, Italian mathematician, 1843–1896. He worked in Milano (Milan), Italy.

Proof: One extracts a subsequence u_n which converges in $L^p(0,T;E_3)$ and therefore it is a Cauchy sequence in $L^p(0,T;E_3)$. From Lemma 24.1 one has $||u_n - u_m||_{E_2} \leq \varepsilon\, ||u_n - u_m||_{E_1} + C(\varepsilon)||u_n - u_m||_{E_3}$, and taking the norms in $L^p(0,T)$ gives

$$||u_n - u_m||_{L^p(0,T;E_2)} \leq \varepsilon\, ||u_n - u_m||_{L^p(0,T;E_1)} + C(\varepsilon)||u_n - u_m||_{L^p(0,T;E_3)}. \tag{24.2}$$

Taking lim sup as n, m tend to ∞ gives $\limsup_{n,m\to\infty} ||u_n - u_m||_{L^p(0,T;E_2)} \leq 2M\varepsilon$, where M is a bound for the sequence in $L^p(0,T;E_1)$, and letting ε tend to 0 shows that u_n is a Cauchy sequence in $L^p(0,T;E_2)$.∎

The conclusion of Lemma 24.1 also occurs in the case $E_1 \subset E_2 \subset E_3$ (without assuming an injection compact) as a consequence of an interpolation inequality, i.e. if there exists $\theta \in (0,1)$ such that $||u||_{E_2} \leq M\, ||u||_{E_1}^{1-\theta}||u||_{E_3}^{\theta}$ for all $u \in E_1$ (there is a result of Jacques-Louis LIONS and Jaak PEETRE about interpolation spaces, that the preceding inequality is equivalent to $(E_1, E_3)_{\theta,1} \subset E_2$ with continuous injection). In that case the conclusion of Lemma 24.1 follows from Young's inequality, and Lemma 24.2 is strengthened, as no assumption of compactness is made.

Lemma 24.3. *Let E_1, E_2, E_3 be three Banach spaces with $E_1 \subset E_2 \subset E_3$ and assume that there exists $\theta \in (0,1)$ such that $||u||_{E_2} \leq M\, ||u||_{E_1}^{1-\theta}||u||_{E_3}^{\theta}$ for all $u \in E_1$. Let $p_1, p_3 \in [1,\infty]$. If a sequence is bounded in $L^{p_1}(0,T;E_1)$ and belongs to a compact of $L^{p_3}(0,T;E_3)$, then it belongs to a compact of $L^{p_2}(0,T;E_2)$ with p_2 defined by $\frac{1}{p_2} = \frac{1-\theta}{p_1} + \frac{\theta}{p_3}$.*

Proof: One has $||u_n - u_m||_{E_2}^{p_2} \leq M\, ||u_n - u_m||_{E_1}^{(1-\theta)p_2}||u_n - u_m||_{E_3}^{\theta p_2}$, and applying Hölder's inequality with the conjugate exponents $\frac{p_1}{(1-\theta)p_2}$ and $\frac{p_3}{\theta p_2}$ gives $||u_n - u_m||_{L^{p_2}(0,T;E_2)} \leq M\, ||u_n - u_m||_{L^{p_1}(0,T;E_1)}^{1-\theta}||u_n - u_m||_{L^{p_3}(0,T;E_3)}^{\theta}$, showing that $\limsup_{n,m\to\infty} ||u_n - u_m||_{L^{p_2}(0,T;E_2)} = 0$ because $\theta < 1$.∎

Notice that there is no hypothesis of compact injection in Lemma 24.3, and one often applies it to the case $E_1 = E_2 = E_3$ in order to change the value of p.

In his lectures, Jacques-Louis LIONS used a compactness result in which a sequence is bounded in $L^p(0,T;E_1)$ while its derivative with respect to t is bounded in $L^p(0,T;E_3)$ and the conclusion is that the sequence belongs to a compact of $L^p(0,T;E_2)$ (the injection of E_1 into E_2 being compact); he used $1 < p < \infty$ and reflexive spaces, and he referred to Jean-Pierre AUBIN,[2] but as I never read the corresponding article, I do not know what each of them did; in his book on nonlinear problems he also refers to Roger

---

[2] Jean-Pierre AUBIN, French mathematician, born in 1939. He works at Université Paris IX-Dauphine, Paris, France.

TEMAM[3] for some variants. I had then heard Pascal MARONI[4] mention that the hypothesis of reflexivity is not necessary, so when I taught the part [23] of my graduate course at University of Wisconsin in 1974–1975, I extended the proof that I had learnt in order to avoid the hypothesis of reflexivity, still using $1 < p < \infty$; a few years later, I remember discussing the case $p = 1$ with François MURAT and Lucio BOCCARDO,[5] but I only wrote the scenario of a proof on the paper tablecloth of a restaurant in Roma (Rome), Italy. Putting different ideas together leads to the following results.

Lemma 24.4. *Let E be a Banach space. Assume that there exists $p \in [1, \infty)$, $\eta > 0$ and a constant M such that a sequence u_n is bounded in $L^p(0, T; E)$ with $\left(\int_0^{T-h} ||u_n(t + h) - u_n(t)||^p \, dt \right)^{1/p} \leq M \, |h|^\eta$ for all $h \in (0, T/2)$. Then u_n is bounded in $L^q(0, T; E)$ with $q < p/(1 - \eta\, p)$ if $\eta < 1/p$, and all $q < \infty$ if $\eta \geq 1/p$.*

Proof: Let $\varphi(t) = ||u_n(t)||$, then $\varphi \in L^1(0, T)$ and $\left(\int_0^{T-h} |\varphi(t + h) - \varphi(t)|^p \, dt \right)^{1/p} \leq M \, |h|^\eta$ for all $h \in (0, T/2)$. Lemma 24.4 is just a part of the analog of the Sobolev embedding theorem 16.1 for Besov[6,7] spaces, and the natural spaces here are not the $L^p(0, T)$ spaces but the Marcinkiewicz[8] spaces ($L^{p,\infty}(0, T)$ in the notations of Lorentz spaces). One may assume that the support of u is included on $(0, T/2)$ by localization, and one then uses nonnegative regularizing sequences $\varrho_\varepsilon(x) = \frac{1}{\varepsilon}\varrho\left(\frac{x}{\varepsilon}\right)$ with support in $(-1, 0)$.

Assume that $u \in L^a(0, T; E)$ and $\int_0^{T-h} ||u(t + h) - u(t)||_E^p \, dt \leq M^p \, |h|^{\eta\, p}$ for all $h < \frac{T}{2}$; then one writes $u = (\varrho_\varepsilon \star u) + (u - (\varrho_\varepsilon \star u))$, and one has $\left(u - (\varrho_\varepsilon \star u) \right)(x) = \frac{1}{\varepsilon} \int \varrho\left(\frac{y}{\varepsilon}\right)\left(u(x) - u(x - y) \right) dy$ because $\int_R \varrho(x) \, dx = 1$, and therefore $||u - (\varrho_\varepsilon \star u)||_{L^p(0, T/2)} \leq \frac{M}{\varepsilon} \int \varrho\left(\frac{y}{\varepsilon}\right) |y|^\eta \, dy = C \, \varepsilon^\eta$. On the other hand, $(\varrho_\varepsilon \star u)(x) = \frac{1}{\varepsilon} \int \varrho\left(\frac{y}{\varepsilon}\right) u(x - y) \, dy$, so $||\varrho_\varepsilon \star u||_{L^\infty(0, T/2)} \leq ||u||_{L^a} ||\varrho_\varepsilon||_{L^{a'}} \leq ||u||_{L^a}\left(\frac{K}{\varepsilon}\right)^{1/a}$, where $K = ||\varrho||_{L^\infty}$.

For $\lambda > 0$, one looks for an estimate of the measure of $\omega_{2\lambda} = \{t \in (0, T/2) \mid |u(t)| \geq 2\lambda\}$, and one chooses $\varepsilon > 0$ such that $||u||_{L^a}\left(\frac{K}{\varepsilon}\right)^{1/a} = \lambda$, so that

[3] Roger TEMAM, Tunisian-born mathematician, born in 1940. He works at Université Paris XI (Paris-Sud), Orsay, France, and at Indiana University, Bloomington, IN.

[4] Pascal MARONI, Swiss-born mathematician, born in 1933. He worked at CNRS (Centre National de la Recherche Scientifique) at Université Paris VI (Pierre et Marie CURIE), Paris, France.

[5] Lucio BOCCARDO, Italian mathematician, born in 1948. He works at Università di Roma "La Sapienza", Roma (Rome), Italy.

[6] Oleg V. BESOV, Russian mathematician, born in 1933. He works at the STEKLOV Institute of Mathematics, Moscow, Russia.

[7] Vladimir Andreevich STEKLOV, Russian mathematician, 1864–1926. He worked in St Petersburg, Russia.

[8] Józef MARCINKIEWICZ, Polish mathematician, 1910–1940. He worked in Wilno, then in Poland, now Vilnius, Lithuania.

$u = v + w$ with $v = \varrho_\varepsilon \star u$ bounded by λ; therefore $\omega_{2\lambda}$ is included in the set where $|w| = |u - (\varrho_\varepsilon \star u)| \geq \lambda$, which gives the estimate $\lambda^p meas(\omega_{2\lambda}) \leq M^p \varepsilon^{\eta p}$, and using the choice of ε, i.e. $K\lambda^{-a}||u||_{L^a}^a$, one obtains the estimate $\lambda^{p+a\eta p} meas(\omega_{2\lambda}) \leq M^p K^{\eta p}||u||_{L^a}^{a\eta p}$.

This estimate shows that $u \in L^b$ for $b < p(1 + a\eta)$. In the case $\eta p < 1$, let $a^* = p(1 + a^*\eta)$, then for $p \leq a < a^*$ one has $a < p(1 + a\eta)$, and starting with $a_1 = p$ one deduces that $u \in L^{a_k}$ with a_k converging to a^*, and therefore one can find a bound for $||u||_{L^q}$ for any $q \in [p, a^*)$ in terms of M and $||u||_{L^p}$ (and q). In the case $\eta p \geq 1$, one has $a < p(1 + a\eta)$ for all a, and starting with $a_1 = p$ one deduces that $u \in L^{a_k}$ with a_k converging to $+\infty$; therefore one can find a bound for $||u||_{L^q}$ for any $q \in [p, +\infty)$ in terms of M and $||u||_{L^p}$ (and q). ∎

Lemma 24.5. *Let E_1 and E_3 be two Banach spaces with $E_1 \subset E_3$, the injection being continuous and compact. Assume that for some $p \in [1, \infty]$ a sequence u_n is bounded in $L^p(0, T; E_1)$, and that there exists $\theta > 0$ and a constant M such that $\left(\int_0^{T-h} ||u_n(t + h) - u_n(t)||_{E_3}^p \, dt\right)^{1/p} \leq M|h|^\theta$ for all $h \in (0, T/2)$, then u_n belongs to a compact set of $L^p(0, T; E_3)$.*

Proof: After localization so that the support of all u_n is in $(0, T/2)$, one chooses $h > 0$ small and one defines $v_n(t) = \frac{1}{h} \int_t^{t+h} u_n(s) \, ds$ and $w_n = u_n - v_n$. As before, $w_n(t) = \frac{1}{h} \int_t^{t+h} (u_n(t) - u_n(s)) \, ds$, giving the bound $||w_n||_{L^p(0,T;E_3)} \leq C|h|^\theta$. For a fixed $h > 0$, v_n takes its values in a bounded set of E_1, and therefore in a compact set of E_3, but as from Lemma 24.4, u_n is bounded in some $L^q(0, T; E_3)$ with $q > 1$, one sees that v_n has its derivative bounded in $L^q(0, T; E_3)$ and is therefore uniformly Hölder continuous with values in E_3; by Ascoli's theorem a subsequence v_m converges uniformly. One deduces that $\limsup_{m,m' \to \infty} ||u_m - u_{m'}||_{L^p(0,T;E_3)} \leq \limsup_{m,m' \to \infty} ||w_m - w_{m'}||_{L^p(0,T;E_3)} \leq 2C|h|^\theta$, and letting h tend to 0 shows that u_m is a Cauchy sequence in $L^p(0, T; E_3)$. ∎

In our application to the Navier–Stokes equation using a special Faedo–Ritz–Galerkin basis, one has u_n bounded in $L^2(0, T; V)$ and in $L^\infty(0, T; H)$, and $\frac{du_n}{dt}$ is bounded in $L^p(0, T; V')$, with $p = 2$ for $N = 2$, $p = 4/3$ for $N = 3$, and $p = 1$ for $N = 4$; moreover, V is continuously and compactly embedded into H (and therefore into V'). One can take $\theta = 1$ and one first deduces that u_n belongs to a compact of $L^p(0, T; V')$; but as it is bounded in $L^\infty(0, T; V')$, it belongs to a compact of $L^q(0, T; V')$ for all $q < \infty$; then it belongs to a compact of $L^2(0, T; H)$, the limitation by 2 being due to the estimate of u_n in $L^2(0, T; V)$, and one can extract subsequences which converge almost everywhere in $\Omega \times (0, T)$. In dimension $N = 2$, using an interpolation inequality, each component of u_n is bounded in $L^4(\Omega \times (0, T))$ and therefore the term $(u_m)_j (u_m)_i$ for which one needs the limit (in order to compute the limit of the term $\langle B(u_m, u_m), e_k \rangle$) is bounded in $L^2(\Omega \times (0, T))$ and converges almost everywhere to $(u_\infty)_j (u_\infty)_i$; in dimension $N = 3$, each component of u_n is bounded in $L^{8/3}(0,T;L^4(\Omega))$ and therefore the term $(u_m)_j(u_m)_i$ is bounded in $L^{4/3}(0,T;L^2(\Omega))$ and converges then almost everywhere to $(u_\infty)_j(u_\infty)_i$. The case $N \geq 4$ uses interior regularity for the e_k.

[Taught on Wednesday March 10, 1999.]

25

Existence of smooth solutions

We have obtained the existence of a weak solution for an abstract formulation of the Navier–Stokes equation; our solution is defined on $(0, T)$, and one may take $T = +\infty$ without much change in the proof. Even in dimension $N = 2$, where we know the solution to be unique, it is useful to know whether or not it is regular when the data are more regular. For $N = 3$ (or $N \geq 4$ for purely mathematical reasons), one may wonder if one can find a strong solution, i.e. a solution having a better regularity so that the solution would be unique, for example, or if the "pressure" would be found in a space of locally integrable functions in $(\mathbf{x}, t)$. Some of the regularity results depend upon the smoothness of the boundary $\partial\Omega$, for example in order to use the fact that $D(A) \subset H^2(\Omega; R^N)$, which is not true for some Lipschitz domains.

Lemma 25.1. *If $N = 2$, if $f \in L^2\big(0, T; L^2(\Omega; R^2)\big) = L^2\big((0, T) \times \Omega; R^2\big)$ and $u_0 \in V$, if Ω is smooth enough so that Poincaré's inequality holds and $D(A) \subset H^2(\Omega; R^2)$, then the solution of $u' + B(u, u) + Au = f$ and $u(0) = u_0$ satisfies $u \in C^0([0, T]; V)$, $u \in L^2\big(0, T; H^2(\Omega; R^2)\big)$, and $u' \in L^2(0, T; H)$.*

Proof: One proves the estimates for the approximation with the special basis, multiplying by $A u_n$, and one needs to bound terms like $(u_n)_j \frac{\partial(u_n)_i}{\partial x_j}$ in $L^2(\Omega)$, for example from a bound of $(u_n)_j$ in $L^\infty(\Omega)$; one may use a bound $||v||_{L^\infty(\Omega)} \leq C \, ||v||_{L^2(\Omega)}^{1/2} ||v||_{H^2(\Omega)}^{1/2} \leq C \, |v|^{1/2} |A v|^{1/2}$ (valid in dimension 2), and one obtains

$$\frac{1}{2} \frac{d(||u_n||^2)}{dt} + |A u_n|^2 \leq |f| \, |A u_n| + |u_n|^{1/2} ||u_n|| \, |A u_n|^{3/2} \\ \leq \varepsilon |A u_n|^2 + C(\varepsilon) |f|^2 + C(\varepsilon) |u_n|^2 ||u_n||^4, \tag{25.1}$$

where one has used Young's inequality $ab \leq |\varepsilon a|^p / p + |b/\varepsilon|^{p'} / p'$, with $p = 4/3, p' = 4$, and one deduces that

$$||u_n(t)||^2 \leq ||u_0||^2 + C \int_0^t |f(s)|^2 \, ds + C \int_0^t \lambda_n(s) ||u_n(s)||^2 \, ds, \\ \text{with } \lambda_n = C \, |u_n|^2 ||u_n||^2, \tag{25.2}$$

from which a uniform bound for $||u_n||$ is deduced by applying Gronwall's inequality, as λ_n is bounded in $L^1(0,T)$, and this implies that $|A\,u_n|$ is bounded in $L^2(0,T)$.■

In the preceding proof, one may use different estimates for the bound in L^∞; for example, trying to get the power of $|A\,u_n|$ as low as possible, one can use $||v||_{L^\infty(\Omega)} \le C(\eta)||v||_{L^2(\Omega)}^{1/(1+\eta)}||v||_{H^{1+\eta}(\Omega)}^{\eta/(1+\eta)}$ (valid in dimension 2), and if one uses $\theta = \eta/(1+\eta) \in (0,1/2)$, it gives a bound for $B(u_n, u_n)$ in $L^2(\Omega; R^2)$ of the form $C(\theta)|u_n|^\theta||u_n||^{2-2\theta}|A\,u_n|^\theta$. The application of Young's inequality, with $p = 2/(1+\theta), p' = 2/(1-\theta)$, gives a term in $\varepsilon\,|A\,u_n|^2 + C(\varepsilon, \theta)|u_n|^{2\theta/(1-\theta)}||u_n||^4$, and therefore one gains on the power of $|u_n|$ but not on the power of $||u_n||$.

Lemma 25.2. *If $N = 3$, if $f \in L^2\big(0,T; L^2(\Omega; R^3)\big) = L^2\big((0,T)\times\Omega; R^3\big)$ and $u_0 \in V$, if Ω is smooth enough so that Poincaré's inequality holds and $D(A) \subset H^2(\Omega; R^3)$, then there exists $T_c \in (0,T]$ depending upon the norms of the data such that there exists a unique solution of $u' + B(u,u) + A\,u = f$ and $u(0) = u_0$ on $[0, T_c]$ which satisfies $u \in C^0([0,T_c]; V)$, $u \in L^2\big(0, T_c; H^2(\Omega; R^3)\big)$, and $u' \in L^2(0, T_c; H)$.*

Proof: Same type of proof as before, but one has $||v||_{L^\infty(\Omega)} \le C\,||v||^{1/2}|A\,v|^{1/2}$ (valid in dimension 3), giving a bound for $B(u_n, u_n)$ in $L^2(\Omega; R^3)$ of the form $||u_n||^{3/2}|A\,u_n|^{1/2}$. The application of Young's inequality, with $p = 4/3, p' = 4$, gives a term in $\varepsilon\,|A\,u_n|^2 + C(\varepsilon)||u_n||^6$, and the exponent is too large to obtain a bound by Gronwall's inequality, and one can only obtain a local bound from the inequality

$$\frac{d(||u_n||^2)}{dt} + \alpha\,|A\,u_n|^2 \le C\,|f|^2 + C\,||u_n||^6. \tag{25.3}$$

After integration, and omission of the term in $|A\,u_n|^2$, one obtains

$$||u_n(t)||^2 \le ||u_0||^2 + C\int_0^t |f(s)|^2\,ds + C\int_0^t ||u_n(s)||^6\,ds \le \\ K + C\int_0^t ||u_n(s)||^6\,ds,\ \text{in}\ (0,T), \tag{25.4}$$

where $K = ||u_0||^2 + C\int_0^T |f(s)|^2\,ds$. If one defines φ by $\varphi(t) = \int_0^t ||u_n(s)||^6\,ds$, then one has $\varphi' \le (K + C\,\varphi)^3$, which implies $\big((K + C\,\varphi)^{-2}\big)' = -2C(K + C\,\varphi)^{-3}\varphi' \ge -2C$ and therefore $\big(K + C\,\varphi(t)\big)^{-2} \ge K^{-2} - 2C\,t$, which is only useful on $[0, T_c]$ if $K^{-2} - 2C\,T_c > 0$, in which case it gives the desired bounds.

To prove uniqueness one assumes that one has two solutions u^1, u^2, one subtracts the equations and one multiplies by $A(u^2 - u^1)$, and one has to estimate the norm in $L^2(\Omega; R^3)$ of $B(u^2, u^2) - B(u^1, u^1)$, which one writes as $B(u^2, u^2 - u^1) + B(u^2 - u^1, u^1)$; the first term can be bounded as $||u^2||_{L^6(\Omega; R^3)}||u^2 - u^1||_{W^{1,3}(\Omega; R^3)}$, bounded by $C\,||u^2||\,||u^2 - u^1||^{1/2}|A\,u^2 - A\,u^1|^{1/2}$; the second term can be bounded as $||u^2 - u^1||_{L^\infty(\Omega; R^3)}||u^1||$, bounded by $C\,||u^1||\,||u^2 - u^1||^{1/2}|A\,u^2 - A\,u^1|^{1/2}$. One then obtains $(||u^2 - u^1||^2)' + 2|A\,u^2 - A\,u^1|^2 \leq C(||u^1|| + ||u^2||)||u^2 - u^1||^{1/2}|A\,u^2 - A\,u^1|^{3/2} \leq |A\,u^2 - A\,u^1|^2 + C(||u^1|| + ||u^2||)^4\,||u^2 - u^1||^2$, and one concludes by using Gronwall's inequality.∎

If the data are small enough, one can take $T_c = T$, but one can even obtain global existence on $[0, \infty)$ if the data are small enough; for simplicity, I consider first the case $f = 0$.

Lemma 25.3. *If $N = 3$, if Ω is smooth enough so that Poincaré's inequality holds and $D(A) \subset H^2(\Omega; R^3)$, then if $u_0 \in V$, and if $|u_0|\,||u_0||$ is small enough, then the solution of $u' + B(u, u) + A\,u = 0$ with $u(0) = u_0$ exists for all $t \in [0, \infty)$ and satisfies $u \in C^0([0, \infty); V)$, $u \in L^2(0, \infty; H^2(\Omega; R^3))$, and $u' \in L^2(0, \infty; H)$.*

Proof: One bounds the norm of $v_j \frac{\partial v_i}{\partial x_j}$ in $L^2(\Omega)$ by $||v_j||_{L^3(\Omega)}||\frac{\partial v_i}{\partial x_j}||_{L^6(\Omega)}$, $||\frac{\partial v_i}{\partial x_j}||_{L^6(\Omega)}$ by $C\,||v||_{H^2(\Omega)}$ and $||v_j||_{L^3(\Omega)}$ by $C\,|v|^{1/2}||v||^{1/2}$, so that

$$|\langle B(u_n, u_n), A\,u_n \rangle| \leq C_0 |u_n|^{1/2}\,||u_n||^{1/2}\,|A\,u_n|^2 \text{ for every } u_n \in D(A). \quad (25.5)$$

Because $f = 0$, one obtains

$$
\begin{aligned}
\tfrac{1}{2}\tfrac{d(|u_n|^2)}{dt} + ||u_n||^2 \leq 0 \\
\tfrac{1}{2}\tfrac{d(||u_n||^2)}{dt} + |A\,u_n|^2 \leq C_0 |u_n|^{1/2}\,||u_n||^{1/2}\,|A\,u_n|^2 \text{ on } (0, T),
\end{aligned}
\quad (25.6)
$$

and therefore as long as $C_0 |u_n|^{1/2}\,||u_n||^{1/2} < 1$, both the quantities $|u_n|$ and $||u_n||$ are nonincreasing and their product is less than its value at time 0; consequently, if the initial data $u_0 \in V$ is chosen so that

$$C_0 |u_0|^{1/2}\,||u_0||^{1/2} < 1, \quad (25.7)$$

then one has $|u_n(t)| \leq |u_0|$ and $||u_n(t)|| \leq ||u_0||$ on $(0, \infty)$, and a global solution exists on $(0, \infty)$.∎

In the case $f \neq 0$, one has $|u_n(t)| \leq |u_0| + \int_0^t |f(s)|\,ds$, but if one wants to assume more than $f \in L^1(0, \infty; L^2(\Omega; R^3))$, one may use Poincaré's inequality $||v||^2 \geq \lambda_1 |v|^2$ for all $v \in H_0^1(\Omega)$, and the inequality $(|u_n|^2)' + 2\lambda_1 |u_n|^2 \leq 2|f|\,|u_n| \leq 2\lambda_1 |u_n|^2 + |f|^2/2\lambda_1$ gives $|u_n(t)|^2 \leq |u_0|^2 + \frac{1}{2\lambda_1}\int_0^t |f(s)|^2\,ds$. If one can enforce the condition $C_0 |u_n|^{1/2}||u_n||^{1/2} \leq 1/2$, then one has $(||u_n||^2)' + |A\,u_n|^2 \leq 2|f|\,|A\,u_n| \leq |A\,u_n|^2 + |f|^2$ and therefore $||u_n||^2 \leq ||u_0||^2 + \int_0^t |f|^2\,dt$, and the condition to enforce is satisfied if one asks that

$$\left(|u_0|^2 + \frac{1}{2\lambda_1}\int_0^\infty |f(s)|^2\,ds\right)\left(||u_0||^2 + \int_0^\infty |f(s)|^2\,ds\right) \leq \frac{1}{16C_0^4}. \quad (25.8)$$

In the case where $f = 0$, instead of putting conditions on $|u_0|\,||u_0||$ one can impose a more natural condition that u_0 be small in the domain of $A^{1/4}$;

this is done by multiplying by $A^{1/2}u_n$, and using the estimate $||v||_{L^3(\Omega;R^3)} \leq C|A^{1/4}v|$, which implies $|(B(u_n, u_n), A^{1/2}u_n)| \leq C_1|A^{1/4}u_n||A^{3/4}u_n|^2$, and therefore if $C_1|A^{1/4}u_0| \leq 1$, then the norm of $|A^{1/4}u_n|$ is nonincreasing and stays $\leq \frac{1}{C_1}$; one easily extends this idea to the case $f \neq 0$.

All these types of inequalities are quite standard, and although I may have improved on details, I learnt most of these techniques in lectures of Jacques-Louis LIONS in the late 1970s; I had taught these techniques in the part [23] of my 1974–1975 course in Madison (the lecture notes having been written by graduate students). After that, I advocated following a little more the physics of the fluid flows in order to get better results, and I still insist that one should not cut the transport term into two parts; I thought that everything important had been found out about these differential inequalities, but I was wrong. In January 1980, Colette GUILLOPÉ[1] showed me some handwritten pages by Ciprian FOIAS, and I took a copy which I looked at during the following month, which I spent at the TATA[2] Institute in Bangalore, India; I made an improvement on FOIAS's original computation, but the idea is his. In the case $N = 3$, taking $f = 0$ in order to simplify, one starts from the already-mentioned differential inequality $(||u_n||^2)' + |A\,u_n|^2 \leq C\,||u_n||^6$, together with $(|u_n|^2)' + 2||u_n||^2 = 0$, which gives the existence on $(0, T)$ of the approximate solution u_n; FOIAS's idea was to divide by $1 + ||u_n||^6$, while my improvement is to divide only by $1 + ||u_n||^4$! One obtains

$$\frac{d}{dt}\left(\arctan(||u_n||^2)\right) + \frac{|A\,u_n|^2}{1+||u_n||^4} = \frac{1}{1+||u_n||^4}\left((||u_n||^2)' + |A\,u_n|^2\right) \leq \frac{C\,||u_n||^6}{1+||u_n||^4} \leq C\,||u_n||^2. \tag{25.9}$$

Integrating from 0 to T (which can be $+\infty$), one obtains

$$\int_0^T \frac{|A\,u_n|^2}{1+||u_n||^4}\,dt \leq \arctan(||u_n(0)||^2) + C\int_0^T ||u_n||^2\,dt \leq \frac{\pi}{2} + \frac{C\,|u_0|^2}{2}. \tag{25.10}$$

One has $\frac{|A\,u_n|^{1/2}}{1+||u_n||}$ bounded in $L^4(0, T)$, but as $||u_n||^{1/2}(1 + ||u_n||)$ is bounded in $L^{4/3}(0, T)$, one has

$$||u_n||^{1/2}|A\,u_n|^{1/2} \text{ bounded in } L^1(0, T), \tag{25.11}$$

from which one obtains

$$u_n \text{ is bounded in } L^1\left(0, T; L^\infty(\Omega; R^3)\right). \tag{25.12}$$

One deduces similar properties for the limit, like $u \in L^1\left(0, T; L^\infty(\Omega; R^3)\right)$, but one should notice how far this estimate is from that which would give well defined curves followed by particles along the flow.

[Taught on Friday March 12, 1999.]

---

[1] Colette GUILLOPÉ, French mathematician, born in 1951. She works at Université de Paris XII-Val de Marne, Créteil, France.

[2] Jamsetji Nusserwanji TATA, Indian industrialist, 1839–1904.

26

Semilinear models

If U_0 is a characteristic velocity and L_0 is a characteristic length of a flow, then the corresponding Reynolds[1] number $\frac{U_0 L_0}{\nu}$ is non dimensional; large ν or small R corresponds to laminar flows, while small ν or large R corresponds to turbulent flows. In the ocean, the characteristic lengths L_0 are large.

As **u** denotes a velocity, it has the dimensions LT^{-1}, where L denotes length and T denotes time, and the kinematic viscosity $\nu = \mu/\varrho$ has the dimensions $L^2 T^{-1}$ (while μ has the dimensions $M L^{-1} T^{-1}$, where M denotes mass). In dimension $N = 3$, the norm $|u|$ has the dimensions $L^{5/2} T^{-1}$ (and therefore as ϱ has the dimensions $M L^{-3}$, $\varrho_0 |u|^2$ has the dimensions $M L^2 T^{-2}$, i.e. energy), the norm $||u||$ has the dimensions $L^{3/2} T^{-1}$ (and $\mu ||u||^2$ has the dimensions $M L^2 T^{-3}$, as energy dissipated per unit of time), and $|u|^{1/2} ||u||^{1/2}$ has the dimensions $L^2 T^{-1}$, as does $||u||_{L^3}$.

The term $\langle B(u,u), A u \rangle$, as $\nu \int u \, \partial u \, \partial^2 u \, d\mathbf{x}$ has the dimensions $L^5 T^{-4}$, while $|u|^{1/2} ||u||^{1/2} |A u|^2$ has the dimensions $L^7 T^{-5}$, and therefore C_1 has the dimensions $L^{-2} T$, i.e. $1/C_1$ has the same dimensions $L^2 T^{-1}$ as ν. As the only parameter in the equation is ν (the term $1/\varrho_0$ in front of $grad\, p$ is hidden, and as ϱ is assumed constant, it is p/ϱ_0 which we have called "pressure"), it is natural to compare norms with that number, but that only makes sense for norms whose dimensions are a power of $L^2 T^{-1}$.

The limitations of the estimates shown before are due in part to the fact that one uses norms which give global information on the solution and not local information; the total kinetic energy at time t is seen by $\varrho_0 |u(t)|^2$ and the energy dissipated by viscosity between time 0 and T is seen by $\mu \int_0^T ||u(t)||^2 \, dt$, but these norms do not reveal if some regions correspond to large velocities or to a large dissipation of energy (as we have assumed that ϱ_0 and μ are independent of temperature, the energy dissipated by viscosity appears in the equation of balance of energy, which is decoupled from the equation of motion that we have been dealing with up to now).

---

[1] Osborne REYNOLDS, Irish-born mathematician, 1842–1912. He worked in Manchester, England, UK.

A different approach, which I initiated in 1979 for a different class of equations, consists in avoiding the semi-group approach where one deals with functional spaces which are functions in x alone and one defines the domain of a nonlinear operator, and one deals instead with functional spaces in (x,t) adapted to the equation. In the class of discrete velocity models in kinetic theory, which are supposed to simplify the Boltzmann equation, there is a particular model attributed to BROADWELL,[2] although Renée GATIGNOL[3] mentions in her book [14] that this kind of model goes back to MAXWELL); in 2 dimensions (x,y) it is

$$
\begin{aligned}
&\frac{\partial u_1}{\partial t} + \frac{\partial u_1}{\partial x} + (\alpha\, u_1 u_2 - \beta\, u_3 u_4) = 0 \text{ in } R^2 \times (0,T), \\
&u_1(x,y,0) = u_{01}(x,y) \text{ in } R^2 \\
&\frac{\partial u_2}{\partial t} - \frac{\partial u_2}{\partial x} + (\alpha\, u_1 u_2 - \beta\, u_3 u_4) = 0 \text{ in } R^2 \times (0,T), \\
&u_2(x,y,0) = u_{02}(x,y) \text{ in } R^2 \\
&\frac{\partial u_3}{\partial t} + \frac{\partial u_3}{\partial y} - (\alpha\, u_1 u_2 - \beta\, u_3 u_4) = 0 \text{ in } R^2 \times (0,T), \qquad (26.1) \\
&u_3(x,y,0) = u_{03}(x,y) \text{ in } R^2 \\
&\frac{\partial u_4}{\partial t} - \frac{\partial u_4}{\partial y} - (\alpha\, u_1 u_2 - \beta\, u_3 u_4) = 0 \text{ in } R^2 \times (0,T), \\
&u_4(x,y,0) = u_{04}(x,y) \text{ in } R^2,
\end{aligned}
$$

where u_1, u_2, u_3, u_4 denote the density of particles at (x,y,t); these particles all have the same mass but their velocities are respectively $(+1,0)$, $(-1,0)$, $(0,+1)$, $(0,-1)$ (Jim GREENBERG[4] defines the unknowns ℓ, r, u, d, for left, right, up, down); α, β are positive parameters related to probability of collisions, which are usually taken equal to $1/\varepsilon$, where ε is related to a mean free path between collisions (BROADWELL was actually interested in the formal fluid limit $\varepsilon \to 0$, and there are plenty of open questions in that direction). Local existence for data in $L^\infty(R^2)$ is standard (locally Lipschitz perturbation of a bounded linear semi-group), and if the data are nonnegative the solution is nonnegative. The model (26.1) conserves mass (density of mass is $u_1 + u_2 + u_3 + u_4$), momentum (density of momentum is $(u_1 - u_2, u_3 - u_4)$), and kinetic energy, which is proportional to mass because all velocities have the same modulus, so that there is no temperature for the "gas" following this model. An analog of the H-theorem of BOLTZMANN holds, and there is an entropy which decreases (density of entropy is $u_1 \log(u_1) + u_2 \log(u_2) + u_3 \log(u_3) + u_4 \log(u_4)$). Takaaki NISHIDA[5]

---

[2] James E. BROADWELL, American engineer. He works at Caltech (California Institute of Technology), Pasadena, CA.

[3] Renée FLANDRIN GATIGNOL, French mathematician. She works at Université Paris VI (Pierre et Marie CURIE), Paris, France.

[4] James M. GREENBERG, American mathematician, born in 1941. For a few years he was Head of the Department of Mathematics (and changed it into a Department of Mathematical Sciences) at CARNEGIE MELLON University, Pittsburgh PA, and he works now at the US Office of Naval Research International Field Office in London, England, UK.

[5] Takaaki NISHIDA, Japanese mathematician, born in 1942. He works at Kyoto University, Kyoto, Japan.

and I (independently) have noticed that there is global existence of a solution for small nonnegative data in $L^2(R^2)$, and it is useful to notice that the $L^2(R^2)$ norm is *invariant by scaling*: if $U = (u_1, u_2, u_3, u_4)$ is a solution, then V defined by $V(x, y, t) = \lambda U(\lambda x, \lambda y, \lambda t)$ is also a solution for any $\lambda > 0$, and the norm of the initial data in $L^2(R^2; R^4)$ is the same for U or V; my proof generalized what I had done for the 1-dimensional case, which I describe now.

If the initial data are independent of y then the solution is independent of y for $t > 0$ (also for $t < 0$ as long as the solution exists, but nonnegativity is only conserved when t increases); if additionally $u_3 = u_4$ at time 0, then it stays true (as long as the solution exists); one considers then the simplified *Broadwell model* (where I have chosen $u = u_1$, $v = u_2$, $w = u_3 = u_4$ and $\alpha = \beta = 1$)

$$u_t + u_x + u\,v - w^2 = 0 \text{ in } R \times (0, \infty), u(x, 0) = u_0(x) \text{ in } R$$
$$v_t - v_x + u\,v - w^2 = 0 \text{ in } R \times (0, \infty), v(x, 0) = v_0(x) \text{ in } R \qquad (26.2)$$
$$w_t - u\,v + w^2 = 0 \text{ in } R \times (0, \infty), w(x, 0) = w_0(x) \text{ in } R.$$

In 1975, in collaboration with Mike CRANDALL[6], we proved global existence (i.e. for $t \in [0, \infty)$) for bounded nonnegative data by using finite propagation speed, the entropy estimate as a compactness argument in L^1 weak, and a crucial result that MIMURA[7] and Takaaki NISHIDA had just published, where they showed that for nonnegative data in $L^1 \cap L^\infty$ with small L^1 norm but arbitrary L^∞ norm, the L^∞ norm is controlled for all t. The generalization of the estimate of MIMURA and NISHIDA is unlikely for n-dimensional problems with $n \geq 2$, so our argument seems restricted to particular 1-dimensional problems of the form (26.3).

It is part of my philosophy that when dealing with partial differential equations from continuum mechanics or physics, it is better to use functional spaces which can be related to some physical properties, and although L^∞ spaces are mathematically convenient for local existence, there does not seem

---

[6] Michael G. CRANDALL, American mathematician, born in 1940. He works at University of California Santa Barbara, Santa Barbara, CA.

[7] Masayasu MIMURA, Japanese mathematician. He works at Hiroshima University, Hiroshima, Japan.

to be any physically relevant idea behind L^∞ estimates.[8,9,10,11] I thought that BMO could be a good substitute, as a BMO semi-norm seems to control the portion of the mass which is out of equilibrium, but on one hand I could not get the attention of my colleague Yves MEYER in order to help me on that question and I never went forward with this idea, and on the other hand I found a new interesting idea, which produced L^∞ estimates as well as information on the asymptotic behavior for a special class of 1-dimensional problems of the form (26.3) satisfying the condition (26.4).

In 1979, I was wondering which discrete velocity models of kinetic theory were stable by weak convergence, because weak convergence is natural for additive quantities like densities of particles (or coefficients of differential forms, which are naturally integrated on manifolds), and in my point of view on continuum mechanics or physics it serves to explain the relation between mesoscopic scales and macroscopic scales for these additive quantities (and other topologies of weak type, like H-convergence, must be used for some other physical quantities), so if one has found effective equations valid at a macroscopic level, they must be stable with respect to these adapted convergences. I had first noticed that the Carleman model is not stable, although it is not a model of kinetic theory as it does not conserve momentum), and then I found that (apart from the affine case) stability with respect to weak convergence occurs for 1-dimensional models of the form

$$\frac{\partial u_i}{\partial t} + C_i \frac{\partial u_i}{\partial x} + \sum_{j,k=1}^{m} A_{ijk} u_j u_k + affine(u) = 0 \text{ in } R \times (0, \infty), \\ u_i(x,0) = u_{0i}(x) \text{ in } R, i = 1, \ldots, m \tag{26.3}$$

if and only if, besides $A_{ijk} = A_{ikj}$ for all i, j, k, the interaction coefficients satisfy the condition

$$\text{for all } j, k, C_j = C_k \text{ implies } A_{ijk} = 0 \text{ for } i = 1, \ldots, m. \tag{26.4}$$

---

[8] Except when the maximum principle holds, as for scalar elliptic or parabolic equations, or for systems with order-preserving properties, like the Carleman model, which is not a good model for kinetic theory as it does not conserve momentum (it can easily be interpreted as a model of self-destruction with conservation of mass). The Carleman model appears in the appendix of a posthumous book [2], edited by Lennart CARLESON and FROSTMAN, and as they mention that they had to finish a few proofs, it is not clear if the model is their own or if they had found this example in CARLEMAN's papers and put it in an appendix because it is not a model of kinetic theory, but an entropy inequality holds for that model. It was first studied mathematically by Ignace KOLODNER, who noticed the order-preserving property.

[9] Lennart CARLESON, Swedish mathematician, born in 1928. He received the Wolf Prize in 1992. He works at the Royal Institute of Technology, Stockholm, Sweden.

[10] Otto FROSTMAN, Swedish mathematician, 1907–1977.

[11] Ignace Izaak KOLODNER, Polish-born mathematician, 1920–1996. For a few years around 1970 he was Head of the Department of Mathematics at CARNEGIE MELLON University, Pittsburgh, PA, and remained there, and was then my colleague after I joined the university in 1987.

I had first discovered this class by applying the div-curl lemma 33.1, a result obtained in 1974 with François MURAT, which we then extended into a more general argument, called compensated compactness at the suggestion of Jacques-Louis LIONS. In the case of gradients, the div-curl lemma 33.1 is proved by a simple integration by parts and a classical compactness argument, but that particular case does not apply in the context of (26.3); our initial proof relies on the use of the Fourier transform and another line of proof had been shown to me in 1975 by Joel ROBBIN,[12] based on Hodge's[13] theory and therefore uses regularity of solutions of elliptic equations (here the Laplacian), while a similar idea lies behind a variant used by François MURAT, based on the use of Fourier multipliers, namely the Mikhlin[14]–Hörmander theorem on $\mathcal{F}L^p$ multipliers. In order to improve my compensated compactness method, I wanted to avoid the use of the Fourier transform in order to get a better understanding of the interaction between the linear differential equations (given by balance equations in continuum mechanics/physics) dealt with in the compensated compactness theorem 38.1, and the nonlinear pointwise relations (given by constitutive relations in continuum mechanics/physics) dealt with in parametrized[15,16,17,18,19,20]/Young[21] measures in order to improve what I had already done by the use of "entropies". I had met Laurence YOUNG on my first visit to United States in 1971 in Madison WI, at a time where my spoken English was so poor that I had been glad to find someone with whom

---

[12] Joel W. ROBBIN, American mathematician, born in 1941. He works at University of Wisconsin, Madison, WI.

[13] William Vallance Douglas HODGE, Scottish mathematician, 1903–1975. He worked at Cambridge, England, UK.

[14] Solomon Grigorevich MIKHLIN, Russian mathematician, 1908–1990. He worked in Leningrad, now St Petersburg, Russia.

[15] I used this term in my 1978 lectures at HERIOT-WATT University, because it was the one that I had heard in the seminar of Robert PALLU DE LA BARRIÈRE at IRIA (Institut de Recherche en Informatique et Automatique) in the late 1960s, where the idea was attributed to GHOUILA-HOURI, and I then I had seen it used in questions of convexity by Henri BERLIOCCHI & Jean-Michel LASRY.

[16] George HERIOT, Scottish goldsmith, 1563–1624.

[17] Robert PALLU DE LA BARRIÈRE, French mathematician, born in 1922. He worked at Université Paris VI (Pierre et Marie CURIE), Paris, France.

[18] Alain GHOUILA-HOURI, French mathematician, –196?.

[19] Henri BERLIOCCHI, French mathematician. He works at Université Paris Nord (Paris XIII), Villetaneuse, France.

[20] Jean-Michel LASRY, French mathematician. He works at Université Paris IX Dauphine, Paris, France.

[21] Laurence Chisholm YOUNG, English-born mathematician, 1905–2000. He worked at University of Wisconsin, Madison, WI. He was the son of William Henry YOUNG and Grace CHISHOLM-YOUNG, both mathematicians who collaborated extensively.

I could speak in French, besides John NOHEL[22] and his wife Vera, but we had not discussed mathematics, and it was much later that I learnt that the idea behind parametrized measures should be attributed to him.

For that particular application of the div-curl lemma 33.1 to systems (26.3) with condition (26.4), I did find a different approach, which gave me a new insight for proving the existence of solutions (local or global) for these special models.

Lemma 26.1. *For $c \in R$, let*

$$V_c = \{u \mid u_t + c\,u_x \in L^1(R^2), u(\cdot, 0) \in L^1(R)\} \tag{26.5}$$

and

$$W_c = \{u \mid |u(x,t)| \le U(x - ct) \text{ a.e., } U \in L^1(R)\}, \tag{26.6}$$

then $V_c \subset W_c$ and if $u \in W_c, v \in W_{c'}$ with $c \ne c'$, then $u\,v$ belongs to $L^1(R^2)$, and

$$|c - c'| \, ||u\,v||_{L^1(R^2)} \le ||u||_{W_c} ||v||_{W_{c'}}. \tag{26.7}$$

Proof: If $u_t + c\,u_x = f \in L^1(R^2)$ and $u(\cdot, 0) = g \in L^1(R)$, then $u(x,t) = g(x - ct) + \int_0^t f(x - cs, t - s)\, ds$ and therefore $|u(x,t)| \le U(x - ct)$ with $U(\sigma) = |g(\sigma)| + \int_R |f(\sigma + cs, s)|\, ds$ and, by Fubini's theorem, $U \in L^1(R)$ with norm $||v||_{L^1(R)} + ||f||_{L^1(R^2)}$ and all the formulas are true almost everywhere. Then if $|u(x,t)| \le U(x - ct)$ and $|v(x,t)| \le V(x - c't)$, one uses the change of variable $\xi = x - ct, \eta = x - c't$ and one has $d\xi\, d\eta = |c - c'| dx\, dt$.∎

Then (at least when the affine functions in (26.3) are 0, as for models in kinetic theory), I used a natural iterative scheme in $V = \prod_i V_{c_i}$, which is a strict contraction in a small ball centered at 0, and this gives global existence for small data in $L^1(R)$ for a system (26.3) satisfying condition (26.4). Existence of solutions has then been proven without trying to define the domain of the nonlinear operator, as I found that all products $u_j u_k$ appearing in the equation belong to $L^1(R^2)$ and therefore Fubini's theorem implies that for almost every t the product $u_j u_k$ belongs to $L^1(R)$.

For the 2-dimensional case, it is u_i^2 which belongs to a space like V_c, and for example the analog of the space W_c is $|u_1(x, y, t)| \le U_1(x - t, y), |u_2(x, y, t)| \le U_2(x + t, y), |u_3(x, y, t)| \le U_3(x, y - t), |u_4(x, y, t)| \le U_4(x, y + t)$, with $U_1, U_2, U_3, U_4 \in L^2(R^2)$; one has then to show that $u_1 u_3 u_4 \in L^1(R^3)$, and this is analogous to the result used in the proof of the Sobolev embedding theorem in a method of Emilio GAGLIARDO or Louis NIRENBERG or Olga LADYZHENSKAYA, as sketched in (16.2) to (16.5).

---

[22] John A. NOHEL, Czech-born mathematician, 1924–1999. He worked at University of Wisconsin, Madison, WI.

I have not found how to use this idea for the Navier–Stokes equation, but there has been some application to Boltzmann's equation or the Fokker[23]–Planck equation[24] by my student Kamel HAMDACHE.[25]. However, if the idea to use an inequality $|f(\mathbf{x}, \mathbf{v}, t)| \leq F(\mathbf{x} - \mathbf{v}\,t, \mathbf{v})$ is clear, it is not obvious how to choose F, and Kamel HAMDACHE extended an initial result of Reinhard ILLNER[26] and SHINBROT,[27] who had taken $F(\boldsymbol{\xi}, \mathbf{v}) = M\,e^{-\alpha\,|\boldsymbol{\xi}|^2}$ for a model of "hard spheres".

I had found the systems (26.3) with condition (26.4) in connection with applying compensated compactness theorem 38.1, but the method for existence relies on something different, which is that one has better bounds for special nonlinearities, and I have proposed to call this topic compensated integrability or compensated regularity. One reason for coining a new name is that in the late 1980s, Raphaël COIFMAN,[28] Pierre-Louis LIONS,[29] Yves MEYER and Stephen SEMMES[30],[31] proved an interesting result involving Hardy spaces, but they also claimed to have improved the compensated compactness theory: however, in my opinion, their result is not about compensated compactness at all (or, at least, not about *what we had called* compensated compacteness), but

---

[23] Adriaan Daniël FOKKER, Dutch physicist and composer, 1887–1972. He worked in Leiden, The Netherlands. He wrote music under the pseudonym Arie DE KLEIN.

[24] There is a growing tendency in some circles to call the Fokker–Planck equation any equation of diffusion with a drift term, even when there are no velocity variables. Here, I mean the equation which I first heard about in the late 1960s in connection with the work of Lars HÖRMANDER on his famous class of hypoelliptic operators, $\frac{\partial f}{\partial t} + \sum_{i=1}^{N} v_i \frac{\partial f}{\partial x_i} - \varepsilon\,\Delta_v f = 0$; later, I learnt of its physical meaning, and the equation has a transport term and is supposed to model grazing collisions, which result in jumps in velocity, and therefore a diffusion term in velocity appears, while jumps in position are highly unrealistic, although widely used by physicists and mathematicians. In the mid 1980s, I learnt that physicists often write a nonlinear Fokker–Planck equation by "imposing" that Maxwellian distributions satisfy it, and I find that it is a typical attitude in the presence of a dogma, here statistical mechanics, that although some of its defects are known, many prefer to ignore them and pass on these errors to another generation of students.

[25] Kamel HAMDACHE, Algerian-born mathematician, born in 1948. He works at École Polytechnique, Palaiseau, France He studied under my supervision for his thesis (1986).

[26] Reinhard ILLNER, German-born mathematician, born in 1950. He works at University of Victoria, British Columbia (Canada).

[27] Marvin SHINBROT, American-born mathematician, 1928–1987. He worked at University of Victoria, British Columbia (Canada).

[28] Ronald Raphaël COIFMAN, Israeli-born mathematician, born in 1941. He works at YALE University, New Haven, CT.

[29] Pierre-Louis LIONS, French mathematician, born in 1956. He received the Fields Medal in 1994. He holds a chair (Equations aux dérivées partielles et applications) at Collège de France, Paris, France. He is the son of Jacques-Louis LIONS.

[30] Stephen W. SEMMES, American mathematician, born in 1962. He works at RICE University, Houston, TX.

[31] William Marsh RICE, American financier and philanthropist, 1816–1900.

actually improves one of my arguments of compensated integrability based on using Lorentz spaces. After that, people in harmonic analysis started using the term compensated compactness for a type of result unrelated to what François MURAT and I had done under that name, and the net effect, unfortunately, has been a considerable increase of the confusion regarding this term.

[Taught on Monday March 15, 1999.]

27

Size of singular sets

At the moment, uniqueness of weak solutions of the 3-dimensional incompressible Navier–Stokes equation is an open problem.[1,2] Jean LERAY had conjectured that there could be point singularities of the equation and he had imagined that they could resemble self-similar solutions, of the form

$$\mathbf{u}(\mathbf{x}, t) = \frac{1}{\sqrt{2a(T - t)}} \mathbf{U} \left(\frac{\mathbf{x}}{\sqrt{2a(T - t)}} \right). \tag{27.1}$$

He also thought that it was related to turbulence, but the prevailing ideas on turbulence now are those imagined much later by KOLMOGOROV, and I think that KOLMOGOROV was wrong too, but Jean LERAY's idea does not fit well with what specialists of fluid dynamics have observed. What I understand about effective properties of mixtures and the effectiveness of microstructures suggests something quite different than Jean LERAY's idea, and somewhat different from KOLMOGOROV's idea: it is a basic result of homogenization (and there is no result of this type that I know of in the theory of Γ-convergence, which a few people still confuse with homogenization), that the effective properties of a mixture depend upon more than average properties (apart from 1-dimensional microstructures, which are not observed in turbulent flows), and the original idea[3,4] of KOLMOGOROV was similar to assuming that a turbulent

---

[1] The course was taught in 1999, before the Clay Millenium Prizes were instituted, and one of these prizes is related to the quite academic questions of uniqueness and regularity for a simplified version of the Navier–Stokes equation in domains without boundary!

[2] Landon CLAY, American investment banker and philanthropist.

[3] In his book [11], Uriel FRISCH argues along slightly different lines than KOLMOGOROV.

[4] Uriel FRISCH, French mathematician, born in 1940. He works at CNRS (Centre National de la Recherche Scientifique) at Observatoire de la Côte d'Azur, Nice, France.

flow can be described with two parameters,[5,6,7] modulo a universal stochastic process, but of course he was thinking of *developed isotropic turbulence*, so that leaves completely open the question of discovering effective equations for the evolution of turbulent flows (which in homogenization theory are treated without any invocation to probabilities, of course!). Recently, Jindřich NEČAS, Michael RUŽICKA[8] and Vladimír ŠVERÁK[9] have shown that the self-similar solutions imagined by Jean LERAY cannot have $U \in L^3(R^3; R^3)$.

Measuring the *Hausdorff dimension* of the *singular set* of a solution has been a way to determine how far it is from being smooth, and at the moment the best result has been obtained by Luis CAFFARELLI[10], Robert KOHN[11] and Louis NIRENBERG; they showed that the 1-dimensional Hausdorff measure of the singular set is 0, and therefore it cannot be a point singularity moving along a nice curve; Michael STRUWE[12] has obtained a similar result for the stationary case in 5 dimensions.

Jean LERAY had already obtained results bounding the Hausdorff dimension of the singular set in t alone; I have not read his argument, but I think that it is based on the already-mentioned differential inequality $(||u||^2)' \leq C\,||u||^6$ as follows. Let $\varphi = ||u||^2$, so we start with the information $\varphi \in L^1(0,T)$ and $\varphi' \leq a\,\varphi^3$; the differential inequality implies $(\varphi^{-2})' \leq -2a$ and therefore $\varphi(t) \leq \varphi(0)\big(1 - 2a\,t\,\varphi(0)^2\big)^{-1/2}$ as long as $1 - 2a\,t\,\varphi(0)^2 > 0$, and therefore the (possible) blow-up time must satisfy $T_c \geq \frac{1}{2a\,\varphi(0)^2}$. One divides $(0,T)$ into N equal intervals $I_1, \ldots, I_N$, of length $\tau = T/N$; if $j < N$ and $\int_{I_j} \varphi(t)\,dt < \sqrt{\tau/4a}$, then there is a point $x_j \in I_j$ such that $\varphi(x_j) < \sqrt{1/4\tau\,a}$; one deduces that the blow-up time after x_j is at least $\frac{1}{2a\,\varphi(x_j)^2} > 2\tau$, and therefore the next interval I_{j+1} is free of singularities. The singular set is then contained in I_1 and the union of all the I_{j+1} for an index j such that $\int_{I_j} \varphi(t)\,dt > \sqrt{\tau/4a}$, but as

[5] Engineers do use the k-epsilon method, but from what I have read in a book [20] by Olivier PIRONNEAU and MOHAMMADI, the method begs to be improved.

[6] Olivier PIRONNEAU, French mathematician, born in 1945. He works at Université Paris VI (Pierre et Marie CURIE), Paris, France.

[7] Bijan MOHAMMADI, Iranian-born mathematician, born in 1964. He works at Université de Montpellier II, Montpellier, France.

[8] Michael RUŽIČKA, German mathematician, born in 1964. He works in Freiburg, Germany.

[9] Vladimír ŠVERÁK, Czech-born mathematician, born in 1959. He works at University of Minnesota Twin Cities, Minneapolis, MN.

[10] Luis Angel CAFFARELLI, Argentine-born mathematician, born in 1948. He works at University of Texas, Austin, TX.

[11] Robert V. KOHN, American mathematician, born in 1953. He works at the COURANT Institute of Mathematical Sciences at New York University, New York, NY.

[12] Michael STRUWE, German-born mathematician, born in 1955. He works at ETH (Eidgenössische Technische Hochschule) in Zürich, Switzerland.

$$\sum_j \sqrt{meas(I_{j+1})} \le 2\sqrt{a} \sum_j \int_{I_j} \varphi(t)\, dt \le 2\sqrt{a}||\varphi||_{L^1(0,T)}, \qquad (27.2)$$

one deduces that the 1/2 Hausdorff dimension of the singular set (in t alone) is finite by letting N tend to infinity; by applying the argument to a family of intervals containing the set where φ takes large values, one deduces that the 1/2 Hausdorff dimension of the singular set of $||u||$ (in t alone) is 0.

Estimating the Hausdorff dimension of the singular set in $(\mathbf{x}, t)$ relies on local regularity results, but Luis CAFFARELLI, Robert KOHN and Louis NIRENBERG used the regularizing effect of the heat kernel, considering the "pressure" as given by an equation; for scaling, instead of balls in the $(\mathbf{x}, t)$ space, they used flat cylinders, scaling in ε in $\mathbf{x}$ and ε^2 in t. Taking the divergence of the equation, one has

$$-\Delta p = \sum_{i,j=1}^3 \frac{\partial u_i}{\partial x_j} \frac{\partial u_j}{\partial x_i}, \qquad (27.3)$$

and there are special results, which I call compensated integrability results, for that equation.

Before reviewing some of that information, I want to describe a different approach, which is based on using pointwise estimates in terms of *maximal functions*, a subject which I had understood from an example used by Lars HEDBERG,[13] reproduced by Haïm BREZIS and Felix BROWDER[14] for a question of truncation (which some justly call the *Hedberg truncation method*).

For a function $f \in L^1_{loc}(R^N)$, the maximal function of f, denoted by $M(f)$ or $M f$, is defined by

$$M f(\mathbf{x}) = \sup_{r>0} \frac{\int_{B(\mathbf{x},r)} |f(\mathbf{y})|\, d\mathbf{y}}{\int_{B(\mathbf{x},r)} d\mathbf{y}}. \qquad (27.4)$$

At every Lebesgue point, and therefore almost everywhere, one has $|f(\mathbf{x})| \le M f(\mathbf{x})$. If $f \in L^\infty$, one has $||M f||_{L^\infty} \le ||f||_{L^\infty}$, but if $f \in L^1(R^N)$ and $f \neq 0$, then $M f \notin L^1(R^N)$. However, using a simple covering argument, one can show that $meas\{\mathbf{x} \mid M f(\mathbf{x}) \ge t\} \le \frac{C ||f||_{L^1}}{t}$ for every $t > 0$, and using an interpolation argument one deduces that for $p > 1$ one has $||M f||_{L^p} \le C_p ||f||_{L^p}$ for every $f \in L^p(R^N)$, and $C_p \to \infty$ as $p \to 1$ (HARDY and LITTLEWOOD,[15,16]

---

[13] Lars Inge HEDBERG, Swedish mathematician, 1935–2005. He worked at Linköping University, Linköping, Sweden.

[14] Felix Earl BROWDER, American mathematician, born in 1928. He works at RUTGERS University, Piscataway, NJ.

[15] John Edensor LITTLEWOOD, English mathematician, 1885–1977. He held the first Rouse BALL professorship at Cambridge, England, UK.

[16] Walter William Rouse BALL, English mathematician, 1850–1925. He worked in Cambridge, England, UK.

WIENER). In 1972, Lars HEDBERG used an inequality giving a pointwise estimate of a convolution product by a radial function in terms of the maximal function, i.e.

$$\text{for a radial function } f, |(f \star g)(\mathbf{x})| \leq C M g(\mathbf{x}). \qquad (27.5)$$

His function f was special ($1/r^\lambda$ for $0 < r < r_0$, 0 for $r > r_0$), and he used a dyadic decomposition, but after studying his proof a few years ago I realized that one could write the inequality above for any f radial and nonincreasing, in which case $C = ||f||_{L^1(R^N)}$, and more generally, assuming f radial and smooth in order to avoid technical details, with $C = ||r\,grad\,f||_{L^1(R^N)}$.

Two years ago, during a meeting dedicated to Jindrich NEČAS in Lisbon, Portugal, I thought of a way to use Lars HEDBERG's argument for estimating various norms of solutions of the heat equation or of Stokes's equation in all the space, and I thought that if one knew how to extend this type of inequality when transport terms are present, it could be quite useful for improving the abstract approach to solutions of the Navier–Stokes equation. I then had an e-mail exchange with Lars HEDBERG in order to learn about the origin of the idea that he had used, and he said that it was his program to show that many classical global inequalities can actually be improved into pointwise inequalities using maximal functions, but he saw the first example of this kind in a result of Lennart CARLESON of 1967, showing that the solution of $\Delta u = 0$ in the unit disc, $u = \varphi$ on the boundary satisfied a bound $|u(\mathbf{x})| \leq C M \varphi\left(\frac{\mathbf{x}}{|\mathbf{x}|}\right)$ for $\mathbf{x} \neq 0$ (maybe with $C = 1$). When I pointed out the inequalities which I had shown for the heat kernel, it reminded him of another result, shown by STEIN in his 1970 book on singular integrals [22], where he shows the idea for the Poisson integrals (same as Lennart CARLESON but for a half space instead of a disc), but STEIN does add a remark showing that my bound $C = ||r\,grad\,f||_{L^1(R^N)}$ is not the good one: he assumes that $|f| \leq \psi$ with ψ radial as I had done, but he only considers ψ nonincreasing and integrable, and he proves $C = \inf_\psi ||\psi||_{L^1(R^N)}$, where the infimum is taken on the radial nonincreasing ψ larger than or equal to $|f|$ almost everywhere (if I had noticed that the bound $||r\,grad\,f||_{L^1(R^N)}$ may decrease while replacing f by a larger function, I would have been automatically led to the bound proven by STEIN).

The proof is easy once one realizes that a radial nonincreasing function is an integral with nonnegative coefficient of characteristic functions of balls centered at 0, or simply that it is a limit in the $L^1(R^N)$ norm of finite combinations with positive coefficients of characteristic functions of balls centered at 0; by linearity it suffices then to prove the result for f being the characteristic function χ of a ball $B(0, \varrho)$ centered at 0, but then the convolution by χ does compute $\int_{B(\mathbf{x},\varrho)} |g(\mathbf{x})|\,d\mathbf{x}$ which is bounded by $M g(\mathbf{x})$ multiplied by the volume of $B(0, \varrho)$.

If one applies the idea to the solution of the heat equation

$$\frac{\partial u}{\partial t} - \Delta u = 0 \text{ in } R^N \times (0, T); \quad u(\mathbf{x}, 0) = v(\mathbf{x}) \text{ in } R^N, \qquad (27.6)$$

then the solution is

$$u(\mathbf{x}, t) = \int_{R^N} E(\mathbf{x} - \mathbf{y}, t) v(\mathbf{y}) \, d\mathbf{y}, \text{ with } E \text{ given by}$$
$$E(\mathbf{z}, t) = C_N t^{-N/2} e^{-|\mathbf{z}|^2/4t} \text{ on } R^N \times (0, \infty), \tag{27.7}$$

and C_N is such that the elementary solution E satisfies $\int_{\mathbf{x} \in R^N} E(\mathbf{x}, t) \, d\mathbf{x} = 1$ for any (or all) $t > 0$. As $E(\cdot, t)$ is radial and decreasing with integral 1, one deduces that

$$|u(\mathbf{x}, t)| \leq M\, v(\mathbf{x}) \text{ a.e. } \mathbf{x} \in R^N, \text{ for all } t > 0. \tag{27.8}$$

As $|u(\cdot, t)| \leq E(\cdot, t) \star |v|$, if χ is the characteristic function of the ball of radius ϱ, then $\chi \star E(\cdot, t)$ is also radial and decreasing and one deduces the more precise inequality

$$M\, u(\mathbf{x}, \cdot) \leq M\, v(\mathbf{x}) \text{ a.e. } \mathbf{x} \in R^N, \text{ for all } t > 0. \tag{27.9}$$

This inequality cannot be deduced from the bounds $\int_{\mathbf{x} \in R^N} \Phi(u(\mathbf{x}, t)) \, d\mathbf{x} \leq \int_{\mathbf{x} \in R^N} \Phi(v(\mathbf{x}, 0)) \, d\mathbf{x}$ for every convex function Φ, or simply the inequalities $||u(\cdot, t)||_{L^p(R^N)} \leq ||v||_{L^p(R^N)}$ for all $p \in [1, \infty]$; the maximal function changes if one replaces v by an equimeasurable function. If one applies the idea to derivatives, then one obtains

$$M(D^\alpha u)(\mathbf{x}, t) \leq C_\alpha t^{-|\alpha|/2} M\, v(\mathbf{x}) \text{ a.e. } \mathbf{x} \in R^N, \text{ for all } t > 0,$$
$$\text{for all derivatives } D^\alpha \text{ of order } |\alpha|, \tag{27.10}$$

and STEIN had noticed the analogous inequality for the Poisson integrals. My remark was that similar results are also true for Stokes's equation, but that the extension of this idea to the Navier–Stokes equation is not straightforward, and another idea must certainly be added (recently, after a discussion with Mariarosaria PADULA,[17] I have thought that the techniques that she has used for studying questions of stability of flows could be of interest in my question).

[Taught on Wednesday March 17, 1999.]

---

[17] Mariarosaria PADULA, Italian mathematician, born in 1949. She works at Università di Ferrara, Ferrara, Italy.

28

Local estimates, compensated integrability

A few years ago, my colleague Victor MIZEL[1] had asked me if I knew about Hedberg's truncation method. At that time I did not know what was meant by that term, and as François MURAT is a good friend of Lars HEDBERG, I had asked him about it, and he had sent me a few pages of an article by Haïm BREZIS and Felix BROWDER who had reproduced Lars HEDBERG's proof in an appendix (they had read it in an article by Jeffrey WEBB,[2] to whom Lars HEDBERG had taught the method); I carried the copy for a while before looking at it, one summer on the beach, and saw what the key ideas were. Later I also looked at a paper of Emilio ACERBI[3] and Nicola FUSCO,[4,5] because I had understood from a comment of Irene FONSECA[6] that there might be similarities between the methods; the key point in their argument is a result of F. C. LIU,[7] which I could derive easily by Lars HEDBERG's trick,

---

[1] Victor J. MIZEL, American mathematician, 1931–2005. He works at CARNEGIE MELLON University, Pittsburgh, PA, and had been my colleague since I joined the university in 1987.

[2] Jeffrey Ronald Leslie WEBB, English mathematician. He works in Glasgow, Scotland, UK.

[3] Emilio Daniele G. ACERBI, Italian mathematician, born in 1955. He works at Università di Parma, Parma, Italy.

[4] Nicola FUSCO, Italian mathematician, born in 1956. He works at Università di Napoli "FEDERICO II", Napoli (Naples), Italy.

[5] FRIEDRICH II (HOHENSTAUFEN), 1194–1250. Holy Roman Emperor (1220–1250). He founded the first European state university in Naples in 1224.

[6] Irene Maria QUINTANILHA COELHO DA FONSECA, Portuguese-born mathematician, born in 1956. She works at CARNEGIE MELLON University, Pittsburgh PA. She was my second wife, from 1985 to 2002, and is the mother of two of my four children, André and Marta.

[7] Fon Che LIU, Chinese mathematician. He works at Academia Sinica, Taipei, Taiwan.

and I had asked my student Sergio GUTIERREZ[8] to work on an extension. Although these results are not directly related to the questions of fluids, I discuss them briefly as they may be useful in order to obtain a clear picture of what the methods are.

Lars HEDBERG's truncation method permits us to approach a function in $u \in W^{m,p}(R^N)$ by a sequence $u_n \in L^\infty(R^N) \cap W^{m,p}(R^N)$ such that $u_n(\mathbf{x})u(\mathbf{x}) \geq 0$ a.e., and as it uses the Calderón–Zygmund theorem one must have $p > 1$ (for $p > \frac{N}{m}$ there is nothing to prove as $W^{m,p}(R^N) \subset C_0(R^N)$). In order to simplify, I show how it works for approaching a function with second derivatives in $L^p(R^N)$ with $1 < p < \frac{N}{2}$. One solves $-\Delta v = |\Delta u|$, which gives $\partial_j \partial_k v \in L^p(R^N)$ for all j, k by the Calderón–Zygmund theorem, and $v \geq |u|$ by the maximum principle (in general one takes convolutions by powers of $\frac{1}{r}$), and one defines u_n by $u_n(\mathbf{x}) = u(\mathbf{x})\varphi\left(\frac{v(\mathbf{x})}{n}\right)$, where φ is smooth and is equal to 1 on $[0, 1/2]$ and 0 on $[1, \infty)$, showing that $|u_n| \leq n$. Then $\partial_j \partial_k u_n = \partial_j \partial_k u\, \varphi\left(\frac{v}{n}\right) + \partial_j u\, \varphi'\left(\frac{v}{n}\right)\frac{1}{n}\partial_k v + \partial_k u\, \varphi'\left(\frac{v}{n}\right)\frac{1}{n}\partial_j v + u\, \varphi''\left(\frac{v}{n}\right)\frac{1}{n^2}\partial_j v \partial_k v + u\, \varphi'\left(\frac{v}{n}\right)\frac{1}{n}\partial_j \partial_k v$, and the first term converges to $\partial_j \partial_k u$ and the last term converges to 0 by the Lebesgue dominated convergence theorem (using $|u| \leq v$); Lars HEDBERG's method is based on the fact that both $\partial_j u$ and $\partial_j v$ are bounded (pointwise) by $C \sqrt{v} \sqrt{M\, \Delta u}$, so that the other terms can also be treated in the same way. Indeed one has $\partial_j u = \partial_j E \star \Delta u$ and $\partial_j v = \partial_j E \star |\Delta u|$, both bounded by $\frac{C}{r^{N-1}} \star |\Delta u|$, which is cut into two parts; the first one is $f \star |\Delta u|$ with $f = \frac{C}{r^{N-1}}$ for $0 < r < \delta$ and 0 for $r > \delta$, and this term is bounded by $C\, \delta\, M\, \Delta u$ by using the argument on convolution with radial functions, and the second is bounded by $\frac{C}{\delta} E \star |\Delta u| = C \frac{v}{\delta}$; then the best δ is chosen (depending upon $\mathbf{x}$).

The estimate of F. C. LIU is about

$$|u(\mathbf{x}) - u(\mathbf{y})| \leq C\, |\mathbf{x} - \mathbf{y}|(M\, |grad\, u|(\mathbf{x}) + M\, |grad\, u|(\mathbf{y})) \text{ for a.e. } \mathbf{x}, \mathbf{y} \in R^N.$$
(28.1)

One starts from $u(\mathbf{x}) - u(\mathbf{y}) = \int_0^1 \left(grad\, u(\mathbf{x} + t(\mathbf{y} - \mathbf{x})\right).\mathbf{y} - \mathbf{x})\, dt$, from which one deduces that

$$\int_{B(\mathbf{x},\varrho)} \frac{|u(\mathbf{y}) - u(\mathbf{x})|}{|\mathbf{x} - \mathbf{y}|}\, d\mathbf{y} \leq \int_0^1 \int_{B(\mathbf{x},\varrho)} \left|grad\, u(\mathbf{x} + t(\mathbf{y} - \mathbf{x}))\right|\, dt\, d\mathbf{y}, \quad (28.2)$$

and one uses the change of variable $\mathbf{z} = -t(\mathbf{y} - \mathbf{x})$; the variable t varies from $|\mathbf{z}|/\varrho$ to 1, and the last integral is $(N-1) \int_{B(\mathbf{x},\varrho)} |grad\, u(\mathbf{x}-\mathbf{z})|(|\mathbf{z}|^{1-N} - 1)\, d\mathbf{z}$, which is a convolution of $|grad\, u|$ by a radial decreasing function and is therefore bounded by $M\, |grad\, u|(\mathbf{x})$ multiplied by the L^1 norm of the radial function, which is the value obtained when one replaces $|grad\, u|$ by 1, i.e. the volume of $B(0, \varrho)$. One then integrates $\frac{|u(\mathbf{y})-u(\mathbf{x_1})|}{|\mathbf{y}-\mathbf{x_1}|} + \frac{|u(\mathbf{y})-u(\mathbf{x_2})|}{|\mathbf{y}-\mathbf{x_2}|}$

---

[8] Sergio Enrique GUTIÉRREZ, Chilean mathematician, born in 1963. He works at Pontificia Universidad Católica de Chile, Santiago, Chile. He was my PhD student (1997) at CARNEGIE MELLON University, Pittsburgh, PA.

on $A = B(\mathbf{x_1}, \varrho) \cap B(\mathbf{x_2}\varrho)$, and it is bounded by the integral on $B = B(\mathbf{x_1}, \varrho) \cup B(\mathbf{x_2}\varrho)$, i.e. by $M\,|grad\,u|(\mathbf{x_1}) + M\,|grad\,u|(\mathbf{x_2})$ multiplied by twice the volume of $B(0, \varrho)$; one chooses $\varrho = |\mathbf{x_1} - \mathbf{x_2}|$ for example and one finds a $\mathbf{y} \in A$ such that $\frac{|u(\mathbf{y}) - u(\mathbf{x_1})|}{|\mathbf{y} - \mathbf{x_1}|} + \frac{|u(\mathbf{y}) - u(\mathbf{x_2})|}{|\mathbf{y} - \mathbf{x_2}|} \leq C(M\,|grad\,u|(\mathbf{x_1}) + M\,|grad\,u|(\mathbf{x_2}))$ and as $|\mathbf{y} - \mathbf{x_1}|, |\mathbf{y} - \mathbf{x_2}| \leq |\mathbf{x_1} - \mathbf{x_2}|$ it implies the desired inequality.

The inequality that Lars HEDBERG used in his truncation method is reminiscent of an inequality due (independently, I believe) to Emilio GAGLIARDO and Louis NIRENBERG, and indeed one can find a pointwise version of the Gagliardo–Nirenberg inequality, as I checked last December, only to discover a week or two after that Patrick GÉRARD[9] had made the same observation; however, his proof is different from mine, relying on a dyadic decomposition in the style of LITTLEWOOD and PALEY, while mine is more elementary and uses a parametrix (a concept introduced by HADAMARD[10]). Let E be the usual elementary solution of $-\Delta$, i.e. $E(\mathbf{x}) = \frac{C_N}{|\mathbf{x}|^{N-2}}$ if $N \geq 3$, or $E(\mathbf{x}) = C\log(|\mathbf{x}|)$ for $N = 2$; let $\varphi \in C_c^\infty(R^N)$ be such that $\varphi(\mathbf{x}) = 1$ for $|\mathbf{x}| \leq 1$ and $\varphi(\mathbf{x}) = 0$ for $|\mathbf{x}| \geq 2$, and for $\alpha > 0$ let us consider the parametrix P_α defined by $P_\alpha(\mathbf{x}) = E(\mathbf{x})\varphi(\frac{\mathbf{x}}{\alpha})$, so that $-\Delta\,P_\alpha = \delta_0 + g$ with $g(\mathbf{x}) = \frac{1}{|\mathbf{x}|^N}\psi(\frac{\mathbf{x}}{\alpha})$, with $\psi \in C_c^\infty(R^N)$. Taking the convolution by $\partial_j u$ gives $\partial_j P_\alpha \star (-\Delta\,u) = \partial_j u + \partial_j g \star u$, and as $||r\,grad(\partial_j P_\alpha)||_{L^1(R^N)} = C\,\alpha$ and $||r\,grad(\partial_j g)||_{L^1(R^N)} = \frac{C}{\alpha}$, one deduces that $|\partial_j u| \leq C\,\alpha\,M\,\Delta u + \frac{C}{\alpha}\,M\,u$, and then taking the best α (depending on $\mathbf{x}$) gives

$$|grad\,u| \leq C\,\sqrt{M\,u}\,\sqrt{M\,\Delta u}. \tag{28.3}$$

I want to finish with some remarks on what I call compensated integrability, although in some cases a term like compensated regularity would be better, the reason for coining these new terms being that some have used the term compensated compactness for describing results which have almost nothing to do with the compensated compactness theorem 38.1 which I had proven with François MURAT, and certainly nothing to do with the compensated compactness method that I had taught at HERIOT-WATT University in the summer of 1978.

In the summer of 1982, at a meeting in Oxford, I heard about a result of WENTE,[11] and back in Paris I derived a proof by interpolation which I mentioned around. If $u, v \in H^1(R^2)$ then one cannot assert that $u_x v_y$ and $u_y v_x$ belong to $H^{-1}(R^2)$ because $L^1(R^2)$ is not embedded in $H^{-1}(R^2)$ as

[9] Patrick GÉRARD, French mathematician, born in 1961. He works at Université Parix XI (Paris-Sud), Orsay, France.

[10] Jacques Salomon HADAMARD, French mathematician, 1865–1963. He held a chair (Mécanique analytique et mécanique céleste) at Collège de France, Paris, France.

[11] Henry C. WENTE, American mathematician, born in 1936. He works in Toledo, OH.

$H^1(R^2)$ is not embedded into $L^\infty(R^2)$, but the difference $u_x v_y - u_y v_x$ belongs to $H^{-1}(R^2)$; it does not follow immediately from writing that quantity as $(u\,v_y)_x - (u\,v_x)_y$ or $(u_x v)_y - (u_y v)_x$, which is the key to the sequential weak lower semicontinuity observed by MORREY[12] and which I heard from John M. BALL (which also follows from the div-curl lemma 33.1 that I had already proven with François MURAT, which we extended later into the compensated compactness theorem 38.1). However, WENTE had not only proven that $u_x v_y - u_y v_x \in H^{-1}(R^2)$ but that if one solves the equation $-\Delta w = u_x v_y - u_y v_x$, then one also has $w \in H^1(R^2) \bigcap C_0(R^N)$.

I do not know what the original proof of WENTE is, but my first proof used interpolation and Lorentz spaces. It would take us too far if I explained what interpolation of Banach spaces is according to the theory developed by Jacques-Louis LIONS and Jaak PEETRE, and how the theory applied to L^1 and L^∞ creates the family of Lorentz spaces; therefore I shall just state the ideas for those readers who know these tools.[13] First one uses the fact that $H^{1/2}(R^2) \subset L^{4,2}(R^2)$, as noticed by Jaak PEETRE (a result already used was $||u||_{L^4(R^2)} \leq C\,||u||_{L^2(R^2)}^{1/2}||grad\,u||_{L^2(R^2)}^{1/2}$, and it is not as precise because a theorem of Jacques-Louis LIONS and Jaak PEETRE asserts that this statement is equivalent to the fact that the interpolation space $\big(H^1(R^2), L^2(R^2)\big)_{1/2,1}$, which is smaller than $H^{1/2}(R^2)$, is included in $L^4(R^2)$, which is bigger than $L^{4,2}(R^2)$). Then one uses the fact that the product of two functions in $L^{4,2}(R^2)$ is in $L^{2,1}(R^2)$. This shows that $B(u,v) = u_x v_y - u_y v_x$ is a sum of derivatives of functions of $L^{2,1}(R^2)$, in the case where $u \in H^{1/2}(R^2)$ and $v \in H^{3/2}(R^2)$ by using the formula $(u\,v_y)_x - (u\,v_x)_y$, or in the case where $u \in H^{3/2}(R^2)$ and $v \in H^{1/2}(R^2)$ by using the formula $(u_x v)_y - (u_y v)y$; by another theorem of Jacques-Louis LIONS and Jaak PEETRE on bilinear mappings the same property is then true for $u,v \in H^1(R^2)$. Then by the Calderón–Zygmund theorem and interpolation, one finds that w has its two partial derivatives in $L^{2,1}(R^2)$ (a smaller space than $L^2(R^2)$), and this implies that $w \in C_0(R^2)$.

In 1984, I described a second method which extends immediately to more general situations similar to those found in applying compensated compactness theorem 38.1 for the quadratic forms which are sequentially weakly continuous; the method uses the Fourier transform and interpolation, but not the Calderón–Zygmund theorem, and the results are slightly different. The example that I had chosen was the equation for the "pressure" in the Navier–Stokes equation in 2 dimensions, but that is similar to the previous example. Using x_1, x_2, instead of x, y, one has $|\xi|^2 \mathcal{F}w(\xi) = \int_{\eta \in R^2} B(\xi - \eta, \eta)\mathcal{F}u(\xi - \eta)\mathcal{F}v(\eta)\,d\eta$, where $B(\zeta, \eta) = \zeta_1\,\eta_2 - \zeta_2\,\eta_1$, but using the fact that $B(\xi, \xi) = 0$ for all ξ, one has $B(\xi - \eta, \eta) = B(\xi, \eta) = B(\xi, \eta - \xi)$, so that one has the two bounds $|B(\xi - \eta, \eta)| \leq C\,|\xi|\,|\eta|$ or

---

[12] Charles Bradfield Jr. MORREY, American mathematician, 1907–1980. He worked at University of California Berkeley, Berkeley, CA.

[13] For those who have not learnt these tools, I have taught this subject in another set of lecture notes.

$\leq\,C\,|\boldsymbol{\xi}|\,|\boldsymbol{\xi}-\boldsymbol{\eta}|$ and therefore $\leq\,C\,|\boldsymbol{\xi}|\,|\boldsymbol{\eta}|^{1/2}|\boldsymbol{\xi}-\boldsymbol{\eta}|^{1/2}$. One deduces that $|\boldsymbol{\xi}|\,|\mathcal{F}w| \leq C\,(|\boldsymbol{\xi}|^{1/2}|\mathcal{F}u|) \star (|\boldsymbol{\xi}|^{1/2}|\mathcal{F}v|)$, but as $|\boldsymbol{\xi}|^{-1/2} \in L^{4,\infty}(R^2)$, one deduces that $\boldsymbol{\xi}\,\mathcal{F}u \in L^2(R^2)$ implies $|\boldsymbol{\xi}|^{1/2}|\mathcal{F}u| \in L^{4/3,2}(R^2)$ and the convolution product of two functions in $L^{4/3,2}(R^2)$ is in $L^{2,1}(R^2)$. In particular $|\boldsymbol{\xi}|\,|\mathcal{F}w| \in L^{2,1}(R^2)$, but I do not know how to compare the informations $grad\,w \in L^{2,1}(R^2;R^2)$ and $|\boldsymbol{\xi}|\mathcal{F}w \in L^{2,1}(R^2;R^2)$. As $|\boldsymbol{\xi}|^{-1} \in L^{2,\infty}(R^2)$, one deduces that $\mathcal{F}w \in L^1(R^2)$, and then $w \in \mathcal{F}L^1(R^2) \subset C_0(R^2)$.

My approach has been slightly improved by Raphaël COIFMAN, Pierre-Louis LIONS, Yves MEYER and Stephen SEMMES, using the Hardy spaces $\mathcal{H}^1$; their result has the advantage of showing that the second derivatives of w belong to $\mathcal{H}^1(R^2)$, and therefore $w \in W^{2,1}(R^2)$; however, contrary to what they have claimed, many applications do not require their improvement and can be obtained by using my second method.

So much for technical details concerning the Navier–Stokes equation. Let us go back to oceanography!

[Taught on Friday March 19, 1999.]

29

Coriolis force

Up to now, I have only mentioned the effect of Coriolis force due to the rotation of the Earth in Lectures 2 and 6; it is small but it does have some effect.

Assume that we have a first frame, called fixed, in which Newton's law of classical mechanics, $force = mass \times acceleration$ applies, and let us see what it implies for the equation in a moving frame. Let $\mathbf{x}(t)$ be the position of a material point in the fixed frame, and let $\boldsymbol{\xi}(t)$ be the position of the same point in the moving frame; let $\mathbf{a}(t)$ be the position of the origin of the moving frame and let $\mathsf{P}(t)$ be the rotation which maps the basis of the initial frame into the basis of the moving frame, so that one has

$$\mathbf{x}(t) = \mathbf{a}(t) + \mathsf{P}(t)\boldsymbol{\xi}(t). \tag{29.1}$$

As $\mathsf{P}(t)^T \mathsf{P}(t) = \mathsf{I}$, if $'$ denotes the derivative with respect to t, one has $\mathsf{P}'(t)^T \mathsf{P}(t) + \mathsf{P}(t)^T \mathsf{P}'(t) = 0$, and so if one defines $\mathsf{B}(t)$ by $\mathsf{P}'(t) = \mathsf{P}(t)\mathsf{B}(t)$, one obtains $\mathsf{B}(t)^T + \mathsf{B}(t) = 0$, and therefore, as we work in R^3, there exists a vector $\boldsymbol{\Omega}(t)$ such that $\mathsf{B}(t)\mathbf{z} = \boldsymbol{\Omega}(t) \times \mathbf{z}$ for every $\mathbf{z} \in R^3$; one deduces

$$\mathbf{x}'(t) = \mathbf{a}'(t) + \mathsf{P}(t)\big[\boldsymbol{\xi}'(t) + \big(\boldsymbol{\Omega}(t) \times \boldsymbol{\xi}(t)\big)\big], \tag{29.2}$$

and

$$\mathbf{x}''(t) = \mathbf{a}''(t) + \mathsf{P}(t)\big[\boldsymbol{\xi}''(t) + \big(\boldsymbol{\Omega}'(t) \times \boldsymbol{\xi}(t)\big) + \big(\boldsymbol{\Omega}(t) \times \boldsymbol{\xi}'(t)\big)\big] + \mathsf{P}(t)\,\boldsymbol{\Omega}(t) \times \big[\boldsymbol{\xi}'(t) + \big(\boldsymbol{\Omega}(t) \times \boldsymbol{\xi}(t)\big)\big]. \tag{29.3}$$

In the case of the rotation of the Earth, one considers that $\boldsymbol{\Omega}'(t) = 0$. The term $2\boldsymbol{\Omega}(t) \times \boldsymbol{\xi}'(t)$ is the Coriolis acceleration (although LAGRANGE had introduced it in 1778–1779 in his studies of tides, while CORIOLIS's work dates from 1835). If one uses the formula $\mathbf{a} \times (\mathbf{b} \times \mathbf{c}) = (\mathbf{a}.\mathbf{c})\,\mathbf{b} - (\mathbf{a}.\mathbf{b})\,\mathbf{c}$, one deduces that $\boldsymbol{\Omega} \times (\boldsymbol{\Omega} \times \boldsymbol{\xi}) = (\boldsymbol{\Omega}.\boldsymbol{\xi})\boldsymbol{\Omega} - |\boldsymbol{\Omega}|^2\boldsymbol{\xi}$, and therefore the term $\boldsymbol{\Omega} \times (\boldsymbol{\Omega} \times \boldsymbol{\xi})$, which is related to the centrifugal acceleration with $\mathbf{a}''(t)$, derives from a potential, which changes slightly the gravitation potential, creating the *geopotential*. For the rotation of the Earth, $|\Omega| = \frac{2\pi}{86400} \approx 3.6 \times 10^{-5}$, so that at the equator the centrifugal acceleration is about 8.3×10^{-3}, less than one thousandth of the acceleration of gravity.

Because the term $\Omega \times (\Omega \times \xi)$ is a gradient, it changes only what p is, and therefore adding the Coriolis term $\Omega \times u$ in the Navier–Stokes equation does not change much in the proofs that we have seen, because this term is orthogonal to u and therefore does not work, and the basic estimates are the same as before.

The Coriolis force depends upon the velocity, in a way that reminds us of electromagnetism, where the *Lorentz force* acting on a charge q moving with velocity v in an electric field $\mathbf{E}$ and magnetic induction field $\mathbf{B}$ is $q(\mathbf{E} + v \times \mathbf{B})$. The analogy goes further and it has been used in connection with MHD (*Magneto-Hydro-Dynamics*), at least by Keith MOFFATT:[1] in MHD the fluid is a *plasma*, which has electrical charges moving around (heavy ions and light electrons), but the forces acting on a neutral fluid are very similar, as we shall see by computing $u \times curl\, u$ in a domain of R^3.

Let ε_{ijk} be the totally antisymmetric tensor, which is 0 if two of the indices i, j, k, are equal, and equal to the signature of the permutation $123 \mapsto ijk$ in other cases, i.e. $\varepsilon_{123} = \varepsilon_{231} = \varepsilon_{312} = 1$ and $\varepsilon_{321} = \varepsilon_{213} = \varepsilon_{132} = -1$. Then the definition of the exterior product of two vectors in R^3 is

$$\mathbf{c} = \mathbf{a} \times \mathbf{b} \text{ means } c_i = \sum_{j,k=1}^{3} \varepsilon_{ijk} a_j b_k, \tag{29.4}$$

and the *curl* of a function u, also denoted by $\nabla \times u$, is defined by

$$(curl\, u)_i = \sum_{j,k=1}^{3} \varepsilon_{ijk} \frac{\partial u_k}{\partial x_j}. \tag{29.5}$$

One has

$$(\mathbf{u} \times curl\, u)_i = \sum_{j,k=1}^{3} \varepsilon_{ijk} u_j \left(\sum_{\ell,m=1}^{3} \varepsilon_{k\ell m} \frac{\partial u_m}{\partial x_\ell} \right) =$$
$$\sum_{j,k=1}^{3} \varepsilon_{ijk} u_j \left(\varepsilon_{kij} \frac{\partial u_j}{\partial x_i} + \varepsilon_{kji} \frac{\partial u_i}{\partial x_j} \right) = \sum_{j,k=1}^{3} \varepsilon_{ijk}^2 u_j \left(\frac{\partial u_j}{\partial x_i} - \frac{\partial u_i}{\partial x_j} \right) = \tag{29.6}$$
$$\sum_{j \neq i} u_j \left(\frac{\partial u_j}{\partial x_i} - \frac{\partial u_i}{\partial x_j} \right) = \sum_{j=1}^{3} u_j \frac{\partial u_j}{\partial x_i} - \sum_{j=1}^{3} u_j \frac{\partial u_i}{\partial x_j},$$

and therefore

$$\sum_{j=1}^{3} u_j \frac{\partial u_i}{\partial x_j} = (\mathbf{u} \times curl(-u))_i + \frac{1}{2} \frac{\partial |u|^2}{\partial x_i}, \tag{29.7}$$

so that the Navier–Stokes equation becomes

$$\frac{\partial u}{\partial t} - \nu \, \Delta u + \mathbf{u} \times curl(-u) + grad\left(\frac{p}{\varrho_0} + \frac{|u|^2}{2} \right) = 0, \; div\, u = 0, \tag{29.8}$$

and Coriolis acceleration just adds 2Ω to $curl(-u)$.

---

[1] Henry Keith MOFFATT, Scottish mathematician, born in 1935. He worked at Cambridge, England, UK.

In the case $\nu = 0$, corresponding to Euler's equation, one sees that a stationary irrotational flow (i.e. satisfying $curl\, u = 0$), corresponds to $\frac{p}{\varrho_0} + \frac{|\mathbf{u}|^2}{2} = constant$ (*Bernoulli's law*); one also sees that in the whole space R^3 helicity ($\mathbf{u}.curl\, u$) is conserved (*curl* is a symmetric operator); this was first observed by Jean-Jacques MOREAU[2] (and also by someone else, whose name I do not remember), and MOFFATT has given an interpretation of this quantity in terms of *linking of vorticity lines* (in order to avoid boundary conditions, the result is considered in the whole space, as I do not care much for unrealistic periodic conditions). As the quantity integrated changes sign, the conservation of helicity has not helped for questions of global existence or smoothness of solutions of the Navier–Stokes equation in 3 dimensions.

[Taught on Monday March 29, 1999. There were no classes during the week March 21–28, which was Spring Break.]

---

[2] Jean-Jacques MOREAU, French mathematician, born in 1923. He worked at Université des Sciences et Techniques de Languedoc (Montpellier II), Montpellier, France.

30

Equation for the vorticity

I want to derive now the equation describing the evolution of vorticity in 2 dimensions, and then in 3 dimensions, as a consequence of Navier–Stokes equation.

In 2 dimensions, the Navier–Stokes equation with zero exterior forces (the gravitational force being included in the pressure term) is

$$
\begin{aligned}
&\frac{\partial u_1}{\partial t} + u_1 \frac{\partial u_1}{\partial x_1} + u_2 \frac{\partial u_1}{\partial x_2} - \nu \, \Delta u_1 + \frac{1}{\varrho_0} \frac{\partial p}{\partial x_1} = 0 \\
&\frac{\partial u_2}{\partial t} + u_1 \frac{\partial u_2}{\partial x_1} + u_2 \frac{\partial u_2}{\partial x_2} - \nu \, \Delta u_2 + \frac{1}{\varrho_0} \frac{\partial p}{\partial x_2} = 0 \\
&div \, u = 0,
\end{aligned}
\tag{30.1}
$$

and the vorticity is the scalar quantity

$$
\omega = \frac{\partial u_2}{\partial x_1} - \frac{\partial u_1}{\partial x_2}.
\tag{30.2}
$$

Applying $-\frac{\partial}{\partial x_2}$ to the first equation, $\frac{\partial}{\partial x_1}$ to the second equation and adding, one finds that the vorticity ω satisfies the equation

$$
\frac{\partial \omega}{\partial t} + u_1 \frac{\partial \omega}{\partial x_1} + u_2 \frac{\partial \omega}{\partial x_2} - \nu \, \Delta \omega = 0,
\tag{30.3}
$$

because the "pressure" disappears and the supplementary terms coming from the first equation are $-\frac{\partial u_1}{\partial x_2} \frac{\partial u_1}{\partial x_1} - \frac{\partial u_2}{\partial x_2} \frac{\partial u_1}{\partial x_2} = -\frac{\partial u_1}{\partial x_2} div \, u$, while the supplementary terms coming from the second equation are $\frac{\partial u_1}{\partial x_1} \frac{\partial u_2}{\partial x_1} + \frac{\partial u_2}{\partial x_1} \frac{\partial u_2}{\partial x_2} = \frac{\partial u_2}{\partial x_1} div \, u$.

In 3 dimensions the computation is a little more involved; the Navier–Stokes equation with zero exterior forces is

$$
\begin{aligned}
&\frac{\partial u_i}{\partial t} + \sum_{j=1}^{3} u_j \frac{\partial u_i}{\partial x_j} - \nu \, \Delta u_i + \frac{1}{\varrho_0} \frac{\partial p}{\partial x_i} = 0 \text{ for } i = 1, 2, 3, \\
&div \, u = 0,
\end{aligned}
\tag{30.4}
$$

and the vorticity is the vector-valued quantity

$$\boldsymbol{\omega} = curl\, u, \text{ i.e. } \omega_i = \sum_{j,k=1}^{3} \varepsilon_{ijk} \frac{\partial u_k}{\partial x_j} \text{ for } i = 1, 2, 3. \qquad (30.5)$$

The equation for $\boldsymbol{\omega}$ is

$$\frac{\partial \omega_i}{\partial t} + \sum_{j=1}^{3} u_j \frac{\partial \omega_i}{\partial x_j} - \sum_{j=1}^{3} \omega_j \frac{\partial u_i}{\partial x_j} - \nu \, \Delta \omega_i = 0 \text{ for } i = 1, 2, 3. \qquad (30.6)$$

Indeed, the "pressure" disappears and the supplementary term in the equation for ω_i is $\sum_{j,k,\ell=1}^{3} \varepsilon_{ijk} \frac{\partial u_\ell}{\partial x_j} \frac{\partial u_k}{\partial x_\ell}$. As only the terms where $\varepsilon_{ijk} \neq 0$ are useful, ℓ must take the value i, j or k, the sum is $\sum_{j,k=1}^{3} \varepsilon_{ijk} \left(\frac{\partial u_i}{\partial x_j} \frac{\partial u_k}{\partial x_i} + \frac{\partial u_j}{\partial x_j} \frac{\partial u_k}{\partial x_j} + \frac{\partial u_k}{\partial x_j} \frac{\partial u_k}{\partial x_k} \right)$, and using $div\, u = 0$ it is $\sum_{j,k=1}^{3} \varepsilon_{ijk} \left(\frac{\partial u_i}{\partial x_j} \frac{\partial u_k}{\partial x_i} - \frac{\partial u_i}{\partial x_i} \frac{\partial u_k}{\partial x_j} \right)$, which one writes $-\omega_i \frac{\partial u_i}{\partial x_i} + \sum_{j,k=1}^{3} \varepsilon_{ijk} \frac{\partial u_i}{\partial x_j} \left(\frac{\partial u_k}{\partial x_i} - \frac{\partial u_i}{\partial x_k} \right)$, and for $j \neq i$ and k being the third index, the term $\frac{\partial u_k}{\partial x_i} - \frac{\partial u_i}{\partial x_k}$ is indeed $-\omega_j$.

Except in the whole space or the unrealistic periodic case, there are no clear boundary conditions for the vorticity (actually, vorticity is created at the boundary).

[Taught on Wednesday March 31, 1999.]

31

Boundary conditions in linearized elasticity

In a talk by Roger LEWANDOWSKI we have seen a model used in oceanography: a horizontal fixed boundary is used to model the interface between ocean and atmosphere and a boundary condition is chosen there, which is supposed to take into account the *turbulent kinetic energy* (TKE) arising in a neighborhood of the interface. Before looking at questions of averaging, I want to discuss the question of which types of boundary conditions are natural.

In studying the stationary Stokes's equation, I have mentioned the approach of considering linearized elasticity and letting the Lamé coefficient λ tend to ∞, which forces the constraint $div\, u = 0$ at the limit, and the limit p of $-\lambda\, div\, u$ plays the role of a pressure. This similitude disappears as soon as one considers the evolution problems, because in linearized elasticity **u** denotes a *displacement* (whose gradient is supposed to be small), while for Stokes's equation **u** denotes a *velocity*; the acceleration involves then a term in $\frac{\partial^2 u}{\partial t^2}$ in the first case and a term in $\frac{\partial u}{\partial t}$ in the second case.

I have initially discussed the homogeneous Dirichlet condition $\mathbf{u} = 0$ on $\partial\Omega$, and one may also consider the case of a nonhomogeneous Dirichlet condition, $u = g$ on $\partial\Omega$: one first chooses a function equal to g on the boundary, and the difference satisfies a homogeneous Dirichlet condition; one must then have characterized the space of traces of functions of $H^1(\Omega)$ (which is $H^{1/2}(\partial\Omega)$ in the good cases), but in the limiting case $\lambda \to \infty$, one needs to add a constraint. If $u \in H^1(\Omega; R^N)$ satisfies $div\, u = 0$, and $u = g$ on $\partial\Omega$, then integrating $div\, u$ in Ω gives $\int_{\partial\Omega} \mathbf{g}.\boldsymbol{\nu}\, d\sigma = 0$, where $\boldsymbol{\nu}$ denotes the exterior normal to Ω; conversely if $g \in H^{1/2}(\partial\Omega; R^N)$ satisfies $\int_{\partial\Omega} \mathbf{g}.\boldsymbol{\nu}\, d\sigma = 0$, then one first chooses $v \in H^1(\Omega; R^N)$ equal to g on the boundary and it remains to add a function $u \in H_0^1(\Omega; R^N)$ with $div\, u = -div\, v$, but as $\int_\Omega div\, v\, d\mathbf{x} = 0$ because of the condition on g, a function u exists if Ω is smooth enough (bounded, with $X(\Omega) = L^2(\Omega)$ for example).

The case of Neumann[1] condition (one talks about a *traction* boundary condition) over all the boundary of Ω is of the form

$$-\sum_{j=1}^{N} \frac{\partial \sigma_{ij}}{\partial x_j} = f_i \text{ in } \Omega$$
$$\sum_{j=1}^{N} \sigma_{ij}\nu_j = g_i \text{ on } \partial\Omega, \tag{31.1}$$

and it requires the *compatibility conditions*

$$\int_\Omega f_i \, dx + \int_{\partial\Omega} g_i \, d\sigma = 0 \text{ for all } i$$
$$\int_\Omega \left(\sum_{j,k=1}^{N} \varepsilon_{ijk} x_j f_k\right) dx + \int_{\partial\Omega} \left(\sum_{j,k=1}^{N} \varepsilon_{ijk} x_j g_k\right) d\sigma = 0 \text{ for all } i, \tag{31.2}$$

which express the fact that the *total force* and the *total torque* acting on $\overline{\Omega}$ are 0. It is important to notice that this follows from the equilibrium equation and the symmetry of the stress tensor, so that it is true for linearized elasticity as well as for the general nonlinear elasticity in the deformed configuration, where the symmetric Cauchy stress tensor appears. Indeed, the variational formulation is

$$\int_\Omega \sum_{i,j=1}^{N} \sigma_{ij} \frac{\partial v_i}{\partial x_j} \, dx = \int_\Omega \sum_{i=1}^{N} f_i v_i \, dx + \int_{\partial\Omega} g_i v_i \, d\sigma \text{ for all } v \in H^1(\Omega; R^3),$$

$$\tag{31.3}$$

and one has $\sum_{i,j=1}^{N} \sigma_{ij} \frac{\partial v_i}{\partial x_j} = \sum_{i,j=1}^{N} \sigma_{ij} \varepsilon_{ij}(v)$ by symmetry of the stress tensor, where as usual $\varepsilon_{ij}(v) = \frac{1}{2}\left(\frac{\partial v_i}{\partial x_j} + \frac{\partial v_j}{\partial x_i}\right)$, and therefore the left side is 0 if v is such that $\varepsilon_{ij}(v) = 0$ for all i, j; this is the case if $v_i = a_i + \sum_{j=1}^{N} M_{ij} x_j$ for all i with M antisymmetric (skew adjoint), and in 3 dimensions it means $M\mathbf{x} = \mathbf{m} \times \mathbf{x}$ for some $\mathbf{m} \in R^3$, and writing that the right side is 0 for all these v gives the necessary conditions on f and g, corresponding to the physical interpretation of total force and total torque. In linearized elasticity, i.e. $\sigma_{ij} = \sum_{k,\ell=1}^{N} C_{ijk\ell} \varepsilon_{k\ell}(u)$ for all i, j, with $C_{ijk\ell} = C_{jik\ell} = C_{ij\ell k}$ for all i, j, k, ℓ, and under the hypothesis of very strong ellipticity (i.e. there exists $\alpha > 0$ such that $\sum_{i,j,k,\ell=1}^{N} C_{ijk\ell} A_{ij} A_{k\ell} \geq \alpha \sum_{i,j=1}^{N} |A_{ij}|^2$ for all symmetric A), the necessary conditions are sufficient if the injection of $H^1(\Omega)$ into $L^2(\Omega)$ is compact and if Korn's inequality holds, as a consequence of the equivalence lemma 13.3. This requires that one identifies all the $v \in H^1(\Omega; R^3)$ satisfying $\varepsilon_{ij}(v) = 0$ for all i, j, and it follows from the identity

$$2\frac{\partial^2 u_i}{\partial x_j \partial x_k} = \frac{\partial}{\partial x_j}\left(\frac{\partial u_i}{\partial x_k} + \frac{\partial u_k}{\partial x_i}\right) - \frac{\partial}{\partial x_i}\left(\frac{\partial u_k}{\partial x_j} + \frac{\partial u_j}{\partial x_k}\right) + \frac{\partial}{\partial x_k}\left(\frac{\partial u_j}{\partial x_i} + \frac{\partial u_i}{\partial x_j}\right) \tag{31.4}$$

for all i, j, k, that $\varepsilon_{ij}(v) = 0$ for all i, j implies that all second derivatives are 0, so that $\mathbf{v}(\mathbf{x}) = \mathbf{a} + M\mathbf{x}$ for all $\mathbf{x}$, and then M must be antisymmetric. The solution u then exists and is defined up to the addition of $\mathbf{a} + \mathbf{m} \times \mathbf{x}$ for $\mathbf{a}, \mathbf{m} \in R^3$,

---

[1] Franz Ernst NEUMANN, German mathematician, 1798–1895. He worked in Königsberg, then in Germany, now Kaliningrad, Russia.

and it must be pointed out that these are not *rigid displacements* but *linearized rigid displacements* (the antisymmetric matrices appear as the tangent space at I to $SO(3)$, the compact manifold of all rotations). If the necessary conditions are not satisfied, there is no stationary solution but the evolution equation does have a solution, and the body moves away, in the direction of a linearized rigid displacement.

Let us imagine now, in the approximation of linearized elasticity, an elastic body with a flat part of its boundary put on an horizontal table, and assume that the system of forces applied to it does not take it away from the table (or consider the purely mathematical problem that the displacement satisfies $u_3 = 0$ on this flat part of the boundary); the body is allowed to slide horizontally on the table, and one expects to have less stringent compatibility conditions, corresponding to the horizontal part of the total force being 0 (there is no friction on the table and so the table will give a vertical reaction which will cancel the vertical component of the total force), and the torque along the x_3 axis must be 0 (the reactions of the table being able to compensate for the rest of the total torque). Mathematically, the condition $u_3 = 0$ on a piece of the boundary sitting in the plane $x_3 = H$ is imposed in the definition of the functional space, and v is constrained to be in this space, so only the elements $\mathbf{a} + \mathbf{m} \times \mathbf{x}$ satisfying this constraint are allowed, i.e. one must choose $a_3 = m_1 = m_2 = 0$, and the necessary conditions corresponding to a_1, a_2 imply that the horizontal part of the total force is 0, while the necessary condition corresponding to m_3 then implies that the total torque around any vertical axis is 0.

Mathematically, one can study nonhomogeneous conditions, like imposing u_3 on a piece of the boundary which is not necessarily flat, and the natural boundary conditions implied by the variational formulation will involve the traction $\mathbf{T}$ defined by $T_i = \sum_{j=1}^{N} \sigma_{ij} \nu_j$ (as for normal traces in $H(div; \Omega)$), and T_1 and T_2 can be imposed, with natural compatibility conditions.

Mathematically, one could also impose the displacements u_1 and u_2 on a piece of the boundary, and the natural boundary condition implied by the variational formulation will involve T_3.

For (Newtonian) fluids, one has

$$\sigma_{ij} = 2\mu\, \varepsilon_{ij} - p\, \delta_{ij} = \mu \left(\frac{\partial u_i}{\partial x_j} + \frac{\partial u_j}{\partial x_i} \right) - p\, \delta_{ij} \text{ for all } i, j, \tag{31.5}$$

so that

$$\sum_{j=1}^{N} \sigma_{ij} \nu_j = \mu \frac{\partial u_i}{\partial n} + \mu \sum_{j=1}^{N} \frac{\partial u_j}{\partial x_i} \nu_j - p\, \nu_i \text{ for all } i. \tag{31.6}$$

For a horizontal boundary, like the fixed interface separating ocean from atmosphere in the model considered by Roger LEWANDOWSKI, one has $\nu_1 = \nu_2 = 0$, and $\nu_3 = 1$ for the ocean and $\nu_3 = -1$ for the atmosphere, so that in the ocean one has $T_1^O = \mu \frac{\partial u_1^O}{\partial x_3} + \mu \frac{\partial u_3^O}{\partial x_1}, T_2^O = \mu \frac{\partial u_2^O}{\partial x_3} + \mu \frac{\partial u_3^O}{\partial x_2}, T_3^O = 2\mu \frac{\partial u_3^O}{\partial x_3} - p^O$, and similarly for the atmosphere $T_1^A = -\mu \frac{\partial u_1^A}{\partial x_3} - \mu \frac{\partial u_3^A}{\partial x_1}, T_2^A = -\mu \frac{\partial u_2^A}{\partial x_3} - \mu \frac{\partial u_3^A}{\partial x_2},$

$T_3^A = -2\mu \frac{\partial u_3^A}{\partial x_3} + p^A$, and it is usually the jumps of these quantities which appear in the variational formulations.

[Taught on Friday April 2, 1999.]

32

Turbulence, homogenization

Modeling of turbulent flows is an important scientific and technological question, and although engineers may say that they are able to control turbulent flows, it is mainly because adaptive control ideas seem to work even in situations where no one knows what the right equations are for describing the phenomena which one wants to control. From a scientific point of view, not so much is understood about turbulence. For what concerns oceanography, some modeling of turbulent flows is necessary in order to describe correctly what goes on at "small scales", remembering that the scales used for the ocean, or the atmosphere, are quite large.

It is quite common to experience the presence of microstructures in some fluid flows, but it is a very arduous task to propose a model that would describe accurately the important effects occurring in these flows.

My first experimental evidence concerns the structure of the interface in front of a rainstorm, as I had observed many times after a hot summer day in the French countryside, long before becoming a mathematician: one knows that a storm is coming, although the air is still, perhaps because the pressure is higher than usual, and then one starts to hear the leaves of the trees moving while the branches stay still; soon after the small branches start to move too, followed by the large branches a little after and the whole trees are in motion when the rain arrives. It clearly suggests that the classical idea of a sharp interface with some partial differential equations being satisfied on each side and with some boundary conditions being imposed on the "interface" might not be so efficient for describing the effects occurring in that living layer, with small vortices on the dry side and large vortices on the wet side.

My second experimental evidence concerns the structure of the wind, as I had observed twenty years ago, on a weekend where I had expected to sail between La Rochelle and Ile de Ré,[1] but the morning had provided us with what one calls *calme plat* in French: there was no wind, and the surface of the

---

[1] Ile is the French word for island, and Ré was indeed an island in those days, but it is now connected to the continent by a bridge.

sea was extremely smooth and only showing a long swell (*houle* in French), which combined with the steady movement sustained by the small engine of the boat to produce a beginning of seasickness; fortunately, it did not last too long, because after a while we saw what one calls *risée* in French (light squall in English): the wind waiting for us; it is an amazing experience to come from the windless side with a smooth sea surface to the place where the wind is, with the surface of the sea all wrinkled with wavelengths of the order of 5 to 10 centimeters, and when one crosses the transition line (which seemed stationary, but it might have been moving at a much slower pace that the boat, which was carried by its small engine), the sails inflated, and sailing started.

At a meeting in 1981 (in New York, NY), I had heard Joe KELLER[2] mention that at one time there had been a lot of articles about the statistical distribution of wavelengths of the waves at the surface of the sea, until one had been able to measure this distribution and it had appeared that all the theories had been wrong, as one had observed much more energy that any theorist had expected in the small capillary waves, those which I had observed as the signature of the wind waiting for us, because it must be mostly in these wavelengths that the ocean and atmosphere exchange energy. In other words, many like to imagine that natural phenomena obey the probabilistic processes or the statistical laws that are already known, and these people usually do not care that the phenomena that they are trying to study are described by complicated systems of partial differential equations for which their standard processes are obviously not adapted.[3,4] At another meeting in 1984 (in Nice, France), Joe KELLER had mentioned the evolution from ideas about 3-dimensional turbulence by KOLMOGOROV, the 2-dimensional turbulence ideas used in meteorology (where the stratification by gravity simplifies the full 3-dimensional aspects), some 1-dimensional ideas that were not so

---

[2] Joseph Bishop KELLER, American mathematician, born in 1923. He received the Wolf Prize for 1996–1997. He works at STANFORD University, Stanford, CA.

[3] When I learnt continuum mechanics with Jean MANDEL, I was quite interested to learn that a 1-dimensional sinusoidal swell travels at a speed depending upon the depth (explaining the breaking of waves near submerged reefs and on beaches), but also that particles do not move much as it goes by (with a noticeable phase speed), and that it carries (linear) momentum but not mass. This is something that one should bear in mind when trying to understand turbulence, that too many people get lost in various questions of statistics which are forced upon the main question, instead of trying to identify the effective equations valid for the important physical quantities like mass, linear momentum, vorticity, pressure, heat flux, and other quantities that are needed for describing the effective constitutive relations; one should not get lost in making a precise description of the microstructures which Nature creates, but try to understand that it creates them for arriving at unusual effects like transporting various physical quantities at various speeds.

[4] Jean MANDEL, French mathematician, 1907–1982. I had him as a teacher, for the course of Continuum Mechanics at École Polytechnique in 1966 in Paris, France.

good, and the zero-dimensional ideas of iterating maps, followed by continually improving numerical simulations in 1, 2 and even 3 dimensions, but he emphasized that something important had been lost on the way: in the 1940s, specialists of turbulence talked about *velocity, pressure, kinetic energy, temperature, heat flux*, while now they talk about statistics without reference to any important physical quantity related to fluids.

The only thing about turbulence that everybody agrees upon is that it is created by *oscillations in the velocity field*, and REYNOLDS might have been the first to notice that if the "average" of u_i is denoted by $\overline{u_i}$, then the average of $u_i u_j$ is $\overline{u_i}\,\overline{u_j} + R_{ij}$, where the symmetric Reynolds tensor R with entries R_{ij} is not necessarily 0. Probabilists like to imagine that all functions in the fluid depend upon a parameter ω belonging to a space endowed with a probability measure, and integration with respect to this probability measure, the expectation, plays the role of the intuitive averaging technique. Some specialists of asymptotic expansions like to plug functions like $u_0(\mathbf{x}) + \varepsilon_n u_1\left(\mathbf{x}, \frac{\mathbf{x}}{\varepsilon_n}\right) + \ldots$ into the system of equations governing fluids, where the functions $u_j(\mathbf{x}, \mathbf{y})$ are periodic in $\mathbf{y}$, the vague idea of averaging becoming the precise technique of averaging in $\mathbf{y}$, and this deterministic approach is sometimes useful, although it is not able to explain some multiple scale effects that turbulent flows are known to show.

For about twenty-five years, I have been developing a mathematical approach to the study of *oscillations* in solutions of partial differential equations, partly in collaboration with François MURAT, and various notions of weak convergence appear in this approach, which definitely has an advantage on all the others, in that it does not postulate anything about oscillations but tries to determine what kind of oscillations are compatible with linear differential balance laws and nonlinear constitutive relations. First, I should point out that I use the term "oscillations" to describe also *concentration effects*, i.e. the meaning used is to consider weakly convergent sequences which are not strongly convergent; however, convergences of a weak type but different from the usual weak convergence are also used. Second, I should point out that the use of sequences is a purely mathematical trick whose object is to identify the correct topology (usually related to some kind of weak convergence) that one should use for various physical or nonphysical quantities (it is similar to the description of R by starting from Cauchy sequences in Q, and once R is understood a real number is not related to a sequence of rationals any more!).

The classical weak convergence appears to be natural for some quantities and not for others, and the notion of differential forms clarifies this question (as I learned from discussions with Joel ROBBIN in 1974–1975, in Madison, WI). In the equation expressing *conservation of mass*, $\frac{\partial \varrho}{\partial t} + div(\varrho\,u) = 0$, the quantities ϱ and $q_i = \varrho\,u_i$ are coefficients of differential forms, but u_i only appears as a quotient of two quantities for which the adapted topology is weak convergence; therefore density and momentum are more easy to handle than velocity. It will be useful then to describe some properties of H-convergence (introduced with François MURAT, and generalizing the notion of G-convergence introduced by Sergio SPAGNOLO, with some ideas from Ennio DE GIORGI), and I shall describe some properties of weakly converging sequences of solutions of equations like $div(A_n \, grad \, u_n) = f$. It will then be natural to consider sequences of operators of the form $\frac{\partial}{\partial t} + \sum_{i=1}^{N} u_i^n \frac{\partial}{\partial x_i}$, and as nothing general is known in the case when the coefficients only converge weakly, I shall describe in detail some special cases.

It is worth mentioning that geometers like to think that they know how to write equations for fluid flows in intrinsic forms, but as long as one does not know how to pass to the limit in weakly convergent sequences of solutions of these equations, one cannot assert that geometers have or have not introduced the correct framework (my guess is that they have not!).

[Taught on Monday April 5, 1999.]

33

G-convergence and H-convergence

In questions of asymptotic expansions (which I first saw in the work of Henri SANCHEZ-PALENCIA), one considers sequences of functions like $v^n(\mathbf{x}) = u_0(\mathbf{x}) + \varepsilon_n u_1\left(\mathbf{x}, \frac{\mathbf{x}}{\varepsilon_n}\right) + \ldots$, where the functions $u_j(\mathbf{x}, \mathbf{y})$ are periodic in $\mathbf{y}$ (and smooth enough in $(\mathbf{x}, \mathbf{y})$), and ε_n tends to 0. If ε_n is a small characteristic length, and if the solution of a physical problem has this form, then if one measured the value of v^n at a few points, quite far apart compared with the characteristic length ε_n, then one would find $u_0(\mathbf{x}) + O(\varepsilon_n)$ (plus some eventual errors due to the measuring process), and one might well believe that the measured solution is u_0, considering the little discrepancies as errors in measurements.

It is usual in physics courses to be told that one term is small and that it will be neglected (it is not always clear if these terms are indeed small, as they may be small in the real world, but if the equation used is not a good model of the physical world the corresponding term might not be so small); having neglected some terms one performs some formal computations with the simplified equation, like taking derivatives, and the first remark, which seems to infuriate physics teachers, is that the derivative of a small term might not be small; actually, in our example one has $\frac{\partial v^n}{\partial x_i} = \frac{\partial u_0}{\partial x_i} + \frac{\partial u_1}{\partial y_i} + O(\varepsilon_n)$, and $\frac{\partial u_1}{\partial y_i}$ is not small when u_1 does depend upon $\mathbf{y}$.

Fortunately, in some cases like linear partial differential equations with smooth coefficients, the procedure can be shown to work, because of the generalized framework of the theory of distributions of Laurent SCHWARTZ for example: if a sequence v^n converges to v^∞, then $\frac{\partial v^n}{\partial x_i}$ converges to $\frac{\partial v^\infty}{\partial x_i}$ for every i, but in this statement it must be realized that the meaning of convergence is not that the differences are uniformly small, as one may have $v^n(\mathbf{x}) = u_0(\mathbf{x}) + O(\varepsilon_n)$, while $\frac{\partial v^n}{\partial x_i} \neq \frac{\partial u_0}{\partial x_i} + O(\varepsilon_n)$, but only $\frac{\partial v^n}{\partial x_i} \neq \frac{\partial u_0}{\partial x_i} + O(1)$ and nevertheless $\frac{\partial v^n}{\partial x_i}$ converges weakly to $\frac{\partial u_0}{\partial x_i}$.

If one wants to avoid the too general framework of the theory of distributions of Laurent SCHWARTZ, one may instead use classical results of functional analysis concerning weak topologies and weak $\star$ topologies for Banach spaces, which appear to be natural for quantities which are integrated against test functions, or integrated on certain sets, and in continuum mechanics it is often the case that such quantities are coefficients of differential forms (and it is probably only for those that these weak convergences should be used).

For example, in the equation of conservation of mass $\frac{\partial \varrho}{\partial t} + \sum_{i=1}^{N} \frac{\partial(\varrho u_i)}{\partial x_i} = 0$, ϱ and ϱu_i, $i = 1, 2, 3$, are the coefficients of a 3-differential form in space-time, namely

$$\omega = \varrho \, dx_1 \wedge dx_2 \wedge dx_3 - \varrho u_1 \, dt \wedge dx_2 \wedge dx_3 - \varrho u_2 \, dt \wedge dx_3 \wedge dx_1 - \varrho u_3 \, dt \wedge dx_1 \wedge dx_2,$$

(33.1)

and as

$$d\omega = \left(\frac{\partial \varrho}{\partial t} + \sum_{i=1}^{3} \frac{\partial(\varrho u_i)}{\partial x_i}\right) dt \wedge dx_1 \wedge dx_2 \wedge dx_3, \qquad (33.2)$$

the equation of conservation of mass is $d\omega = 0$. One must notice that the components u_i of the velocity field are not themselves coefficients of differential forms (and the weak convergence is not adapted for them), and it is the *momentum* which is the correct *physical quantity*, which has an additive character. The velocity is not mentioned when one deals with conservation of *electric charge*, and it is written as $\frac{\partial \varrho}{\partial t} + div \, j = 0$, and one does not even bother to define a velocity as $\frac{j}{\varrho}$, because it would usually be meaningless, because the electric charge is transported by light electrons and by heavy ions, and an average velocity would be of little use (it is often better to think of two interacting populations, one of electrons and one of ions, eventually having different temperatures).

H-convergence gives an example where a quantity which is not a coefficient of a differential form needs a different type of weak topology, which I have introduced with François MURAT, generalizing to nonsymmetric operators the earlier notion of G-convergence, introduced by Sergio SPAGNOLO. It also comes with a quite different point of view, and although I am using the framework of differential forms, my motivation is quite different from that of geometers, and it is worth describing the differences in points of view.

The *exterior calculus* is purely *algebraic*: one considers the p-linear alternated forms on a finite-dimensional vector space E, i.e. f is multilinear and satisfies $f(e_{s(1)}, \ldots, e_{s(p)}) = \varepsilon(s) f(e_1, \ldots, e_p)$ for all $e_1, \ldots, e_p \in E$ and all permutations s of p elements, where $\varepsilon(s)$ is the signature of the permutation s. One then defines the *exterior product* $\wedge$: if f is a p-linear alternated form and g is a q-linear alternated form, then $f \wedge g$ is the $(p + q)$-linear alternated form defined by $(f \wedge g)(e_1, \ldots, e_{p+q}) = \frac{1}{p!q!} \sum_s \varepsilon(s) f(e_{s(1)}, \ldots, e_{s(p)}) g(e_{s(p+1)}, \ldots, e_{s(p+q)})$, where s runs through the permutations of $p+q$ elements; one checks easily that $g \wedge f = (-1)^{pq} f \wedge g$. The exterior product is associative.

A *differential form* of order p, or a p-form, on an open set Ω of E, is a (smooth enough) mapping from Ω into the space of p-linear alternated forms;

a 0-form is a function, and the derivative of a function is a 1-form. Then one defines the *exterior derivative* d, which maps p-forms into $(p+1)$-forms, with the rules that $d(f \wedge g) = df \wedge g + (-1)^p f \wedge dg$ if f is a p-form, and $df = \sum_{i=1}^{N} \frac{\partial f}{\partial x_i} dx_i$ if f is a function. One shows that $d \circ d = 0$, and *Poincaré's lemma* asserts that if $df = 0$ then locally $f = dh$ for a $(p-1)$-form h (asking for global results leads to questions of algebraic topology).

One can *restrict* a differential form to a *submanifold* by considering its action only on vectors tangent to the submanifold, and actually one can develop all the theory of differential forms on abstract manifolds (not necessarily orientable), with or without boundary, and one proves *Stokes's formula* $\int_\Omega d\omega = \int_{\partial\Omega} \omega$. As a student, before learning this framework, I was taught about formulas attributed to GREEN, STOKES, and OSTROGRADSKI,[1] where *curl* and *div* appear: a vector field V in R^3 can be attached to a 1-form $\omega(V) = V_1 dx_1 + V_2 dx_2 + V_3 dx_3$ but also to a 2-form $\pi(V) = V_1 dx_2 \wedge dx_3 + V_2 dx_3 \wedge dx_1 + V_3 dx_1 \wedge dx_2$ (but this uses the Euclidean[2] structure, so that vectors and covectors are identified), and $d\omega(V) = \pi(curl\,V)$ and $d\pi(V) = (div\,V) dx_1 \wedge dx_2 \wedge dx_3$ (one usually suppresses the $\wedge$ and one replaces $dx_1 \wedge dx_2 \wedge dx_3$ by $d\mathbf{x}$). If *curl* applies to 1-forms in all dimensions, *div* applies to $(N-1)$-forms in dimension N.

I suppose that all this beautiful theory was developed by POINCARÉ and E. CARTAN,[3] but I have also heard the name of PFAFF[4] being mentioned.

If differential forms are natural for geometers, because they are the right objects which transform well under *change of variables*, the reason why I am using them is different: they are *adapted to weak convergence*. In the early 1970s, I worked with François MURAT on questions that were not yet called homogenization, and I understood from reading some work of Henri SANCHEZ-PALENCIA (who was using asymptotic expansions for problems with periodic microstructures), that what we had done is related to effective properties of mixtures (we also discovered that Sergio SPAGNOLO had solved earlier the first step of our program): we were considering a sequence of elliptic problems $-div(A_n\,grad\,u_n) = f$ in Ω, together with some natural boundary conditions, and with A_n converging weakly we extracted a subsequence u_m converging weakly in $H^1(\Omega)$ to u_∞, but the limit of $A_m\,grad\,u_m$ could not be defined easily, so we introduced an adapted notion (later called H-convergence, and generalizing the G-convergence introduced by Sergio SPAGNOLO). Using notation from electrostatics, with $E_n = -grad\,u_n$ and $D_n = A_n E_n$, I considered the weak convergence natural for the electric field E_n, interpreting its weak limit E_∞ as a macroscopic field, and similarly for the polarization field D_n

---

[1] Mikhail Vasilevich OSTROGRADSKI, Russian mathematician, 1801–1862. He worked in St Petersburg, Russia.

[2] EUCLID, mathematician, 325 BCE – 265 BCE. He worked in Alexandria, Egypt.

[3] Elie Joseph CARTAN, French mathematician, 1869–1951. He worked in Paris, France.

[4] Johann Friedrich PFAFF, German mathematician, 1765–1825. He worked in Halle, Germany.

converging weakly to D^∞, but the right question concerning the limit for A_n is to be able to relate D_∞ to E_∞, and this concerns a different physical process where there is no averaging of A_n: one creates a macroscopic field E_∞ by choosing correctly f (which is ϱ in electrostatics) and one measures the weak limit D_∞ and this gives a partial information on a tensor A^{eff} such that $\mathbf{D_\infty} = \mathsf{A}^{\text{eff}}\mathbf{E_\infty}$ a.e. in Ω; therefore one does not "measure" A^{eff} by computing averages, but one *identifies* A^{eff} from averages of the electric and polarization fields. We also discovered the div-curl lemma 33.1, and during the year 1974–1975 which I spent at University of Wisconsin (Madison, WI), Joel ROBBIN explained to me that our result is quite clear when expressed in the framework of differential forms if one uses Hodge's decomposition, and he also taught me how to write Maxwell's equation using differential forms. In the fall of 1975, I met John BALL and learnt from him about the sequential weak continuity of Jacobians (which I thought he had proven, but understood later that MORREY had done that in the 1950s, and RESHETNYAK[5] more recently), and I could derive easily these results from the div-curl lemma 33.1 in 2 or 3 dimensions; the general framework of compensated compactness theorem 38.1 appeared two years after, again with participation of François MURAT.

Lemma 33.1. *(Div-curl lemma, 1974, François MURAT & Luc TARTAR[6]) If $\Omega \subset R^N$, if $E_n \rightharpoonup E_\infty$ in $L^2_{loc}(\Omega; R^N)$ weak, $D_n \rightharpoonup D_\infty$ in $L^2_{loc}(\Omega; R^N)$ weak, $\operatorname{div} D_n \to \operatorname{div} D_\infty$ in $H^{-1}_{loc}(\Omega)$ strong, and $\operatorname{curl} E_n \to \operatorname{curl} E_\infty$ in $H^{-1}_{loc}(\Omega; X)$ strong (where X has the right dimension), then $E_n.D_n$ converges to $E_\infty.D_\infty$ in the sense of measures (i.e. integrated against test functions in $C_c(\Omega)$).*

In the case where $E_n = \operatorname{grad} u_n$, the proof of Lemma 33.1 reduces to an integration by parts, using the compactness of the injection of $H^1_{loc}(\Omega)$ into $L^2_{loc}(\Omega)$ by writing $\mathbf{E_n.D_n} = -(\operatorname{grad} u_n).\mathbf{D_n} = -\operatorname{div}(u_n D_n) + u_n \operatorname{div} D_n$, which one integrates against $\varphi \in C^\infty_c(\Omega)$.

Lemma 33.1 is a particular case of Lemma 40.1, a consequence of compensated compactness theorem 38.1, which considers a general framework of linear differential equations with constant coefficients, $U^n \rightharpoonup U^\infty$ in $L^2_{loc}(\Omega; R^p)$ weak and $\sum_{j=1}^p \sum_{k=1}^N A_{ijk} \frac{\partial U_j^n}{\partial x_k} \to f_i$ in $H^{-1}_{loc}(\Omega)$ strong for $i = 1, \ldots, q$, and identifies the possible limits (in the sense of measures) of all quadratic functions of U^n.

The compensated compactness method, which I developed afterwards, adds the use of "entropies" in order to deduce which Young measures could be associated with the sequence U^n, assumed also to satisfy the constraints $\mathbf{U^n(x)} \in K$ a.e. $\mathbf{x} \in \Omega$ (which correspond to constitutive relations, while the

---

[5] Yuriĭ Grigor'evich RESHETNYAK, Russian mathematician, born in 1930. He worked at the SOBOLEV Institute of Mathematics, Novosibirsk, Russia.

[6] Luc Charles TARTAR, French born mathematician, born in 1946. Since 1987, I have been working at CARNEGIE MELLON University, Pittsburgh, PA.

differential equations correspond to balance equations for problems in continuum mechanics).

If one did not know about differential forms, one would discover them by looking at sequences U^n which converge strongly to U^∞ in $L^2_{loc}(\Omega; R^p)$ but only weakly in $H^1_{loc}(\Omega; R^p)$, and wonder if one could compute the limit of some functions of $grad\, U^n$; indeed compensated compactness theorem 38.1 (or its consequence, lemma 40.1) reveals that $dU^n_i \wedge dU^n_j$ converges to $dU^\infty_i \wedge dU^\infty_j$ in the sense of measures. More generally, let f^n be a sequence of p-forms converging to f^∞ in $L^2_{loc}(\Omega)$ weak (for its coefficients) and such that df^n has its coefficients staying in a compact of $H^{-1}_{loc}(\Omega)$ strong, and let g^n be a sequence of q-forms converging to g^∞ in $L^2_{loc}(\Omega)$ weak (for its coefficients) and such that dg^n has its coefficients staying in a compact of $H^{-1}_{loc}(\Omega)$ strong, then lemma 40.1 shows that $df^n \wedge dg^n$ converges to $df^\infty \wedge dg^\infty$ in the sense of measures (of course one has better convergences if one improves the hypotheses). For example, when applied to Maxwell's equation, compensated compactness theorem 38.1 gives three independent quadratic quantities which are sequentially weakly continuous, and with the framework of differential forms, I will show in Lecture 40 that they come from exterior products of differential forms having good exterior derivatives (so that Lemma 40.1 applies).

After reiteration, using the entropy conditions following from the formula for $d(f \wedge g)$, one recovers MORREY's result about Jacobians (apart from a wrong hypothesis on which value of p one should choose when asking that each component of U^n converges weakly in $W^{1,p}$, if one uses only the Hilbert setting of compensated compactness theorem 38.1 instead of the adapted variant using multipliers in $\mathcal{F}L^p$, valid for bilinear mappings under a supplementary hypothesis of constant rank).

The compensated compactness framework is of course more general than MORREY's ideas like quasiconvexity or polyconvexity, and can be used for any system that one encounters in continuum mechanics. Of course, the compensated compactness method must sometimes be used in conjunction with the ideas of H-convergence that I developed with François MURAT for homogenization, but the compensated compactness method certainly still needs to be improved, although no one has come up yet with an idea that helps toward understanding more about nonlinear partial differential equations of continuum mechanics or physics.

[Taught on Wednesday April 7, 1999.]

34

One-dimensional homogenization, Young measures

In 1 dimension, one can solve explicitly all homogenization problems by computing various weak limits, and the same is true in more than 1 dimension when the oscillating coefficients only depend upon one variable. The general case of a diffusion equation was solved by François MURAT in the early 1970s; then I learnt in 1975 about a computation by MCCONNELL[1] of the general case for linearized elasticity, and I derived the general approach shown below a few years after. In 1979, having been asked how to compute the effective properties of a material layering steel and rubber, I also explained how to carry out the computations in a nonlinear setting, although there is no general theory of indexhomogenizationhomogenization for nonlinear elasticity (despite the claims of those who mistake Γ-convergence for homogenization, this is still the situation today).

The basic idea is an application of the div-curl lemma 33.1: if $\Omega \subset R^N$ and $D^n \rightharpoonup D^\infty$ in $L^2_{loc}(\Omega; R^N)$ weak with $div\, D^n$ staying in a compact of $H^{-1}_{loc}(\Omega)$ strong, then D^n *does not oscillate in* x_1, i.e. whenever f_n depends only upon x_1 and $f_n \rightharpoonup f_\infty$ in $L^2_{loc}(\Omega)$ weak, one has $D^n_1 f_n \rightharpoonup D^\infty_1 f_\infty$ in the sense of measures (the precise definition that a sequence is not oscillating in x_1 says that the corresponding H-measures do not charge the point $\mathbf{e}^1$ of the unit sphere); of course this follows from the fact that $\mathbf{E^n} = f_n \mathbf{e}^1$ is a gradient.

For a diffusion equation, if $\mathbf{E^n} = grad\, u_n$ and $\mathbf{D^n} = A^n \mathbf{E^n}$ satisfies $div\, D^n \rightarrow f$ in $H^{-1}_{loc}(\Omega)$ strong, with A^n only depending upon x_1, one remarks that D^n_1 does not oscillate in x_1 as well as $E^n_2, \ldots, E^n_N$, because of the equation $curl\, E^n = 0$. From the components of $\mathbf{E^n}$ and $\mathbf{D^n}$, one creates a *good vector* $\mathbf{G^n}$ whose components are $D^n_1, E^n_2, \ldots, E^n_N$, and an *oscillating vector* $\mathbf{O^n}$ whose components are $E^n_1, D^n_2, \ldots, D^n_N$, and one has $\mathbf{O^n} = \Phi(A^n)\mathbf{G^n}$, where $\Phi(A^n)$ is obtained by algebraic computations from A^n, and these computations (which start by eliminating E^n_1 in the equation giving D^n_1) only require that A^n_{11} stay away from 0; as A^n depends only upon x_1, so does $\Phi(A^n)$ and one can pass to the limit in $\Phi(A^n)\mathbf{G^n}$, so that $\mathbf{O^\infty} = $ [weak limit

---

[1] William H. MCCONNELL, American mathematician.

$\Phi(\mathsf{A}^n)]\mathsf{G}^\infty$, i.e. $\Phi(\mathsf{A}^{\text{eff}})$ is the weak limit of $\Phi(\mathsf{A}^n)$. For linearized elasticity, the good vector uses the components σ_{i1} (and σ_{1i} which is equal to σ_{i1} because the Cauchy stress tensor is used), and the ε_{ij} for $i, j \geq 2$, while the oscillating vector uses the other components; starting from $\sigma_{ij} = \sum_{k,\ell=1}^N C_{ijk\ell}\varepsilon_{k\ell}$, the algebraic computations only require that the *acoustic tensor* $\mathsf{A}(\mathbf{e}^1)$ be invertible (one defines $\mathsf{A}_{ik}(\xi) = \sum_{j,\ell=1}^N C_{ijk\ell}\xi_j\xi_\ell$). For nonlinear elasticity, the good vector uses the components σ_{i1} (but not σ_{1i} which is different from σ_{i1} because the Piola–Kirchhoff stress tensor is used), and the $\frac{\partial u_i}{\partial x_j}$ for $j \geq 2$, while the oscillating vector uses the other components (in the case of hyperelasticity where there is a stored energy function W, the computations only require a uniform rank-one convexity for W).

If one considers now the general problem of homogenization, and I recall that I do not imply any restriction to periodic structures like so many do when using this term (maybe because they have not heard of the general framework that I developed with François MURAT in the early 1970s), one imposes a uniform ellipticity condition, which for the diffusion case is that there exists $0 < \alpha \leq \beta < \infty$ such that $(\mathsf{A}^n(\mathbf{x})\xi.\xi) \geq \alpha\,|\xi|^2$ and $(\mathsf{A}^n(\mathbf{x})\xi.\xi) \geq \frac{1}{\beta}|\mathsf{A}^n(\mathbf{x})\xi|^2$ for all $\xi \in R^N$ and a.e. $\mathbf{x} \in \Omega$ (if A^n is symmetric, it means that $\alpha\mathsf{I} \leq \mathsf{A}^n \leq \beta\mathsf{I}$ almost everywhere).

In the G-convergence approach, developed in the late 1960s by Sergio SPAGNOLO, one considers only symmetric A^n and one extracts a subsequence such that for every $f \in H^{-1}(\Omega)$ the solution $u_m \in H_0^1(\Omega)$ of the equation $-div(\mathsf{A}^m grad\, u_m) = f$ converges weakly to u_∞, and one shows that there exists A^{eff} (symmetric with $\alpha\mathsf{I} \leq \mathsf{A}^{\text{eff}} \leq \beta\mathsf{I}$ almost everywhere) such that $-div(\mathsf{A}^{\text{eff}} grad\, u_\infty) = f$ (this is the convergence of the Green kernels, and explains the choice of the prefix G).

In the H-convergence approach, which I developed in the early 1970s with François MURAT without knowing at the time what Sergio SPAGNOLO had already done, one can consider nonsymmetric A^n and one extracts a subsequence such that for every $f \in H^{-1}(\Omega)$ the solution $u_m \in H_0^1(\Omega)$ of the equation $-div(\mathsf{A}^m grad\, u_m) = f$ converges weakly to u_∞, but also $\mathsf{A}^m grad\, u_m$ converges weakly to D^∞, and one shows that there exists A^{eff} (with $(\mathsf{A}^{\text{eff}}(\mathbf{x})\xi.\xi) \geq \alpha\,|\xi|^2$ and $(\mathsf{A}^{\text{eff}}(\mathbf{x})\xi.\xi) \geq \frac{1}{\beta}|\mathsf{A}^{\text{eff}}(\mathbf{x})\xi|^2$ for all $\xi \in R^N$ and a.e. $\mathbf{x} \in \Omega$) such that $D^\infty = \mathsf{A}^{\text{eff}} grad\, u_\infty$, implying $-div(\mathsf{A}^{\text{eff}} grad\, u_\infty) = f$ (it is equivalent to G-convergence in the symmetric case, and the choice of the prefix H, chosen in the late 1960s, reminds us of the term homogenization, introduced by Ivo BABUŠKA).

It is important to realize that A^{eff} cannot be computed using the Young measures associated with the sequence A^n in dimensions ≥ 2; using Young measures is the mathematical way of dealing with *one-point statistics* which physicists use in their probabilistic framework, and therefore the preceding statement says that one cannot deduce the effective properties of a mixture by using only the proportions of the different constituents used. If one mixes two isotropic materials with different conductivity (or permittivity) α, β, it means that $A^n = (\chi_n \alpha + (1-\chi_n)\beta)I$, and if $\chi_n \rightharpoonup \theta$ in $L^\infty(\Omega)$ weak $\star$, then $\theta(\mathbf{x})$ is the local proportion of the first material near $\mathbf{x}$; the Young measure in this case is $\nu_{\mathbf{x}} = \theta(\mathbf{x})\delta_{\alpha I} + (1-\theta(\mathbf{x}))\delta_{\beta I}$. One can construct layers in x_1 or layers in x_2 for two sequences having the same Young measure by taking θ constant, but the effective properties are different: if a_+ is the arithmetic average $\theta\alpha + (1-\theta)\beta$, and a_- is the harmonic average $\left(\frac{\theta}{\alpha} + \frac{1-\theta}{\beta}\right)^{-1}$, then layering in the direction x_j corresponds to A^{eff} being diagonal with $A_{ii}^{eff} = a_-\delta_{ij} + a_+(1-\delta_{ij})$ for all i. It is a little more technical to construct sequences for which A^{eff} is of the form γI and show that the value γ can be different for two sequences using the same proportions, but Antonio MARINO[2] and Sergio SPAGNOLO had proven that before François MURAT and I had noticed it in the early 1970s, but I only characterized the correct interval for γ in 1980, giving the first proof that the bounds guessed by Zvi HASHIN[3] and SHTRIKMAN[4,5] are right.

Nevertheless, physicists do write formulas for effective properties of mixtures in terms of proportions, which can then only be approximations, or bounds, and in the late 1980s I developed the tool of H-measures initially for explaining why some formulas guessed by physicists are good approximations in situations where the properties of the constituents are very similar. There are other situations where physicists might be right, because they only observe the result of an evolution, like for mixtures of gases or liquids, and it might be that the evolution dissipates energy and ends up at a stable equilibrium which they can compute, but there are no good mathematical methods yet for studying the evolution of mixtures, and ascertaining the validity of the physicists' arguments.

However, some mathematicians have wrongly claimed that Young measures are the right objects for studying microstructures. I was the first to introduce parametrized/Young measures in questions of partial differential equations in the late 1970s, in connection with my compensated compactness method (which I described in my 1978 lectures at HERIOT-WATT University), the reason being that I used them for describing the limits of sequences constrained (in a pointwise way) by constitutive relations. I had shown the importance of characterizing which Young measures are compatible with a given set of linear differential equations and a nonlinear constitutive relation, but I had not emphasized what had been known since the early 1970s in homogenization, that although they are useful, these objects are not the right ones for doing applications to continuum mechanics except for 1-dimensional microstructures. Those who make such wrong claims often only talk about

[2] Antonio MARINO, Italian mathematician. He works at Universitá di Pisa, Pisa, Italy.

[3] Zvi HASHIN, Israeli physicist. He works in Tel Aviv, Israel.

[4] Shmuel SHTRIKMAN, Israeli physicist. He works at the WEIZMANN Institute, Rehovot, Israel.

[5] Chaim WEIZMANN, Russian-born chemist, 1874–1952. He was the first President of Israel, 1949–1952.

gradient Young measures, forgetting to mention that I had introduced that notion for a general system because the laws of continuum mechanics cannot be expressed using only gradients, and forgetting that my framework did emphasize both balance relations and constitutive relations, while they seem to ignore constitutive relations; probably this relates to a widespread disease of pretending to work on elasticity without ever mentioning stress.

Young measures were introduced in the 1930s by Laurence C. YOUNG; I learnt about these measures as parametrized measures in seminars on control theory in the late 1960s (without attribution to YOUNG) and I first used them under that name (see footnote 15, Lecture 26).

Lemma 34.1. *For a sequence of measurable functions U^n on $\Omega \subset R^N$, taking values in a closed bounded set $K \subset R^p$, there is a subsequence and a (weakly) measurable family $\nu_{\mathbf{x}}$ of probability measures on K such that for every continuous function φ on K the subsequence $\varphi(U_m)$ converges in $L^\infty(\Omega)$ weak $\star$ to a limit ℓ_φ such that $\ell_\varphi(\mathbf{x}) = \langle \nu_{\mathbf{x}}, \varphi \rangle$ for a.e. $\mathbf{x} \in \Omega$.*

If K is unbounded, one may lose information at infinity, and one may use a compactification of K (or various ones, as pointed out by Erik BALDER[6]), and this corresponds to concentration effects; notice that the compensated compactness theorem 38.1, or the theory of H-measures which generalizes it, can deal with oscillations and concentration effects simultaneously. Losing mass at infinity was a classical question in problems of theoretical physics, but observing concentration effects in minimizing sequences was also a well known fact for geometers, and when Pierre-Louis LIONS studied these questions he was not aware of all the earlier results, and he mentioned to me that Raghu VARADHAN[7] had told him that part of his method had been already used by LEVY;[8] his choice of calling his approach the concentration-compactness method has confused a few people because of the similarity in name with my compensated compactness method, which uses microlocal ideas and is therefore of a very different nature than his method.

[Taught on Friday April 9, 1999.]

---

[6] Erik J. BALDER, Dutch mathematician. He works in Utrecht, The Netherlands.
[7] Srinivasa Raghu S. VARADHAN, Indian-born mathematician. He works at the COURANT Institute of Mathematical Sciences at New York University, New York, NY.
[8] Paul Pierre LÉVY, French mathematician, 1886–1971. He worked in Paris, France.

35

Nonlocal effects I

For questions of oceanography, we have to face the extremely difficult problem of passing to the limit in the Navier–Stokes equation in situations where the Reynolds number gets large; if one maintains the size of the domain and the size of the velocities it corresponds to letting the kinematic viscosity ν tend to 0. We have seen that for ϱ and $\varrho\,u$, the density of mass and the density of momentum, the weak convergence is well adapted as they are coefficients of differential forms, but as turbulence is related to situations where the velocity u fluctuates, we must understand how to average **u**, in the sense of finding the *correct topology* adapted to that quantity. The velocity u appears inside the differential operator

$$\frac{D}{Dt} = \frac{\partial}{\partial t} + \sum_{i=1}^{3} u_i \frac{\partial}{\partial x_i}, \qquad (35.1)$$

which is a transport operator, and what is transported is *mass, linear momentum, angular momentum, temperature, salinity, pollutants*, etc. It is natural to ask the same question as was solved for equations of the form $div\big(A^n\, grad\, u_n\big) = f$ for equations of the form $\frac{\partial v^n}{\partial t} + \sum_{i=1}^{3} u_i^n \frac{\partial v^n}{\partial x_i} = f$, where u^n is now a given oscillating field and the solution v^n may represent any of the transported quantities. If v^n is a scalar quantity, we are considering a first-order operator, which is *hyperbolic*, but it can also be considered as a *degenerate elliptic operator* and this could give some hope that the results for elliptic operators would extend to degenerate cases, but this extension is not straightforward, as we shall see on simple examples, because *nonlocal effects* appear.

I had first learnt that *memory effects* may appear by homogenization in the work of Henri SANCHEZ-PALENCIA (using asymptotic expansions in a periodic setting), for questions like viscoelasticity or electricity, corresponding to the fact that some coefficients like permittivity depend upon frequency (which is the physicist's way of dealing with *pseudo-differential operators*), and then

Jacques-Louis LIONS created an academic example where the effective equation is given by a pseudo-differential operator. I started thinking about this question in 1980 because I had guessed that these effects are the main reason behind the strange rules of *absorption and emission* invented by physicists, and I looked at the following much-simplified model:

$$\frac{\partial u_n(x,t)}{\partial t} + a_n(x)u_n(x,t) = f(x,t) \text{ in } \Omega \times (0,T); \; u_n(x,0) = v(x), \quad (35.2)$$

where a_n takes values between α and β and converges to a_∞ in $L^\infty(\Omega)$ weak $\star$ (the Young measure of a_n contains all the information that we shall need, and Ω may be any set endowed with a nonnegative Radon measure without atoms). I guessed that the limiting equation would have a convolution term

$$\frac{\partial u_\infty(x,t)}{\partial t} + a_\infty(x)u_\infty(x,t) - \int_0^t K(x, t-s)u_\infty(x,s)\,ds = f(x,t) \text{ in } \Omega \times (0,T),$$
$$u_\infty(x,0) = v(x) \text{ in } \Omega,$$

$$(35.3)$$

and I expected $K \geq 0$ for reasons related to the maximum principle. If one defines $B_n(x,t) = e^{-t\,a_n(x)}$, then one has $u_n(x,t) = B_n(x,t)v(x) + \int_0^t B_n(x,t-s)f(x,s)\,ds$, and if $B_n \rightharpoonup B_\infty$ in $L^\infty(\Omega \times (0,T))$ weak $\star$, then u_∞ satisfies the same equation with B_n replaced by B_∞; as $B_\infty(x,t) \neq e^{-t\,a_\infty(x)}$ except if a_n converges strongly to a_∞ (in $L^1_{loc}(\Omega)$ for example), one cannot have $K = 0$, and this gives the simplest example of a sequence of semi-groups whose limit is not a semi-group. Of course, the kernel K must satisfy the equation $\frac{\partial B_\infty(x,t)}{\partial t} + a_\infty(x)B_\infty(x,t) = \int_0^t K(x, t-s)B_\infty(x,s)\,ds$ for a.e. $x \in \Omega$; this was the first approach that I had taken, after trying the approach by Laplace transform which I shall follow now.

One defines the Laplace transform of a function g defined on $(0,\infty)$ by $\mathcal{L}g(p) = \int_0^\infty g(t)e^{-pt}\,dt$, and usually the Laplace transform is holomorphic in some half space $\Re p > \gamma$, and the theory has been extended by Laurent SCHWARTZ to some distributions; the important fact is that $\mathcal{L}(g \star h) = \mathcal{L}g\,\mathcal{L}h$, and $\mathcal{L}\frac{dg}{dt} = p\mathcal{L}g + g(0)$. One has

$$\big(p + a_n(x)\big)\mathcal{L}u_n(x,p) = \mathcal{L}f(x,p) + v(x), \quad (35.4)$$

and

$$\big(p + a_\infty(x) - \mathcal{L}K(x,p)\big)\mathcal{L}u_\infty(x,p) = \mathcal{L}f(x,p) + v(x), \quad (35.5)$$

so that K is characterized by

$$p + a_\infty - \mathcal{L}K(\cdot, p) = \left(\text{weak limit } \frac{1}{p + a_n}\right)^{-1}. \quad (35.6)$$

Of course, using the Young measure associated with the sequence a_n, the weak limit of $\frac{1}{p+a_n}$ is $F(x,p) = \int_{[\alpha,\beta]} \frac{d\nu_x(a)}{p+a}$, and the key to the formula for K is a property about functions $G(z)$ of a complex variable which satisfy $\Im G(z) > 0$

when $\Im z > 0$; I had first heard of it in talks by David BERGMAN,[1] and the idea is attributed to various authors such as HERGLOTZ,[2] NEVANLINNA,[3] PICK[4] or STIELTJES.[5] Suppose as a simplification that G is defined in the complex plane except for a bounded closed interval I on the real axis, and satisfies $\Im G(z)\Im z > 0$ for $\Im z \neq 0$ (as is the case here), then there exists $A \geq 0$, $B \in \mathbb{R}$ and a nonnegative Radon measure μ with support in I such that

$$G(z) = A z + B + \int_I \frac{d\mu(\lambda)}{\lambda - z} \text{ for all } z \notin I. \tag{35.7}$$

Using a Taylor[6] expansion near $p = \infty$ gives $A = 1$ and $B = a_\infty(x) = \int_I a \, d\nu_x(a)$, and one deduces that

$$\frac{1}{\int_I \frac{d\nu_x(a)}{p+a}} = p + a_\infty(x) + \int_{[-\beta,-\alpha]} \frac{d\mu_x(\lambda)}{\lambda - p} \text{ for all } p \notin [-\beta, -\alpha], \text{ a.e. } x \in \Omega, \tag{35.8}$$

so the inverse Laplace transform is then easily performed and gives

$$K(x,t) = \int_{-I} e^{\lambda t} \, d\mu_x(\lambda). \tag{35.9}$$

If a_n takes only k different values, then ν_x is a combination of at most k Dirac masses, and μ_x is a combination of at most $(k-1)$ Dirac masses which are the roots of a polynomial for which there is no simple formula in general.

In the preceding example, we found a solution u_∞ and we looked then for an equation that it satisfies, and this is a weird idea for a mathematician, but the reason why the equation obtained is natural is that the operator $\frac{d}{dt} + a_n$ is linear and commutes with translation in t, and a theorem of Laurent SCHWARTZ says that every linear (continuous) operator which commutes with translation is a convolution operator (with a distribution kernel), and the only kernel that works here is $\frac{d\delta_0}{dt} + a_\infty \delta_0 - K$; we shall use this argument again below, but in nonlinear settings the situation is not so clear.

Although the preceding example is not of great interest from a physical point of view (one could use it for a mixture of materials decaying at different rates for example), it shows something important from a philosophical point

---

[1] David J. BERGMAN, Israeli physicist. He works in Tel Aviv, Israel.
[2] Gustav HERGLOTZ, Austrian-born mathematician, 1881–1953. He worked in Göttingen, Germany.
[3] Rolf Herman NEVANLINNA, Finnish mathematician, 1895–1980. He worked in Helsinki, Finland.
[4] Georg PICK, Austrian-born mathematician, 1859–1942. He worked in Prague, Czech Republic.
[5] Thomas Jan STIELTJES, Dutch-born mathematician, 1856–1894. He worked in Toulouse, France.
[6] Brook TAYLOR, English mathematician, 1685–1731. He worked in London, England, UK.

of view: *the memory effect term is not related to any probabilistic argument!* It is actually possible to *invent a probabilistic game*, with particles absorbed and particles emitted, which will create the equation that we have found, but there is absolutely no reason other than ideology to give a better status to the probabilistic approach than to any other way of considering the preceding equation (probability is a part of analysis, but from the point of view of analysis without probability, integral equations with smooth kernels are treated as mere perturbations, in semi-group theory for example).

The model explains qualitatively something about *irreversibility*. One may start from an equation for which one can reverse time and a limiting process may make an irreversible equation appear: diffusion equations arrive naturally in certain situations by letting the velocity of light c tend to ∞, but the equation that one starts from has already incorporated a modeling of scattering which is not reversible. A more puzzling question is asked by people who start from a finite-dimensional Hamiltonian system and let the number of degrees of freedom tend to ∞, as numerical simulations show that something like entropy increases, but the system is reversible and the same occurs for the reversed equation. The answer provided by the example is that one might have to consider memory effects in order to describe well what is going on, and an observer using time in a backward way will then do the same analysis and get an integral term from t to ∞ instead in his equations; it is when one wants to get rid of the nonlocal effects and only use partial differential equations that the problems occur. In the previous example, one can approach $d\mu$ by a finite combination of Dirac masses and transform the equation obtained into a system of differential equations.

The method which I have shown above was applied by my former student Kamel HAMDACHE (with Youcef AMIRAT[7] and Abdelhamid ZIANI[8]) to a question which is more relevant to the questions of fluids that we are interested in (but from a pedagogical point of view I prefer to start with the simpler problem which was done first); their motivation was flows in porous media, and they considered

$$\frac{\partial u_n}{\partial t} + a_n(y)\frac{\partial u_n}{\partial x} = f(x,y,t) \text{ in } R \times \Omega \times (0,T); \ u_n(x,y,0) = v(x,y) \text{ in } R \times \Omega.$$
$$(35.10)$$

The method is essentially the same, using the Laplace transform in t, but also the Fourier transform in x, due to the fact that the partial differential operator that we are dealing with commutes with translations in t but also in x; this gives

$$\left(p + 2i\pi\xi\, a_n(y)\right)\mathcal{L}\mathcal{F}(\xi,y,p) = \mathcal{L}\mathcal{F}f(\xi,y,p) + \mathcal{F}v(\xi,y), \quad (35.11)$$

---

[7] Youcef AMIRAT, Algerian-born mathematician, born in 1949. He works at Université de Clermont-Ferrand II (Blaise PASCAL), Aubière, France.

[8] Abdelhamid ZIANI, Algerian-born mathematician, 1949–2004. He worked at Université de Nantes, Nantes, France.

and if one uses the Young measures ν_y associated with a subsequence, one needs the weak $\star$ limit of $\frac{1}{p+2i\pi\xi\,a_n(y)}$, which is $\int_I \frac{d\nu_y(a)}{p+2i\pi\xi\,a} = \frac{1}{2i\pi\xi}\int_I \frac{d\nu_y(a)}{q+a}$ where $q = \frac{p}{2i\pi\xi}$, and the same formula (35.8) can be used, so one can perform the inverse Laplace–Fourier transform easily and one obtains the only convolution equation in (x,t) (independent of f and v) that the limit solution may satisfy,

$$\frac{\partial u_\infty}{\partial t} + a_\infty(y)\frac{\partial u_\infty}{\partial x} - \int_0^t \int_{[-\beta,-\alpha]} \frac{\partial^2 u_\infty(x+\lambda(t-s),y,t-s)}{\partial x^2}\,d\mu_y(\lambda)\,ds = f(x,y,t) \text{ in } R \times \Omega \times (0,T), \tag{35.12}$$

with $u_\infty(x,y,0) = v(x,y)$ in $R \times \Omega$. Notice that the second derivatives are not computed at the point (x,y,t) but on lines approaching the point with a velocity $-\lambda$, with a weight depending upon λ; of course, the equation obtained has the *finite propagation speed property* and Youcef AMIRAT, Kamel HAMDACHE and Abdelhamid ZIANI checked that this is true for any nonnegative measure $d\mu_y$ with bounded support (the ones coming from the formula have a constraint on their mass for example); they also proposed a way to look at this equation as a possibly infinite hyperbolic system, by using the auxiliary functions

$$\varphi(x,y,t;V) = \int_0^t \frac{\partial u_\infty(x-V(t-s),y,t-s)}{\partial x}\,ds, \tag{35.13}$$

for $V \in [\alpha,\beta]$, so that

$$\frac{\partial\varphi(x,y,t;V)}{\partial t} + V\frac{\partial\varphi(x,y,t;V)}{\partial x} = \frac{\partial u_\infty(x,y,t)}{\partial x} \text{ in } R \times \Omega \times (0,T), \tag{35.14}$$

and the equation becomes

$$\frac{\partial u_\infty}{\partial t} + a_\infty(y)\frac{\partial u_\infty}{\partial x} - \frac{\partial}{\partial x}\left(\int_{[\alpha,\beta]} \varphi(x,y,t;V)\,d\mu_y(-V)\right) = f(x,y,t) \text{ in } R \times \Omega \times (0,T), \tag{35.15}$$

with the initial conditions $u(x,y,0) = v(x,y)$ and $\varphi(x,y,0;V) = 0$ in $R \times \Omega$ for $V \in [\alpha,\beta]$.

This example suggests that if a general transport operator with oscillating coefficients is used, one may expect *nonlocal effects*, but as we lose the commutation property, we need to find other methods of proofs. In the example, the coefficients are divergence free, so that one could write the equation in conservation form, and the transport operator applied to the coefficients gives 0; for fluids the coefficients are the components of u, which is divergence free, but the transport operator applied to u does not give 0, as the gradient of the "pressure" and the viscous term appear.

[Taught on Monday April 12, 1999.]

36

Nonlocal effects II

Physicists often describe some properties of matter at a given frequency, and in the relations that they obtain, the frequency often occurs explicitly; they usually do not bother to explain what could be a general equation valid for all solutions. In linear cases, one can usually give a meaning to these computations, and something like pseudo-differential operators, or nonlocal effects do appear, but not much is understood for nonlinear equations. For example, if one considers *Maxwell's equation*

$$div\,\mathbf{D} = \varrho; \quad -\frac{\partial \mathbf{D}}{\partial t} + curl\,\mathbf{H} = \mathbf{j}$$
$$div\,\mathbf{B} = 0; \quad \frac{\partial \mathbf{B}}{\partial t} + curl\,\mathbf{E} = 0, \tag{36.1}$$

one usually assumes that there are relations $\mathbf{D} = \varepsilon(\mathbf{x})\mathbf{E}$ between the *polarization field* $\mathbf{D}$ and the *electric field* $\mathbf{E}$ (ε is the *dielectric permittivity*), and $\mathbf{B} = \mu(\mathbf{x})\mathbf{H}$ between the *induction field* $\mathbf{B}$ and the *magnetic field* $\mathbf{H}$ (μ is the *magnetic susceptibility*).[1] One often adds Ohm's[2] law $\mathbf{j} = \sigma(\mathbf{x})\mathbf{E}$ (σ is the

---

[1] In the *vacuum*, the dielectric permittivity ε_0 and the magnetic susceptibility μ_0 are related to the *velocity of light* c by $\varepsilon_0\mu_0 c^2 = 1$, and this is obtained by looking at plane waves propagating in the direction of a unit vector $\boldsymbol{\xi}$, i.e. looking for solutions depending only upon $\mathbf{x}.\boldsymbol{\xi}$ and t, and finding that each component of the fields satisfies a scalar wave equation with characteristic velocity c. As was noticed by POINCARÉ, Maxwell's equation is invariant by action of the *Lorentz group*, and such a property is also true of *Dirac's equation* which *couples light and matter* (with Planck's constant h appearing in the coupling, of course), but not of *Schrödinger's equation* (which should be understood as a *simplification where one has let c tend to ∞*). As was pointed out to me by Joel ROBBIN around 1975, it is not clear how the equations with variable electric permittivity and magnetic susceptibility, which are not invariant by the Lorentz group, could be deduced from a more basic equation, which should be invariant by action of the Lorentz group!

[2] Georg Simon OHM, German mathematician, 1789–1854. He worked in München (Munich), Germany.

conductivity and its inverse is the *resistivity*), and in that case one forgets about the equation $div \mathbf{D} = \varrho$, which is automatically satisfied if it is true at time 0 because of the relation $\frac{\partial \varrho}{\partial t} + div \, \mathbf{j} = 0$ (which is then used for computing ϱ once one has found $\mathbf{E}$ and therefore $\mathbf{j}$).

The physicists' point of view is to look at solutions of the form $\mathbf{B}(\mathbf{x}, t) = e^{i\omega t}\mathbf{b}(\mathbf{x})$, $\mathbf{D}(\mathbf{x}, t) = e^{i\omega t}\mathbf{d}(\mathbf{x})$, $\mathbf{E}(\mathbf{x}, t) = e^{i\omega t}\mathbf{e}(\mathbf{x})$, $\mathbf{H}(\mathbf{x}, t) = e^{i\omega t}\mathbf{h}(\mathbf{x})$, so that, for example, one has $(\sigma + i\omega \varepsilon)\mathbf{e} + curl \, \mathbf{h} = 0$ and one sees a complex conductivity $\sigma + i\omega \varepsilon$ appear. The mathematicians' point of view is to use the Laplace transform, and the same equation becomes $(\sigma + p\varepsilon)\mathcal{L}\mathbf{E} + curl(\mathcal{L}\mathbf{H}) = \varepsilon \, \mathbf{E}(\cdot, 0)$. Whatever the point of view used, if one considers a mixture of such materials, homogenization usually creates coefficients which depend upon ω or p in a nonpolynomial way, but in the second case one can look for a convolution equation for linking $\mathbf{D}$ and $\mathbf{j}$ to $\mathbf{E}$ and its history (of course, one imposes the *principle of causality*, i.e. nonlocal effects must only *use the past and not the future*), and this was done using asymptotic expansions in a periodic framework by Henri SANCHEZ-PALENCIA. The physicists say that ε depends upon the frequency ω, but the mathematicians go further and try to identify a memory kernel for a convolution equation valid for all solutions and not only for those of the form $e^{i\omega t}f(x)$, and this is what I have done on the model examples. However, physicists do use the same approach for nonlinear problems, but mathematicians do not have a general theory for these cases, as far as I know, and one may start following the approach shown below, but I have not solved the bookkeeping problem and the convergence problem.

One could in principle use *pseudo-differential operators*, which Joseph KOHN and Louis NIRENBERG had introduced for developing a calculus that one can use for expressing the solutions of *elliptic equations*, or *Fourier integral operators*, which Lars HÖRMANDER had developed for similar questions for *hyperbolic equations*, but these theories have unfortunately been developed only with smooth coefficients, and this is a serious handicap even for linear problems originating in continuum mechanics or physics.

It is not known how to extend to more realistic questions of fluid dynamics the results obtained for the models that I have shown, but it is useful to derive the same results with different methods for which there is more hope for an extension; one of these methods, which physicists often use, is a perturbation method. I consider now a time-dependent model problem

$$\frac{\partial u_n}{\partial t} + a_n(x, t)u_n = f(x, t) \text{ in } \Omega \times (0, T); \quad u_n(x, 0) = v(x) \text{ in } \Omega. \quad (36.2)$$

Under the assumption that a_n is globally Lipschitz in t, this was first considered by my student Luisa MASCARENHAS[3], who used a time discretization, but the following method is more easy to apply, and although I assumed

---

[3] Maria Luisa MARTINS MACEDO FARIA MASCARENHAS, Portuguese mathematician, born in 1950. She works at Universidade de Lisboa, Lisbon, Portugal. She studied for her thesis under my supervision (1983).

equicontinuity in t, the result seems valid without such an assumption (it simplifies in that having extracted a subsequence such that $a_n(x,t)a_n(x,s)$ converges in $L^\infty(\Omega)$ weak $\star$ for s,t belonging to a countable dense set of Ω, it is then true for all $s,t \in (0,T)$, but only some integrals of the limits are really needed). Assuming that $a_n \rightharpoonup a_\infty$ in $L^\infty(\Omega)$ weak $\star$, one defines $b_n = a_n - a_\infty$ and one considers for a parameter γ the equation

$$\frac{\partial U^n(x,t;\gamma)}{\partial t} + \big(a_\infty(x,t) + \gamma\,b_n(x,t)\big)U^n(x,t;\gamma) = f(x,t) \text{ in } \Omega \times (0,T),$$
$$U^n(x,0;\gamma) = v(x) \text{ in } \Omega,$$

$$(36.3)$$

so that the preceding problem corresponds to $\gamma = 1$. Obviously U^n is analytic in γ and we can consider the Taylor expansion at $\gamma = 0$,

$$U^n(x,t;\gamma) = \sum_{k=0}^\infty \gamma^k V_k^n(x,t) \text{ in } \Omega \times (0,T), \qquad (36.4)$$

and one finds immediately that V_0^n is independent of n, and a solution of

$$\frac{\partial V_0(x,t)}{\partial t} + a_\infty(x,t)V_0(x,t) = f(x,t) \text{ in } \Omega \times (0,T); \; V_0(x,0) = v(x) \text{ in } \Omega,$$

$$(36.5)$$

and that for $k \geq 1$, V_k^n is a solution of

$$\frac{\partial V_k^n(x,t)}{\partial t} + a_\infty(x,t)V_k^n(x,t) + b_n(x,t)V_{k-1}^n(x,t) = 0 \text{ in } \Omega \times (0,T), \quad (36.6)$$
$$V_k^n(x,0) = 0 \text{ in } \Omega.$$

As $b_n \rightharpoonup 0$ in $L^\infty(\Omega)$ weak $\star$, one sees that V_1^n converges weakly to 0, but one needs the limit of $b_n V_1^n$ in order to compute the limit of V_2^n, and more generally one needs the explicit form of each V_k^n for $k \geq 1$,

$$V_k^n(x,t) = -\int_0^t e^{-\int_s^t a_\infty(x,\sigma)\,d\sigma}\,b_n(x,s)V_{k-1}^n(x,s)\,ds \text{ for } (x,t) \in \Omega \times (0,T),$$

$$(36.7)$$

so that if one defines $R(x,s,t) = exp\big(-\int_s^t a_\infty(x,\sigma)\,d\sigma\big)$, one has $V_1^n(x,t) = -\int_0^t R(x,s,t)b_n(x,s)V_0(x,s)\,ds$ and therefore the limit of $b_n V_1^n$ involves limits of $b_n(x,t)b_n(x,s)$ and has the form $\int_0^t C(x,t,s)V_0(x,s)\,ds$. One can deal similarly with the following terms, and integral terms having appeared naturally, one may look for a kernel having the analytic form

$$K(x,t,s,\gamma) = \sum_{k=2}^\infty \gamma^k K_k(x,t,s), \qquad (36.8)$$

and it is not difficult to obtain bounds for the functions V_k^n and K_k; these bounds show that the Taylor expansions written have an infinite radius of convergence, and one can take $\gamma = 1$ safely. The formula obtained for the kernel is quite different from the one which was obtained by using the representation formula for Pick functions.

In principle one could do the same type of expansions for some nonlinear problems, but the bookkeeping is quite arduous (and FEYNMANN seems to have introduced his famous diagrams for a similar purpose), and the convergence questions are not so clear (and it is for similar reasons that physicists like to use Padé[4] approximants, or other ways to sum divergent series).

[Taught on Wednesday April 14, 1999.]

---

[4] Henri Eugène PADÉ, French mathematician, 1863–1953. He taught in secondary schools and then was a rector, in Besançon, Dijon, Aix-Marseille, France.

37

A model problem

In the preceding analysis, I was studying questions of homogenization for first-order differential equations with oscillating coefficients, but the reality of fluid dynamics is a little different, in particular because of viscosity and pressure. Of course, the questions that we would like to understand are related to small viscosity, and as this problem is far from being understood now, it is useful to derive simpler models retaining as much as possible of the qualitative properties that we are interested in.

I started in this direction in 1976, and my analysis was based on the fact that the nonlinear term in the Navier–Stokes equation may be written as $\mathbf{u} \times curl(-\mathbf{u}) + grad\left(\frac{|\mathbf{u}|^2}{2}\right)$: I knew from electromagnetism that force terms in $\mathbf{u} \times \mathbf{b}$ have the effect of making particles turn, and as I had heard that vorticity is important in turbulent flows, I decided to replace $curl(-\mathbf{u})$ with a given oscillating function in order to study its effect. Not knowing what to expect, I decided to start with the stationary case, and I first used the formal method of asymptotic expansions in a periodic setting, so that my problem was

$$-\nu \Delta \mathbf{u}_\varepsilon + \mathbf{u}_\varepsilon \times \frac{1}{\varepsilon}\mathbf{b}\left(\frac{\mathbf{x}}{\varepsilon}\right) + grad\,p_\varepsilon = \mathbf{f}, \ div\,\mathbf{u}_\varepsilon = 0 \text{ in } \Omega; \ u \in H_0^1(\Omega; R^3),$$
(37.1)

for a periodic vector field **b**. I did the formal computations with Michel FORTIN,[1,2,3] who was visiting Orsay that year and was sharing my office. The first thing that we noticed was that the average of **b** had to be 0 or the whole fluid would turn very fast; in that case we derived an equation satisfied

---

[1] Michel FORTIN, Canadian mathematician. He works at Université LAVAL, Québec, Québec (Canada).

[2] Blessed François DE (MONTMORENCY) LAVAL, French-born bishop, 1623–1708. He was the first Roman Catholic bishop in Canada, archbishop of Québec, Québec (Canada), beatified in 1980 by Pope JOHN-PAUL II.

[3] Karol Józef WOJTYLA, Polish-born Pope, 1920–2004. Elected Pope in 1978, he chose the name JOHN-PAUL II.

by the first term of the formal expansion. Using then my *method of oscillating test functions* in homogenization, I easily proved the formal result that we had obtained. Although the force is perpendicular to the velocity $\mathbf{u}_\varepsilon$ and therefore does not work, it induces oscillations in $grad\,\mathbf{u}_\varepsilon$ and therefore more energy is dissipated by viscosity (per unit of time, as we are looking at a stationary problem), but the interesting feature is that the added dissipation which appears in the limiting equation is not quadratic in $grad\,\mathbf{u}$, but quadratic in $\mathbf{u}$, contrary to a quite general belief about *turbulent viscosity*. It is easy to avoid periodicity hypotheses and to consider terms of the form $\mathbf{u}_\varepsilon \times curl\,\mathbf{v}_\varepsilon$ with $\mathbf{v}_\varepsilon$ converging weakly, but I noticed something else when I wrote it down for a meeting in Minneapolis, MN in 1984, that there is a quadratic effect in a strength parameter λ; the new problem was

$$-\nu\,\Delta\,\mathbf{u}_n + \mathbf{u}_n \times curl(\mathbf{v}_0 + \lambda\,\mathbf{v}_n) + grad\,p_n = \mathbf{f},\ div\,\mathbf{u}_n = 0 \text{ in } \Omega, \quad (37.2)$$

with $\mathbf{v}_0 \in L^3(\Omega; R^3)$ and $\mathbf{v}_n \rightharpoonup 0$ in $L^3(\Omega; R^3)$ weak. I did not impose boundary conditions but I assumed that $\mathbf{u}_n \rightharpoonup \mathbf{u}_\infty$ in $H^1(\Omega; R^3)$ weak (it is a classical requirement in homogenization that if one wants to speak about the effective properties of a mixture one should obtain a result which is independent of the boundary conditions; if one does not do this, one can only talk about global properties of the mixture together with the container). I showed that there exists a symmetric nonnegative matrix M, depending only upon a subsequence of v_n that one may have to extract (and ν), such that $\mathbf{u}_\infty$ satisfies the equation

$$-\nu\,\Delta\,\mathbf{u}_\infty + \mathbf{u}_\infty \times curl\,\mathbf{v}_0 + \lambda^2 M\,\mathbf{u}_\infty + grad\,p_\infty = \mathbf{f},\ div\,\mathbf{u}_\infty = 0 \text{ in } \Omega, \quad (37.3)$$

and of course a more precise convergence result is

$$\mathbf{u}_n \times curl\,\mathbf{v}_n \rightharpoonup \lambda\,M\,\mathbf{u}_\infty \text{ in } H^{-1}_{loc}(\Omega; R^3) \text{ weak}, \quad (37.4)$$

and

$$\nu\,|grad\,\mathbf{u}_n|^2 \rightharpoonup \nu\,|grad\,\mathbf{u}_\infty|^2 + \lambda^2\,(M\,\mathbf{u}_\infty.\mathbf{u}_\infty) \text{ in the sense of measures.} \quad (37.5)$$

The way M is defined follows my approach to homogenization, but the quadratic dependence in λ and a particular formula in the case where $div\,v_n = 0$ was my first hint about the possibility of defining *H-measures*, and after I was led to introduce H-measures for the different purpose of "small amplitude" homogenization, I checked that M could indeed be computed from the H-measures associated with the sequence $\mathbf{v}_n$ (and ν).

One extracts a subsequence from $\mathbf{v}_n$ and one constructs M in the following way: for $\mathbf{k} \in R^3$ one solves

$$-\nu\,\Delta\,\mathbf{w}_n + \mathbf{k} \times curl\,\mathbf{v}_n + grad\,q_n = 0,\ div\,\mathbf{w}_n = 0 \text{ in } \Omega, \quad (37.6)$$

adding boundary conditions which imply that $\mathbf{w}_n \rightharpoonup 0$ in $H^1(\Omega; R^3)$ weak (Dirichlet conditions, or periodic conditions in the case where $\mathbf{v}_n$ is defined in a periodic way, for example); one can indeed extract a subsequence such that this occurs for three independent vectors $\mathbf{k}$, and one defines M by

$$\mathbf{w}_n \times curl\,\mathbf{v}_n \rightharpoonup M\,\mathbf{k} \text{ in } H_{loc}^{-1}(\Omega; R^3) \text{ weak.} \tag{37.7}$$

The first remark is that $\mathbf{u} \mapsto \mathbf{u} \times curl\,\mathbf{v}$ maps continuously $H^1(\Omega; R^3)$ into $H^{-1}(\Omega; R^3)$ if $\mathbf{v} \in L^3(\Omega; R^3)$, if the boundary of Ω is smooth, using the Sobolev embedding theorem $H^1(\Omega) \subset L^6(\Omega)$. Indeed, if $z, \varphi \in H^1(\Omega)$, then $z\varphi \in W^{1,3/2}(\Omega)$; if $\mathbf{f} \in H^{-1}(\Omega; R^3)$ one finds that, after eventually adding a constant, p_n is bounded in $L_{loc}^2(\Omega)$, and one can assume that $p_n \rightharpoonup p_\infty$ in $L_{loc}^2(\Omega)$ weak. Similarly, in the problem for $\mathbf{w}_n$ one can assume that $q_n \rightharpoonup 0$ in $L_{loc}^2(\Omega)$ weak. Using elliptic regularity theory (and the Calderón–Zygmund theorem), $grad\,\mathbf{w}_n$ is bounded in $L_{loc}^3(\Omega; R^{3\times3})$ and therefore $\mathbf{w}_n \to 0$ in $L_{loc}^p(\Omega; R^3)$ strong for every $p < \infty$; one then has a better convergence for $\mathbf{w}_n \times curl\,\mathbf{v}_n$, which converges to $M\,\mathbf{k}$ in $H_{loc}^{-1}(\Omega; R^3)$ strong, because of writing products of the form $(w_n)_i \frac{\partial (v_n)_j}{\partial x_k}$ as $\frac{\partial[(w_n)_i (v_n)_j]}{\partial x_k} - \frac{\partial (w_n)_i}{\partial x_k}(v_n)_j$ and terms like $(w_n)_i (v_n)_j$ converge strongly to 0 in $L_{loc}^2(\Omega)$ and terms like $\frac{\partial (w_n)_i}{\partial x_k}(v_n)_j$ are bounded in $L_{loc}^{3/2}(\Omega)$ and converge strongly to 0 in $L_{loc}^q(\Omega)$ for every $q > 3/2$ and therefore in $H_{loc}^{-1}(\Omega)$ strong. One applies my method of oscillating test functions, multiplying the equation for $\mathbf{u}_n$ by $\varphi\,\mathbf{w}_n$ and the equation for $\mathbf{w}_n$ by $\varphi\,\mathbf{u}_n$, with $\varphi \in C_c^1(\Omega)$, and noticing that $div(\varphi\,\mathbf{w}_n) = (grad\,\varphi).\mathbf{w}_n$ and $div(\varphi\,\mathbf{u}_n) = (grad\,\varphi).\mathbf{u}_n$ and therefore the estimates on the "pressures" p_n and q_n are needed. One assumes that $\mathbf{u}_n \times curl\,\mathbf{v}_n \rightharpoonup \mathbf{g}$ in $H^{-1}(\Omega; R^3)$ weak and one wants to identify $\mathbf{g}$; one finds that

$$\lim_{n\to\infty} \int_\Omega \nu\,\varphi\big(grad\,\mathbf{u}_n.grad\,\mathbf{w}_n\big)\,d\mathbf{x} + \lambda\langle \varphi\,\mathbf{u}_n \times curl\,\mathbf{v}_n, \mathbf{w}_n\rangle = 0, \tag{37.8}$$

and

$$\lim_{n\to\infty} \int_\Omega \nu\,\varphi\big(grad\,\mathbf{u}_n.\mathbf{w}_n\big)\,d\mathbf{x} = \langle \varphi\,\mathbf{g}, \mathbf{k}\rangle, \tag{37.9}$$

but as

$$\langle \varphi\,\mathbf{u}_n \times curl\,\mathbf{v}_n, \mathbf{w}_n\rangle = -\langle \varphi\,\mathbf{w}_n \times curl\,\mathbf{v}_n, \mathbf{u}_n\rangle \to -\langle \varphi\,M\,\mathbf{k}, \mathbf{u}_\infty\rangle, \tag{37.10}$$

one has shown that

$$\mathbf{g} = \lambda\,M^T\,\mathbf{u}_\infty. \tag{37.11}$$

The fact that M is symmetric follows easily by the same method, $\mathbf{w}_n'$ being the solution for $\mathbf{k}'$, multiplying the equation for $\mathbf{w}_n'$ by $\varphi\,\mathbf{w}_n$, the equation for $\mathbf{w}_n$ by $\varphi\,\mathbf{w}_n'$, and comparing. The limit of $\nu|\,grad\,\mathbf{u}_n|^2$ is obtained by multiplying the equation for $\mathbf{u}_n$ by $\varphi\,\mathbf{u}_n$.

In the case where $div\,\mathbf{v}_n = 0$, which is the case for fluid dynamics, one can take

$$\mathbf{w}_n = \mathbf{k}.grad\,\mathbf{z}_n, \tag{37.12}$$

where $\mathbf{z}_n$ solves

$$-\nu \, \Delta \mathbf{z}_n = \mathbf{v}_n, \qquad\qquad (37.13)$$

and therefore

$$\nu \sum_{\ell,m=1}^{3} \frac{\partial^2 (z_n)_\ell}{\partial x_i \partial x_m} \frac{\partial^2 (z_n)_\ell}{\partial x_j \partial x_m} \rightharpoonup \mathsf{M}_{ij} \text{ in } L^{3/2}(\Omega) \text{ weak.} \qquad (37.14)$$

[Taught on Monday April 19, 1999. There were no classes on Friday April 16, which was Spring Carnival.]

38

Compensated compactness I

The same analysis can be done for the evolution problem if one can obtain a bound for the "pressure", so I initially did it for the whole space; Wolf VON WAHL[1] later told me that he had shown by semi-group methods that one can obtain estimates on the "pressure" for any smooth bounded open set. In my original proof for the evolution case, I had not seen how to prove that the matrix M corresponding to the added dissipation is symmetric; two years ago, I worked with Chun LIU[2] and Konstantina TRIVISA[3] on extending the formula using H-measures to the evolution case, and we first checked the symmetry, but then we noticed that one needs a new variant of H-measures, with a parabolic scaling, i.e. instead of identifying rays $s\,\boldsymbol{\xi}$ through a nonzero $\boldsymbol{\xi}$ with $s > 0$, one identifies curves $(s\,\boldsymbol{\xi}, s^2\tau)$ through a nonzero $(\boldsymbol{\xi}, \tau)$.

Of course, one should not lose sight of the reasons why the preceding models were chosen and the previous computations were done: the initial purpose was to understand what is the adapted weak type topology for the velocity, appearing as coefficients of a transport equation, with or without viscosity. Starting from a model without viscosity, we saw that various nonlocal terms could appear, and this analysis could also be useful for correcting the defects of the Navier–Stokes equation, but the class to consider should then at least contain some equations with memory effects. Starting with a model with viscosity but magnifying the oscillations possible for the velocity field, we have seen some lower order terms appear, and it could have some analogy with

---

[1] Wolf VON WAHL, German mathematician. He works in Bayreuth, Germany.

[2] Chun LIU, Chinese-born mathematician, born in 1966. He works at Pennsylvania State University, University Park, PA. He was a postdoctoral associate at the Center for Nonlinear Analysis at CARNEGIE MELLON University, Pittsburgh, PA.

[3] Konstantina TRIVISA, Greek-born mathematician, born in 1969. She works at University of Maryland, College Park MD. She was a postdoctoral associate at the Center for Nonlinear Analysis at CARNEGIE MELLON University, Pittsburgh, PA.

the framework using *affine connections* which geometers have advocated. Obviously one should improve the model, but it is worth mentioning that the formula using H-measures which gives an explicit form for M has a $\frac{1}{\nu}$ in front of an integral on the unit sphere, and although it is tempting to rescale the equation, one should remember that turbulence is supposed to show an infinite number of length scales, and that an object like H-measures which mixes different frequencies cannot reasonably help describe that.

Before describing the tool of H-measures and the various formulas that one can deduce from it, it is worth starting with the compensated compactness theory, which I had developed with François MURAT in the late 1970s. I make a distinction between the basic compensated compactness theorem 38.1, and the compensated compactness method, which I developed afterward, and which is more general, and takes into account the constitutive relations and some "entropies". The basic theorem – which I sometimes call the quadratic theorem of compensated compactness theory – is the following.

Theorem 38.1. *(Compensated compactness, 1976, François* MURAT *& Luc* TARTAR*) Assume that*

$$U^n \rightharpoonup U^\infty \text{ in } L^2_{loc}(\Omega; R^p) \text{ weak}$$
$$\sum_{j=1}^{p} \sum_{k=1}^{N} A_{ijk} \frac{\partial U_j^n}{\partial x_k} \text{ stays in a compact of } H^{-1}_{loc}(\Omega) \text{ for } i = 1, \ldots, q.$$
$$(38.1)$$

Define the two characteristic sets V *and* Λ *by*

$$V = \left\{ (\lambda, \xi) \in R^p \times (R^N \setminus 0) \mid \sum_{j=1}^{p} \sum_{k=1}^{N} A_{ijk} \lambda_j \xi_k = 0 \text{ for } i = 1, \ldots, q \right\}$$
$$\Lambda = \left\{ \lambda \in R^p \mid \text{ there exists } \xi \in R^N \setminus 0, (\lambda, \xi) \in V \right\}.$$
$$(38.2)$$

Let Q *be a real quadratic form on* R^p; *then (38.1) implies*

$$Q(U^n) \rightharpoonup Q(U^\infty) + \nu \text{ in the sense of measures implies } \nu \geq 0, \qquad (38.3)$$

if and only if Q *satisfies*

$$Q(\lambda) \geq 0 \text{ for all } \lambda \in \Lambda. \qquad (38.4)$$

In particular (38.1) implies $Q(U^n) \rightharpoonup Q(U^\infty)$ *in the sense of measures if and only if* $Q(\lambda) = 0$ *for all* $\lambda \in \Lambda$.

I have already mentioned the div-curl lemma 33.1, found with François MURAT in 1974 in connection with homogenization: it is the particular case where $U = (\mathbf{E}, \mathbf{D})$ with the list of differential information corresponding to $div \, \mathbf{D}$ and the components of $curl \, \mathbf{E}$; then $V = \{(\mathbf{E}, \mathbf{D}, \xi) \mid \xi \perp \mathbf{D}, \xi \parallel \mathbf{E}\}$ (where $\mathbf{a} \parallel \mathbf{b}$ means $\mathbf{a}$ parallel to $\mathbf{b}$), and $\Lambda = \{(\mathbf{E}, \mathbf{D}) \mid \mathbf{E}.\mathbf{D} = 0\}$; then the quadratic form Q_0 defined by $Q_0(\mathbf{E}, \mathbf{D}) = \mathbf{E}.\mathbf{D}$ is 0 on Λ, and therefore by applying the theorem to $\pm Q_0$ one deduces that $\mathbf{E}^n.\mathbf{D}^n$ converges to $\mathbf{E}^\infty.\mathbf{D}^\infty$

in the sense of measures. Our proof of Theorem 38.1 followed the one that we had found for the div-curl lemma 33.1, using the Fourier transform and Plancherel's formula.

I spent the year 1974–1975 in Madison, and Joel ROBBIN taught me how to translate some of my results into the language of differential forms, and he showed me a new proof of the div-curl lemma 33.1 using Hodge's theory. We did not notice that the same method was giving the properties of sequential weak continuity of Jacobian determinants, which MORREY had actually proven in the 1950s, and RESHETNYAK more recently, and I only learnt these results from John BALL in the fall of 1975, mistakenly believing that he had proven them.

Later, Jacques-Louis LIONS asked François MURAT to extend our div-curl lemma 33.1, coining the term compensated compactness for this type of result, as it looks like a compactness argument because one can deduce the weak limit of a nonlinear quantity, but it is the result of a compensation effect; he gave him an article by SCHULENBERGER[4] and WILCOX[5] which he thought related, and François MURAT first considered a bilinear setting, i.e. $U = (V, W)$ with some differential list for V and a differential list for W, and then he looked at the bilinear forms $B(V, W)$ which are sequentially weakly continuous. Then, I pointed out to him that the splitting of U and the restriction to bilinear forms is not natural, and therefore François MURAT proved the above theorem in the case where $Q(\lambda) = 0$ for all $\lambda \in \Lambda$ (deducing that $\nu = 0$ in this case). However, his proof was a little different than the one that we had followed for our div-curl lemma 33.1, probably because he had also extended our div-curl lemma 33.1 itself to an (L^p, L^q) setting (for which he used the Mikhlin–Hörmander theorem on $\mathcal{F}L^p$ multipliers, and one needs to check the smoothness of the multiplier), and a similar approach forced him to impose a hypothesis of constant rank: if for each $\boldsymbol{\xi} \in R^N \setminus 0$ one defines $\Lambda_{\boldsymbol{\xi}} = \{\lambda \in R^p \mid \sum_{j=1}^{p} \sum_{k=1}^{N} A_{ijk} \lambda_j \xi_k = 0 \text{ for } i = 1, \ldots, q\}$, so that Λ is the union of the subspaces $\Lambda_{\boldsymbol{\xi}}$, the constant rank hypothesis imposes that the dimension of $\Lambda_{\boldsymbol{\xi}}$ is independent of $\boldsymbol{\xi}$. Then, I extended the result as shown above, and it has the following consequence: if $U_i^n U_j^n \rightharpoonup U_i^\infty U_j^\infty + R_{ij}$ in the sense of measures for $i, j = 1, \ldots, p$, then this defines a symmetric matrix R whose entries may be Radon measures: if all R_{ij} are integrable functions, one has

$$R(\mathbf{x}) \text{ belongs to the closed convex hull of } \{\lambda \otimes \lambda \mid \lambda \in \Lambda\} \text{ a.e. } \mathbf{x} \in \Omega. \quad (38.5)$$

---

[4] John R. SCHULENBERGER, American mathematician. He works at Texas Tech University, Lubbock, TX.

[5] Calvin Hayden WILCOX, American mathematician. He works at University of Utah, Salt Lake City, UT.

The general case is similar, once one uses the *Radon–Nikodým*[6],[7] *theorem*: if $\tau = \sum_{k=1}^{p} R_{kk}$, then $R_{ij} = \varrho_{ij}\tau$ with the functions ϱ_{ij} being τ-integrable for $i, j = 1, \ldots, p$, and it is $\varrho(x)$ which belongs to the convex hull of the elements of the form $\lambda \otimes \lambda$ for $\lambda \in \Lambda$; this property holds τ-almost everywhere. A point in the convex hull of a set K is the center of mass of a probability measure with support on K, and the theory of H-measures, which I developed in the late 1980s, extends the compensated compactness theorem 38.1 and gives explicitly a way to describe these probability measures; the interest is that the H-measures are measures in $(\mathbf{x}, \boldsymbol{\xi})$ and that they permit us to extend the compensated compactness theorem 38.1 to the case of differential equations with variable coefficients, and in some situations the H-measures satisfy partial differential equations in $(\mathbf{x}, \boldsymbol{\xi})$. There are, however, parts of the compensated compactness method which still require improvements, as there is not yet a characterization of which pairs (Young measures/ H-measures) can be created, although I have obtained some information in this direction with François MURAT.

[Taught on Wednesday April 21, 1999.]

---

[6] Otton Marcin NIKODÝM, Polish-born mathematician, 1887–1974. He worked at KENYON College, Gambier, OH.

[7] George KENYON, second Baron KENYON, British statesman, 1776–1855.

39

Compensated compactness II

My analysis of problems of continuum mechanics in the 1970s was that there is a dichotomy between the *constitutive relations* which are possibly *nonlinear pointwise constraints* of the form

$$U(\mathbf{x}) \in K \text{ a.e. } \mathbf{x} \in \Omega, \tag{39.1}$$

with K eventually depending upon $\mathbf{x}$ (with possible oscillations requiring techniques from homogenization), and the *balance equations* which are *linear differential constraints* of the form

$$\sum_{j=1}^{p} \sum_{k=1}^{N} A_{ijk} \frac{\partial U_j}{\partial x_k} = f_i \text{ in } \Omega, \text{ for } i = 1, \ldots, q. \tag{39.2}$$

Perhaps because I had learnt some continuum mechanics as a student, I knew that elasticity meant

$$\varrho(x) \frac{\partial^2 u_i}{\partial t^2} - \sum_{j=1}^{N} \frac{\partial \sigma_{ij}}{\partial x_j} = f_i \text{ in } \Omega \text{ for } i = 1, \ldots, N, \tag{39.3}$$

and in that case U would contain the components of the momentum $\varrho \frac{\partial u}{\partial t}$, the strain ∇u, the stress σ (as I do not remember hearing the term Piola–Kirchhoff stress from my student days, it might be that I was told mostly the Eulerian point of view, where the symmetric Cauchy stress appears), and the constitutive relations relate σ and ∇u, while the list of balance equations contains the equilibrium equation above and the compatibility conditions related to using gradients. (Then in 1977, I added "entropies" to that description, developing the compensated compactness method; it may be useful to point out that "entropies" have nothing to do with the fact that one considers an evolution equation, or that one is interested in hyperbolic problems, as it is just a name (chosen by Peter LAX) for designing supplementary differential equations which are consequences of those already written for smooth solutions; geometers refer to special ones as Casimirs.)

Of course, it is a handicap that the compensated compactness theorem 38.1 cannot handle variable coefficients, but H-measures do not suffer from this defect: each first-order partial differential equation written in conservative form $\sum_{j=1}^{p} \sum_{k=1}^{N} \frac{\partial (A_{jk}(x)U_j^n)}{\partial x_k} \to f$ in $H_{loc}^{-1}(\Omega)$ strong, with the coefficients A_{jk} being continuous, can be seen by H-measures (and the *localization principle* implies $\sum_{j=1}^{p} \sum_{k=1}^{N} \xi_k A_{jk}(x)\mu^{j\ell} = 0$ for every ℓ).

Young measures cannot take into account partial differential equations, but it might be because I used them in order to describe some constraints that they must satisfy as a consequence of the compensated compactness theorem 38.1, that some may have misunderstood their role. Suppose, for example, that a sequence U^n is bounded in L^∞, corresponds to a Young measure $\nu_x, x \in \Omega \subset R^N$, and satisfies a partial differential equation with constant coefficient $\sum_{j=1}^{p} \sum_{k=1}^{N} A_{jk} \frac{\partial U_j^n}{\partial x_k} = 0$; decompose R^N as a union of cubes of size $1/n$ and for each of these cubes choose a rigid displacement mapping the cube onto itself and transport the values of U^n accordingly, and let V^n be any of the new functions obtained this way. It is not difficult to check that V^n corresponds to the same Young measure $\nu_x, x \in \Omega$, as U^n; however, V^n is unlikely to solve the same partial differential equation as U^n, and therefore Young measures cannot reveal if the sequence that they analyze satisfies or not a given partial differential equation. A different way to express the same idea is to notice that in defining Young measures the only important property of Ω is to be endowed with a nonnegative measure without atoms, like the Lebesgue measure, and therefore the structure of a differentiable manifold is not seen by Young measure. However, what the compensated compactness theorem 38.1 says can be expressed in terms of Young measures, as it says that if Q is quadratic and satisfies $Q(\lambda) \geq 0$ for all $\lambda \in \Lambda$, then the limit of $Q(U^n)$ is $\langle \nu_x, Q \rangle$, while the limit of U^n is $\langle \nu_x, id \rangle$ (where id is the identity mapping), and therefore one has the following inequality:

$$\langle \nu_x, Q \rangle \geq Q(\langle \nu_x, id \rangle) \text{ a.e. } x \in \Omega, \tag{39.4}$$

reminiscent of *Jensen's inequality*, which says that if one replaces Q by any convex function the preceding inequality is true.

If $\Lambda = \{0\}$, one is looking at an elliptic system or some overdetermined system; the ellipticity of the system corresponds to saying that for every $\xi \neq 0$, the linear mapping $U \mapsto V$ with $V_i = \sum_{j=1}^{p} \sum_{k=1}^{N} A_{ijk} U_j \xi_k$ for $i = 1, \ldots, q$, is invertible (so that $q = p$); if it is not the case then one necessarily has $q > p$, but that by itself is not enough to imply that $\Lambda = \{0\}$. In the case $\Lambda = \{0\}$ one has $U^n \to U^\infty$ in $L_{loc}^2(\Omega; R^p)$ strong, because one can take Q positive definite, and as Q is 0 on Λ one finds that $Q(U^n) \rightharpoonup Q(U^\infty)$ in the sense of measures; if B is the symmetric bilinear form associated with Q, one then has $Q(U^n - U^\infty) = Q(U^n) - 2B(U^n, U^\infty) + Q(U^\infty)$, so that $Q(U^n - U^\infty)$ tends to 0 in the sense of measures. The case $\Lambda = \{0\}$ corresponds then to using a *compactness argument*.

The case $\Lambda = R^p$ corresponds to using a *convexity argument*, as $Q \geq 0$ is equivalent to Q convex for quadratic forms; this happens if there is no differential equation ($q = 0$), but also for some list of differential equations that are not constraining enough: if U^n consists of the list of k vector fields whose divergence is controlled, then one has $\Lambda = R^p$ if $k < N$, but $\Lambda \neq R^p$ if $k \geq N$.

In the case of the div-curl lemma 33.1, the information about *curl E* gives $\xi_i E_j - \xi_j E_i = 0$ for all i, j, i.e. $\boldsymbol{\xi}$ parallel to $\mathbf{E}$, and the information about *div D* gives $\boldsymbol{\xi}$ orthogonal to $\mathbf{D}$, so that Λ is the set of $(\mathbf{E}, \mathbf{D})$ with $\mathbf{D}$ orthogonal to $\mathbf{E}$. This case is related to the *monotonicity method*.

The case of Maxwell's equation is more intricate; the dual variable is $(\tau, \boldsymbol{\xi})$ with $|\tau| + |\boldsymbol{\xi}| \neq 0$ and the equations are $\boldsymbol{\xi}.\mathbf{D} = 0$, $-\tau \mathbf{D} + \boldsymbol{\xi} \times \mathbf{H} = 0$, $\boldsymbol{\xi}.\mathbf{B} = 0$, $\tau \mathbf{B} + \boldsymbol{\xi} \times \mathbf{E} = 0$. If $\tau \neq 0$, one may assume $\tau = 1$, i.e. one has $\mathbf{D} = \boldsymbol{\xi} \times \mathbf{H}$ and $\mathbf{B} = -\boldsymbol{\xi} \times \mathbf{E}$, so that $\mathbf{D}$ is orthogonal to $\mathbf{H}$, $\mathbf{B}$ is orthogonal to $\mathbf{E}$, and $\mathbf{E}.\mathbf{D} = \mathbf{B}.\mathbf{H}$; if $\tau = 0$, then $\mathbf{H}$ and $\mathbf{E}$ must be parallel to $\boldsymbol{\xi}$ and orthogonal to $\mathbf{B}$ and $\mathbf{D}$, and one still has $\mathbf{E}.\mathbf{B} = \mathbf{D}.\mathbf{H} = \mathbf{E}.\mathbf{D} - \mathbf{B}.\mathbf{H} = 0$; one can check that this is exactly the description of Λ. We shall see later why these quantities are natural, using the framework of differential forms.

It is easy to see that the condition on Q in the theorem is necessary. More generally, let F be a (continuous) function on R^p: if one wants that for all $U^n \rightharpoonup U^\infty$ in $L^\infty(\Omega; R^p)$ weak $\star$, satisfying the equations with $f_i = 0$ for $i = 1, \ldots, q$, and such that $F(U^n) \rightharpoonup F(U^\infty) + \nu$ in $L^\infty(\Omega)$ weak $\star$, one can deduce that $\nu \geq 0$, then it is necessary that F be convex in every direction of Λ, i.e. for every $a \in R^p$ and every $\lambda \in \Lambda$ the mapping $t \mapsto F(a + t\lambda)$ must be convex in $t \in R$ (this is not always sufficient, but theorem 38.1 says that it is sufficient if the function F is quadratic). Indeed, for $(\lambda, \boldsymbol{\xi}) \in V$ the function U defined by $U(\mathbf{x}) = a + \lambda \varphi(\boldsymbol{\xi}.\mathbf{x})$ satisfies $\sum_{j=1}^{p} \sum_{k=1}^{N} A_{ijk} \frac{\partial U_j}{\partial x_k} = (\sum_{j=1}^{p} \sum_{k=1}^{N} A_{ijk} \lambda_j \xi_k) \varphi'(\boldsymbol{\xi}.\mathbf{x}) = 0$, and this stays true if φ is replaced by a characteristic function χ, so that U takes only the values a and $b = a + \lambda$; if one uses a sequence of characteristic functions χ_n converging to $\theta \in (0, 1)$ in $L^\infty(R)$ weak $\star$, then $U^n \rightharpoonup U^\infty$ in $L^\infty(\Omega; R^p)$ weak $\star$, with $U^\infty = a + \theta \lambda = (1 - \theta)a + \theta b$ and because $F(U^n) = (1 - \chi_n)F(a) + \chi_n F(b)$ one has $F(U^n) \rightharpoonup V^\infty$ in $L^\infty(\Omega)$ weak $\star$, with $V^\infty = (1 - \theta)F(a) + \theta F(b)$, and the result follows easily. If one wants to deduce that $F(U^n) \rightharpoonup F(U^\infty)$ in $L^\infty(\Omega)$ weak $\star$, it is necessary that F be affine in every direction of Λ; this is sufficient for quadratic functions but not always sufficient in general. In the case where Λ spans all R^p, F must be a combination of multilinear forms and in particular it should be a polynomial of degree at most p, but there are *other necessary conditions* which imply that the degree can be at most N.

[Taught on Friday April 23, 1999.]

40

Differential forms

We have seen that there are three linearly independent quadratic forms which are 0 on the characteristic set Λ for Maxwell's equation, $\mathbf{E}.\mathbf{B}$, $\mathbf{H}.\mathbf{D}$ and $\mathbf{E}.\mathbf{D} - \mathbf{H}.\mathbf{B}$; Λ is actually the intersection of the zero sets of these three quadratic forms. In discussing the interpretation of these quantities in terms of differential forms, it is worth recalling the analogy between Maxwell's equation and the equation for fluids. The transport term $\frac{\partial \mathbf{u}}{\partial t} + \mathbf{u}.\nabla\,\mathbf{u}$ can be written as $\frac{\partial \mathbf{u}}{\partial t} + \mathbf{u} \times curl(-\mathbf{u}) + grad\left(\frac{|\mathbf{u}|^2}{2}\right)$, and there may also exist a Coriolis term $\mathbf{u} \times 2\mathbf{\Omega}$ so that a part of it has the same form as a part of the Lorentz force $\mathbf{E} + \mathbf{u} \times \mathbf{B}$ that one encounters in electromagnetism; if one denotes $\mathbf{e} = \frac{\partial \mathbf{u}}{\partial t} + grad\left(\Phi + \frac{p}{\varrho_0} + \frac{|\mathbf{u}|^2}{2}\right)$ and $\mathbf{b} = curl(-\mathbf{u}) + 2\mathbf{\Omega}$, then one has $div\,\mathbf{b} = 0$ and $\frac{\partial \mathbf{b}}{\partial t} + curl\,\mathbf{e} = 0$, in the case where the viscosity is 0 and there is no exterior force other than those related to the geopotential Φ. In electromagnetism one introduces the scalar and vector potentials $V, \mathbf{A}$, such that $\mathbf{B} = -curl\,\mathbf{A}$ and $\mathbf{E} = \frac{\partial \mathbf{A}}{\partial t} - grad\,V$, and therefore one has similar relations if one defines $\mathbf{a} = \mathbf{u} - \mathbf{u}^*$ with $\mathbf{u}^*$ being a velocity field such that $curl\,\mathbf{u}^* = 2\mathbf{\Omega}$, and $v = -\Phi - \frac{p}{\varrho_0} - \frac{|\mathbf{u}|^2}{2}$.

In terms of differential forms, let us define the 1-form α by

$$\alpha = V\,dt + \sum_{j=1}^{3} A_j\,dx_j, \tag{40.1}$$

and the 2-form $\beta = d\alpha$, i.e.

$$\beta = \sum_{j=1}^{3} \frac{\partial V}{\partial x_j}\,dx_j \wedge dt + \sum_{j=1}^{3} \frac{\partial A_j}{\partial t}\,dt \wedge dx_j + \sum_{j,k=1}^{3} \frac{\partial A_j}{\partial x_k}\,dx_k \wedge dx_j$$
$$= \sum_{i=1}^{3} E_i\,dt \wedge dx_i - B_1\,dx_2 \wedge dx_3 - B_2\,dx_3 \wedge dx_1 - B_3\,dx_1 \wedge dx_2. \tag{40.2}$$

Then, the equations $div\,B = 0$ and $\frac{\partial B}{\partial t} + curl\,E = 0$ simply mean $d\beta = 0$. Similarly, let us define the 2-form γ by

$$\gamma = \sum_{i=1}^{3} H_i \, dt \wedge dx_i + D_1 \, dx_2 \wedge dx_3 + D_2 \, dx_3 \wedge dx_1 + D_3 \, dx_1 \wedge dx_2, \quad (40.3)$$

and the 3-form δ by

$$\delta = \varrho \, dx_1 \wedge dx_2 \wedge dx_3 - j_1 \, dt \wedge dx_2 \wedge dx_3 - j_2 \, dt \wedge dx_3 \wedge dx_1 - j_3 \, dt \wedge dx_1 \wedge dx_2, \quad (40.4)$$

so that the equations $div \, D = \varrho$ and $-\frac{\partial D}{\partial t} + curl \, H = \mathbf{j}$ simply mean $d\gamma = \delta$, because

$$d\gamma = \sum_{i,j=1}^{3} \frac{\partial H_i}{\partial x_j} \, dx_j \wedge dt \wedge dx_i + \frac{\partial D_1}{\partial t} \, dt \wedge dx_2 \wedge dx_3 +$$
$$+ \frac{\partial D_2}{\partial t} \, dt \wedge dx_3 \wedge dx_1 + \frac{\partial D_3}{\partial t} \, dt \wedge dx_1 \wedge dx_2 + div \, D \, dx_1 \wedge dx_2 \wedge dx_3. \quad (40.5)$$

Conservation of charge is $d\delta = 0$, and therefore by *Poincaré's lemma* there must exist a 2-form γ with $d\gamma = \delta$; naming the six coefficients of γ leads to the introduction of the components of $\mathbf{D}$ and $\mathbf{H}$. Studying the movement of charged particles leads us to *discover experimentally*[1] the *Lorentz force* $\mathbf{E} + \mathbf{u} \times \mathbf{B}$, and the components of $\mathbf{E}$ and $\mathbf{B}$ appear to be the coefficients of an exact 2-form β; this therefore leads by Poincaré's lemma to the introduction of the 1-form α. (I learnt about this interpretation of Maxwell's equation in terms of differential forms from Joel ROBBIN in 1975, and then I heard Laurent SCHWARTZ mention it in a talk, where he considered the vacuum with $\varepsilon_0 = \mu_0 = 1$, and instead of $d\gamma = \delta$ he wrote $d^*\beta = 0$; it is actually important not to identify $\mathbf{E}$ and $\mathbf{D}$ or $\mathbf{H}$ and $\mathbf{B}$, even if that is possible in the vacuum, because in the presence of matter they play different roles.)

Now we can identify what the quantities $\mathbf{E.B}$, $\mathbf{H.D}$ and $\mathbf{E.D} - \mathbf{H.B}$ mean in terms of the 2-forms β and γ introduced in (40.2) and (40.3), because one has

$$\beta \wedge \beta = -2(\mathbf{E.B}) \, dt \wedge dx_1 \wedge dx_2 \wedge dx_3$$
$$\gamma \wedge \gamma = 2(\mathbf{H.D}) \, dt \wedge dx_1 \wedge dx_2 \wedge dx_3 \quad (40.6)$$
$$\beta \wedge \gamma = (\mathbf{E.D} - \mathbf{H.B}) \, dt \wedge dx_1 \wedge dx_2 \wedge dx_3.$$

Of course, the compensated compactness theorem 38.1 says that the situation encountered for Maxwell's equation is general.

Lemma 40.1. *If a sequence of p-forms a^n converges weakly to a^∞ and a sequence of q-forms b^n converges weakly to b^∞ (in the sense that the coefficients of a^n and b^n converge in $L^2_{loc}(\Omega)$ weak) and if $d a^n$ and $d b^n$ have their coefficients staying in compact sets of $H^{-1}_{loc}(\Omega)$ (strong), then $a^n \wedge b^n$ converges weakly to $a^\infty \wedge b^\infty$ (i.e. the coefficients converge in the sense of measures).*

---

[1] Maxwell's equation is about electromagnetism and in particular about electromagnetic waves, which include *light*, but it is not about *matter*, whose density of charge appears as ϱ and density of current appears as j, which are data for Maxwell's equation. How "macroscopic charged particles" move in an electromagnetic field is not written in Maxwell's equation, and the Lorentz force must be added to Maxwell's equation.

Proof: In order to apply compensated compactness theorem 38.1, one must compute what the characteristic set Λ is. If for $\xi \neq 0$ one defines the (alternated) linear form $\eta = \sum_{i=1}^{3} \xi_i \, dx_i$, then one sees easily that $\Lambda = \{a, b \mid a$ is an alternated p-linear form with $a \wedge \eta = 0$, and b is an alternated q-linear form with $b \wedge \eta = 0\}$. It is an elementary result that $a \wedge \eta = 0$ if and only if $a = \eta \wedge c$ for some alternated $(p-1)$-linear form, and as $b \wedge \eta = 0$ if and only if $b = \eta \wedge d$ for some alternated $(q-1)$-linear form, one deduces that $\eta \wedge \eta = 0$ implies $a \wedge b = 0$ on Λ.■

In order to reiterate the result, it is useful to notice that $d(a^n \wedge b^n) = (d\,a^n) \wedge b^n + (-1)^p a^n \wedge (d\,b^n)$. It is also useful to notice that the preceding result holds if the coefficients of a^n converge in $L^r_{loc}(\Omega)$ weak with the coefficients of $d\,a^n$ staying in a compact of $W^{-1,r}_{loc}(\Omega)$ (strong), if the convergence of b^n converges in $L^s_{loc}(\Omega)$ weak with the coefficients of $d\,b^n$ staying in a compact of $W^{-1,s}_{loc}(\Omega)$ (strong), with $1 < r, s < \infty$ and $\frac{1}{r} + \frac{1}{s} \leq 1$. François MURAT proved this improvement of our div-curl lemma 33.1 by using the Mikhlin–Hörmander theorem on Fourier multipliers (which requires a constant rank hypothesis, satisfied here), but one could use a proof involving the Calderón–Zygmund theorem instead, but the general case can also be proven along the lines proposed earlier by Joel ROBBIN, using Hodge's theory.

One finds as a particular case the result of MORREY and RESHETNYAK that Jacobian determinants are sequentially weakly continuous: if $a^n_j = d\,u^n_j$ converges to $a^\infty_j = d\,u^\infty_j$ for $j = 1, \ldots, k$ (with obvious constraints on the values of $p_j, j = 1, \ldots, k$, if the convergence of the coefficients of a^n_j holds in $L^{p_j}_{loc}(\Omega)$ weak), then $a^n_1 \wedge \ldots \wedge a^n_k$ converges weakly to $a^\infty_1 \wedge \ldots \wedge a^\infty_k$. This result is more easy to prove because the 1-forms used are exact, and a simple proof can be obtained by integration by parts, as the formula $d(a \wedge b) = (d\,a) \wedge b - a \wedge (d\,b)$ means here

$$\frac{\partial u}{\partial x_j}\frac{\partial v}{\partial x_k} - \frac{\partial u}{\partial x_k}\frac{\partial v}{\partial x_j} = \frac{\partial}{\partial x_j}\left(u\frac{\partial v}{\partial x_k}\right) - \frac{\partial}{\partial x_k}\left(u\frac{\partial v}{\partial x_j}\right). \qquad (40.7)$$

Using the Sobolev embedding theorem and formulas analogous to (40.7), one can deduce that Jacobian determinants of size k are not only defined for functions in $L^k_{loc}(\Omega)$ but also for $L^p_{loc}(\Omega)$ with $p \geq \frac{kN}{N+1}$, and the sequential weak continuity holds if $p > \frac{kN}{N+1}$. One can of course take the functions in various spaces $W^{1,p_j}_{loc}(\Omega)$, with a corresponding relation for the exponents p_j, but another improvement is sometimes useful and involves Hardy spaces, an idea[2,3] due to Pierre-Louis LIONS, with a basic theorem obtained with Raphaël COIFMAN, Yves MEYER and Stephen SEMMES.

---

[2] Motivated by a result of compensated integrability of Stefan MÜLLER.

[3] Stefan MÜLLER, German mathematician. He works at the Max PLANCK Institute for Mathematics in the Sciences, Leipzig, Germany.

In the div-curl lemma 33.1, one cannot replace the weak convergence in the sense of measures by a weak convergence in $L^1_{loc}(\Omega)$ (for $N \geq 2$, of course).

Lemma 40.2. *Let Ω be a smooth bounded open set of R^N, and let ω be a nonempty open set whose closure is contained in Ω. Then there exists sequences E^n and D^n converging weakly to 0 in $L^2(\Omega; R^N)$, with $curl\, E^n = 0$ and $div\, D^n = 0$, but with $\lim_{n\to\infty} \int_\omega (\mathbf{E^n.D^n})\, dx > 0$.*

Proof: The compensated compactness theorem 38.1 says that for $\varphi \in C_c(\Omega)$ one has $\int_\Omega \varphi(\mathbf{E^n.D^n})\, dx \to 0$, so the counter-example that I construct shows that one does not have the same result if $\varphi = \chi_\omega$, the characteristic function of ω. One then chooses a sequence f_n converging weakly to 0 in $H^{1/2}(\partial\Omega)$ but not strongly (it is here that the hypothesis $N \geq 2$ is used, so that $H^{1/2}(\partial\Omega)$ is indeed an infinite-dimensional Hilbert space); one solves $-\Delta u_n = 0$ in ω with the trace of u_n on $\partial\omega$ being f_n, and the sequence u_n then converges weakly to 0 in $H^1(\omega)$ but not strongly. One uses a linear continuous extension P from $H^1(\omega)$ into $H^1_0(\Omega)$ and one takes $E^n = grad(P\, u_n)$. One also solves the equation $-\Delta v_n = 0$ in $\Omega \setminus \overline{\omega}$ with $v_n = 0$ on $\partial\Omega$ and $\frac{\partial v_n}{\partial\nu} = \frac{\partial u_n}{\partial\nu}$ on $\partial\omega$, where ν is the normal to $\partial\omega$, which in variational formulation means $\int_{\Omega\setminus\overline{\omega}}(grad\, v_n.grad\, w)\, dx = -\int_\omega (grad\, u_n.grad\, w)\, dx$ for every $w \in H^1_0(\Omega)$, and one takes $D^n = grad\, u_n$ in ω and $D^n = grad\, v_n$ in $\Omega\setminus\omega$. One concludes by noticing that $(\mathbf{E^n.D^n}) = |grad(u_n)|^2$ and that u_n does not converge strongly to 0 in $H^1(\omega)$. ∎

The case of Jacobian determinants gives an example where there are some polynomials of degree more than 2 which are sequentially weakly continuous (and $\Lambda \neq 0$, of course). We have seen that a necessary condition that a (continuous) real function F on R^p be such that $F(U^n) \rightharpoonup F(U^\infty)$ in $L^\infty(\Omega)$ weak $\star$ for all sequences U^n converging to U^∞ in $L^\infty(\Omega; R^p)$ weak $\star$ and satisfying the equations $\sum_{j=1}^p \sum_{k=1}^N A_{ijk}\frac{\partial U^n_j}{\partial x_k} = 0$ for $i = 1, \ldots, q$, is that F must be affine in all directions of Λ, but there are in general *other necessary conditions*. In the case where Λ spans R^p, the preceding necessary condition implies that F is a combination of multilinear forms, of degree at most p, while the new conditions will imply that the degree is at most N. The basic idea can be shown on the following example, which I encountered while studying oscillating sequences of solutions of discrete kinetic velocity models like the Broadwell model. Let $N = 2, p = 3$ and the list of equations be $\frac{\partial U^n_1}{\partial x_1} = 0, \frac{\partial U^n_2}{\partial x_2} = 0, \frac{\partial U^n_3}{\partial x_1} + \frac{\partial U^n_3}{\partial x_2} = 0$, so that the characteristic set Λ is defined by $\xi_1 U_1 = \xi_2 U_2 = (\xi_1 + \xi_2)U_3 = 0$, and as ξ must be different from 0, one finds the three axes. The only candidates for sequential weak continuity are then $U_1 U_2, U_1 U_3, U_2 U_3$, and $U_1 U_2 U_3$; the first three are quadratic and therefore they are sequentially weakly continuous, but the fourth one is not, and this is seen by taking the sequence $U^n_1(x) = \cos(n\, x_2), U^n_2(x) = \cos(n\, x_1), U^n_3(x) = \cos(n(x_1 - x_2))$, so that U^n converges weakly to 0 but $U^n_1 U^n_2 U^n_3(x) = \cos^2(n\, x_1)\cos^2(n\, x_2) - \sin(2n\, x_1)\sin(2n\, x_2)/4$, and therefore $U^n_1 U^n_2 U^n_3$ converges weakly to $1/4$.

In the general case, the new necessary conditions use the characteristic set $\mathcal{V} = \{(\lambda,\boldsymbol{\xi}) \in R^p \times (R^N \setminus 0) \mid \sum_{j=1}^{p} \sum_{k=1}^{N} A_{ijk}\lambda_j\xi_k = 0 \text{ for } i = 1,\ldots,q\}$, and not only its projection Λ; taking combinations of m functions of the form $\lambda f(n\boldsymbol{\xi}.\mathbf{x})$ with $(\lambda,\boldsymbol{\xi}) \in \mathcal{V}$, one finds that if $(\lambda^1,\boldsymbol{\xi}^1),\ldots,(\lambda^m,\boldsymbol{\xi}^m) \in \mathcal{V}$ with $rank(\boldsymbol{\xi}^1,\ldots,\boldsymbol{\xi}^m) \leq m-1$, then $F^{(m)}(a)(\lambda^1,\ldots,\lambda^m) = 0$ for all $a \in R^p$ (the case $m = 2$ corresponds to the previously used necessary condition).

Coming back to Maxwell's equation, there are other sequentially weakly continuous quantities if one uses the 1-form $\alpha = V\,dt + \sum_{i=1}^{3} A_i\,dx_1$, as for example, $\alpha \wedge \beta = -(\mathbf{A}.\mathbf{B})\,dx_1 \wedge dx_2 \wedge dx_3 - C_1\,dt \wedge dx_2 \wedge dx_3 - C_2\,dt \wedge dx_3 \wedge dx_1 - C_3\,dt \wedge dx_1 \wedge dx_2$, where $\mathbf{C} = V\,\mathbf{B} + \mathbf{A} \times \mathbf{E}$. In the case of fluids with viscosity $\nu = 0$, $\mathbf{A}$ is replaced by $\mathbf{u} - \mathbf{u}^*$, $\mathbf{B}$ by $curl(-\mathbf{u}) + 2\Omega$, and $\mathbf{A}.\mathbf{B}$ by $(\mathbf{u} - \mathbf{u}^*).(curl(-\mathbf{u}) + 2\Omega)$, i.e. the helicity in the case where the Coriolis force is neglected. This shows then that helicity is a robust quantity, not too sensitive to oscillations, and therefore useful even in turbulent flows.

[Taught on Monday April 26, 1999.]

41

The compensated compactness method

It is not always easy to characterize the quadratic forms which are nonnegative on a characteristic cone Λ, but in the case where $\Lambda = \{(\mathbf{E}, \mathbf{D}) \in R^N \times R^N \mid \mathbf{E}.\mathbf{D} = 0\}$, $Q \geq 0$ on Λ is equivalent to the existence of $c \in R$ such that $Q(\mathbf{E}, \mathbf{D}) + c\,\mathbf{E}.\mathbf{D} \geq 0$ for all $(\mathbf{E}, \mathbf{D}) \in R^N \times R^N$; after I had mentioned this result to Joel ROBBIN (in 1974/75), he pointed out to me the natural generalization of this result.

Lemma 41.1. *(Joel ROBBIN) If Q_0 is a nondefinite nondegenerate quadratic form on R^p and Q is a quadratic form on R^p such that $Q_0(\boldsymbol{\lambda}) = 0$ implies $Q(\boldsymbol{\lambda}) \geq 0$, then there exists $c \in R$ such that $Q + cQ_0$ is nonnegative everywhere, i.e. convex.*

Proof: One first defines α to be the minimum of $Q(\boldsymbol{\lambda})$ for all $\boldsymbol{\lambda} \in R^p$ satisfying $Q_0(\boldsymbol{\lambda}) = 0$ and $|\boldsymbol{\lambda}| = 1$ (and α exists because Q_0 is assumed to be nondefinite, so the set is not empty, and compact); as $Q(\boldsymbol{\lambda}) - \alpha\,|\boldsymbol{\lambda}|^2$ satisfies the hypothesis, one may assume that $\alpha = 0$.

One chooses a Euclidean structure, and a unit vector $\mathbf{e_1}$ such that $Q_0(\mathbf{e_1}) = 0$ and $Q(\mathbf{e_1}) = 0$. Because Q_0 is nondegenerate, one has $Q_0'(\mathbf{e_1}) \neq 0$, so $Q_0'(\mathbf{e_1}) = 2\beta\,\mathbf{e_2}$ for a unit vector $\mathbf{e_2}$ and $\beta \neq 0$, and $\mathbf{e_2}$ must be orthogonal to $\mathbf{e_1}$ because $2\beta(\mathbf{e_2}.\mathbf{e_1}) = (Q_0'(\mathbf{e_1}).\mathbf{e_1}) = 2Q_0(\mathbf{e_1}) = 0$. One completes $\mathbf{e_1}, \mathbf{e_2}$ in order to obtain an orthonormal basis.

Q_0 and Q have no term in x_1^2, and they have the form $Q_0(\mathbf{x}) = 2\beta\,x_1 x_2 + R_0(x_2, \ldots, x_p)$ with R_0 quadratic, $Q(\mathbf{x}) = 2x_1 L(x_2, \ldots, x_p) + R(x_2, \ldots, x_p)$ with L linear and R quadratic. For every $(y_2, \ldots, y_p)$ with $y_2 \neq 0$, one chooses $y_1 = -\frac{R_0(y_2, \ldots, y_p)}{2\beta\,y_2}$ so that $Q_0(\mathbf{y}) = 0$ and therefore $Q(\mathbf{y}) \geq 0$, i.e. $R(y_2, \ldots, y_p) - \frac{R_0(y_2, \ldots, y_p)L(y_2, \ldots, y_p)}{\beta\,y_2} \geq 0$, so by letting y_2 tend to 0 one must have $R_0(0, y_3, \ldots, y_p)L(0, y_3, \ldots, y_p) = 0$, for all $y_3, \ldots, y_p$. One must have have either $R_0(0, y_3, \ldots, y_p) = 0$, for all $y_3, \ldots, y_p$ or $L(0, y_3, \ldots, y_p) = 0$, for all $y_3, \ldots, y_p$ because the ring of polynomials in $m \geq 0$ variables is an integral domain (as the ring of polynomials with coefficients in an integral domain is itself an integral domain).

If one has $L(0, y_3, \ldots, y_p) = 0$ for all $y_3, \ldots, y_p$, there exists $\gamma \in R$ such that $L(y_2, y_3, \ldots, y_p) = \gamma y_2$ for all $y_2, \ldots, y_p$, and if one has $R_0(0, y_3, \ldots, y_p) = 0$ for all $y_3, \ldots, y_p$, then $R_0(y_2, \ldots, y_p)$ is divisible by y_2, but as Q_0 is nondegenerate, it can only happen if $p = 2$, in which case L can only be of the form γy_2). This shows that $R(y_2, \ldots, y_p) - \gamma R_0(y_2, \ldots, y_p)/\beta \geq 0$ when $y_2 \neq 0$ and by continuity for all $y_2, \ldots, y_p$, and therefore $Q - \gamma Q_0/\beta \geq 0$.■

I think that Denis SERRE[1] proved something similar, a few years afterward. As far as I know, no one has found in the general case a simple characterization of the set of quadratic forms which are nonnegative on Λ, or at least the extreme rays of this convex cone; as I shall explain later, one must introduce a list of entropies before doing that.

It is worth looking at simple problems in order to explain what the ideas of the compensated compactness method are. In N-dimensional elasticity (with $N = 2$ or 3 for applications), one must deal with a strain tensor $\mathsf{F} = \nabla u$ (in the evolution case, the derivatives with respect to t, i.e. the velocities, must be added to the list), and the Piola–Kirchhoff stress tensor σ. The constitutive relations relate the stress σ to F; in the hyperelastic case one often prefers to deal with an energy functional. The list of differential information consists of the compatibility conditions for gradients, and Newton's law of mechanics, the equilibrium equation in the stationary case. The characteristic set $\mathcal{V}$ is the set of $\mathsf{F}, \sigma, \boldsymbol{\xi}$, with $\boldsymbol{\xi} \in R^N \setminus 0$, $\mathsf{F} = \mathbf{a} \otimes \boldsymbol{\xi}$ for some $\mathbf{a} \in R^N$ and $\sigma \boldsymbol{\xi} = 0$; the characteristic set Λ is then the set of F, σ, such that the rank of F is less than or equal than 1, σ is singular, and $\sigma \mathsf{F}^T = 0$; the list of quantities which are sequentially weakly continuous includes the subdeterminants extracted from F, but also all the components of $\sigma \mathsf{F}^T$, and if $N = 2$ the determinant of σ. In the hyperelastic case, i.e. if there exists a real function W such that $\sigma_{ij} = \frac{\partial W}{\partial F_{ij}}$ for all $i, j = 1, \ldots, N$, one can derive new equations for smooth solutions, and these go under the general term of "entropies", chosen by Peter LAX: in the stationary case, without exterior forces, one has

$$\sum_{j=1}^{3} \frac{\partial(\sigma_{ij} F_{ik})}{\partial x_j} = F_{ik} \sum_{j=1}^{3} \frac{\partial \sigma_{ij}}{\partial x_j} + \sum_{j=1}^{3} \frac{\partial W}{\partial F_{ij}} \frac{\partial^2 u_i}{\partial x_j \partial x_k} = \frac{\partial(W(\mathsf{F}))}{\partial x_k} \quad \text{for all } i, k.$$

(41.1)

One can reiterate the application of the quadratic theorem after introducing as new components for $\mathbf{U}$ the quantities which appear in the new conserved quantities; I present the idea in the next example, dealing with solutions of the wave equation.

It is an important effect in fluids that waves can transport momentum and energy without transporting mass; one can learn something important about

---

[1] Denis SERRE, French mathematician, born in 1954. He works at École Normale Supérieure, Lyon, France.

this question by looking at weakly converging sequences of solutions of linear wave equations (there is not so much proven at the moment for semi-linear or quasi-linear cases). For simplicity, let us consider a scalar wave equation with constant coefficients $\varrho_0, a > 0$,

$$\varrho_0 \frac{\partial^2 u}{\partial t^2} - a\,\Delta u = f \text{ in } R^N \times (0,T); \quad u(\cdot,0) = v \text{ in } R^N; \quad \frac{\partial u}{\partial t}(\cdot,0) = w \text{ in } R^N,$$

(41.2)

which can arise in various ways, often with $f = 0$, in general after having linearized a nonlinear system near a trivial solution; for example, u can be a vertical displacement, or a variation in pressure. Using methods of functional analysis, similar to the ones we used for the abstract framework for Stokes's equation, for example, one can show that if $v \in H^1(R^N)$, $w \in L^2(R^N)$, and $f \in L^1(0,T;L^2(R^N))$, then there is a unique solution $u \in C^0([0,T];H^1(R^N)) \bigcap C^1([0,T];L^2(R^N))$, and a very important property of the preceding equation, apart from describing an isotropic medium where information travels at velocity $\sqrt{\frac{a}{\varrho_0}}$, is the *balance of energy* (conservation if $f = 0$)

$$\frac{\partial}{\partial t}\left(\frac{\varrho_0}{2}\left|\frac{\partial u}{\partial t}\right|^2 + \frac{a}{2}\sum_{j=1}^{3}\left|\frac{\partial u}{\partial x_j}\right|^2\right) - a\sum_{j=1}^{3}\frac{\partial}{\partial x_j}\left(\frac{\partial u}{\partial x_j}\frac{\partial u}{\partial t}\right) = f\frac{\partial u}{\partial t} \text{ in } R^N \times (0,T).$$

(41.3)

The density of energy $\frac{\varrho_0}{2}\left|\frac{\partial u}{\partial t}\right|^2 + \frac{a}{2}\sum_{j=1}^{3}\left|\frac{\partial u}{\partial x_j}\right|^2$ is the sum of the density of *kinetic energy* and of the density of *potential energy* (related to the elastic properties or to the compressibility properties of the medium, according to the interpretation given to u); if $f = 0$, the integral of the density of energy is constant, equal to the total energy of the entire medium, and therefore equal to its value at time 0, $\int_{\mathbf{x} \in R^N}\left(\frac{\varrho_0}{2}|w|^2 + \frac{a}{2}|grad\,v|^2\right) d\mathbf{x}$.

If one considers a sequence v_n converging to v_∞ in $H^1(R^N)$ weak, and w_n converging to w_∞ in $L^2(R^N)$ weak, and one of these sequences does not converge strongly, the limit of $\int_{\mathbf{x} \in R^N}\left(\frac{\varrho_0}{2}|(u_n)_t|^2 + \frac{a}{2}|grad\,u_n|^2\right) d\mathbf{x}$ will be strictly greater than $\int_{\mathbf{x} \in R^N}\left(\frac{\varrho_0}{2}|(u_\infty)_t|^2 + \frac{a}{2}|grad\,u_\infty|^2\right) d\mathbf{x}$, and it means than part of the initial energy is stored at high frequencies, and it is important to know where this missing energy goes.

If one observes the ocean from a plane, the surface of the ocean looks flat, and it is similar to having $v_\infty = w_\infty = 0$, but there is some energy and momentum moving around, near the surface.

The compensated compactness theorem 38.1 gives interesting but incomplete information about this energy traveling around at high frequencies, *equipartition of energy*: it is the *action* $\frac{\varrho_0}{2}\left|\frac{\partial u}{\partial t}\right|^2 - \frac{a}{2}\sum_{j=1}^{3}\left|\frac{\partial u}{\partial x_j}\right|^2$ which is sequentially weakly continuous. This result does not reveal what the density of energy transported by the high frequencies is, but it shows that half of it is in kinetic form and half of it is in potential form. The proof is just an application of the div-curl lemma 33.1, where E^n is the full gradient of u_n, i.e. in all the variables $(t, \mathbf{x})$, and D^n is $\left(\varrho_0 \frac{\partial u_n}{\partial t}, -a \frac{\partial u_n}{\partial x_1}, \ldots, -a \frac{\partial u_n}{\partial x_N}\right)$, whose divergence is f.

One can describe where the energy goes with a mathematical tool which I introduced in the late 1980s, H-measures (I introduced H-measures for a different purpose, and then I proved a general *propagation theorem for H-measures*; the same objects were also introduced independently by Patrick GÉRARD for still another purpose (compactness by averaging); taking into account the problem of initial data for the wave equation was done by Gilles FRANCFORT[2] and François MURAT, with the technical help of Patrick GÉRARD).

If the gradient of u_n in all the variables $(t, \mathbf{x})$ stays bounded in L^3_{loc}, one can use the div-curl lemma 33.1 with D^n being replaced by $\left(\frac{\varrho_0}{2}\left|\frac{\partial u_n}{\partial t}\right|^2 + \frac{a}{2}\sum_{j=1}^3\left|\frac{\partial u_n}{\partial x_j}\right|^2, -a\frac{\partial u_n}{\partial x_1}\frac{\partial u_n}{\partial t}, \ldots, -a\frac{\partial u_n}{\partial x_N}\frac{\partial u_n}{\partial t}\right)$, and that case gives $(\mathbf{E^n.D^n}) = \frac{\partial u_n}{\partial t}\left(\frac{\varrho_0}{2}\left|\frac{\partial u_n}{\partial t}\right|^2 - \frac{a}{2}\sum_{j=1}^3\left|\frac{\partial u_n}{\partial x_j}\right|^2\right)$, i.e. $\frac{\partial u_n}{\partial t}$ $action_n$; one deduces that if $u_\infty = 0$, then not only $action_n$ converges weakly to 0, but also $\frac{\partial u_n}{\partial t}$ $action_n$ converges weakly to 0 (and actually $\frac{\partial u_n}{\partial x_j}$ $action_n$ also converges weakly to 0 for every j, as a consequence of other balance equations, expressing conservation of linear momentum).

One should not conclude that one has found a new sequentially weakly continuous function, because if $u_\infty \neq 0$, then E^∞ is the full gradient of u_∞, but D^∞ is not necessarily the corresponding quantity associated with u_∞, because the components of D^n are quadratic quantities which are not from the list of sequentially weakly continuous functions (which only contains the action); actually the functions $\frac{\partial u}{\partial t}$ $action$ (as well as $\frac{\partial u}{\partial x_j}$ $action$) are such that their weak limits can be computed from the full gradient of u_∞ together with the H-measure associated with the full gradient of $u_n - u_\infty$.

I have almost described the idea of the compensated compactness method: one looks for "entropies", the consequence of the differential equations and eventually also of the nonlinear constitutive relations, and one applies the compensated compactness theorem 38.1 to that extended system; an obvious difficulty is that one may find an enormous system, and one may have to make a choice of which "entropies" to use, but there is another difficulty that can be described in the example of *Burgers's equation* (9.3).

One considers a weakly converging sequence u_n of solutions of the equation $\frac{\partial u_n}{\partial t} + u_n \frac{\partial u_n}{\partial x} = 0$, written in conservative form

$$\frac{\partial u_n}{\partial t} + \frac{1}{2}\frac{\partial (u_n)^2}{\partial x} = 0. \tag{41.4}$$

If one has (after eventually extracting a subsequence)

$$(u_n)^k \rightharpoonup U_k \text{ in } L^\infty_{loc} \text{ weak } \star \text{ for } k \geq 1, \tag{41.5}$$

[2] Gilles FRANCFORT, French mathematician, born in 1957. He works at Université Paris XIII (Paris-Nord), Villetaneuse, France.

one deduces that

$$\frac{\partial U_1}{\partial t} + \frac{1}{2}\frac{\partial U_2}{\partial x} = 0,\tag{41.6}$$

but the compensated compactness theorem 38.1 does not help, apart from $U_2 \geq U_1^2$ which one deduces from a convexity argument (by taking the weak $\star$ limit of $(u_n - U_1)^2 \geq 0$). If one now assumes that u_n is smooth enough so that one can multiply the equation by u_n and writes the result in conservation form

$$\frac{1}{2}\frac{\partial (u_n)^2}{\partial t} + \frac{1}{3}\frac{\partial (u_n)^3}{\partial x} = 0,\tag{41.7}$$

then one can use the div-curl lemma 33.1 and one deduces that

$$u_n\frac{(u_n)^3}{3} - \frac{(u_n)^2}{2}\frac{(u_n)^2}{2} \rightharpoonup U_1\frac{U_3}{3} - \frac{U_2}{2}\frac{U_2}{2} \text{ in } L_{loc}^\infty \text{ weak } \star,\tag{41.8}$$

and therefore

$$U_4 = 4U_1U_3 - 3U_2^2.\tag{41.9}$$

Because $(u_n - U_1)^4 \geq 0$ gives at the limit

$$U_4 - 4U_1U_3 + 6U_1^2U_2 - 3U_1^4 \geq 0,\tag{41.10}$$

one deduces $-3(U_2 - U_1^2)^2 \geq 0$, and therefore $U_2 = U_1^2$, which implies that $u_n \to U_1$ in L_{loc}^2 strong (and therefore in L_{loc}^p strong for every $p < \infty$).

One sees that the use of "entropies" together with the compensated compactness theorem 38.1 has created some kind of ellipticity for an enlarged system, which forbids oscillations, and that it is the main idea of the compensated compactness method; however, one must complement it with a technical remark, because the sequence u_n is not usually smooth enough: if one starts with oscillations in the initial data, the derivative in x at time 0 must be large and negative somewhere and shocks will therefore appear after a very short time. Actually, as a consequence of an explicit formula valid for $\frac{\partial u}{\partial t} + \frac{\partial [f(u)]}{\partial x} = 0$ with f convex (used by Peter LAX, but Sergei GODUNOV[3] also claims priority), or as a consequence of my argument (valid for more general f), these shocks will interact and decay rapidly; this occurs in such a short time that for every $\tau > 0$ the sequences converges strongly in $t \geq \tau$. To do this, one must restrict attention to physically realistic sequences of approximating solutions, those which satisfy an *entropy condition* like

$$\frac{1}{2}\frac{\partial (u_n)^2}{\partial t} + \frac{1}{3}\frac{\partial (u_n)^3}{\partial x} \leq 0 \text{ in the sense of measures}\tag{41.11}$$

(as noticed by Eberhard HOPF and KRUZHKOV in the scalar case, and extended to systems of conservation laws by Peter LAX), and although measures are not necessarily in H_{loc}^{-1}, here the ones which appear do stay in a compact

---

[3] Sergei Konstantinovich GODUNOV, Russian mathematician, born in 1929. He works at the SOBOLEV Institute of Mathematics, Novosibirsk, Russia.

of H_{loc}^{-1} strong, as a consequence of a result that François MURAT had proven for another purpose: if a sequence is bounded in $W_{loc}^{-1,p}(\Omega)$ and stays bounded in the space of measures, then it stays in a compact of $W_{loc}^{-1,q}(\Omega)$ strong if $1 \leq q < p$.

Of course, it remains to apply these ideas to realistic questions of fluids. I had asked my student Luisa MASCARENHAS to work on the case where one considers the gradient of a divergence free velocity field, in the case where the potential energy is a function of $F + F^T$, and she obtained some results in this direction in the early 1980s; I had in mind to improve some earlier results by Olga LADYZHENSKAYA and by Shmuel KANIEL,[4] but Dan JOSEPH then mentioned to me a defect of this kind of model, and I therefore did not pursue that direction. I was not aware at the time of models coming from oceanography, and there are obviously some potential applications of the methods which I had developed in the late 1970s for these models, but instead of describing models I have made the choice of describing the mathematical tools available for analyzing these models, and it remains to describe quickly what H-measures are, as it would be quite unrealistic to ignore the question of propagation of momentum or energy by waves (and one should not postulate too much about these waves).

[Taught on Wednesday April 28, 1999.]

---

[4] Shmuel KANIEL, Israeli mathematician. He works at The Hebrew University, Jerusalem, Israel.

42

H-measures and variants

In the late 1970s, I was trying to improve Young measures by adding a direction variable ξ in order to prove propagation results.[1] I had mentioned the question to George PAPANICOLAOU[2] and he had suggested to me the *Wigner*[3] *transform*, but I could not find a way to use it (later Pierre-Louis LIONS and Thierry PAUL[4] used it in order to give a different definition of the *semiclassical measures* introduced by Patrick GÉRARD, but these objects use one *characteristic length* and my vision of the physical world being one where there are plenty of different scales, this was not what I was looking for).

I tried to use the limit of functions $F\left(\mathbf{x}, u_n, \frac{grad\, u_n}{|grad\, u_n|}\right)$, but if u_n is not smooth and one approaches it by smooth functions, the limit depends upon the approximating sequence, and I did not pursue that idea.

I thought of using homogenization for various elliptic systems with coefficients being general functions of u_n, because I consider that homogenization as I have developed it with François MURAT is only a small piece of what it should actually be, which is a *nonlinear microlocal theory*, with maybe a connection with a question of mathematics of which I only have a quite vague idea, the theory of sheaves as it developed after the original ideas of Jean

---

[1] Lars HÖRMANDER had proven results of "propagation of singularities" where bicharacteristic rays play a role, but I knew that singularities are not the important objects to consider in continuum mechanics / physics, and that the propagation of oscillations was the key question, and I wanted to know if they were following the same rules. It was only after I had proven results of propagation for my H-measures that Mike CRANDALL pointed out that what Lars HÖRMANDER was really following along bicharacteristic rays is *microlocal regularity*.

[2] George C. PAPANICOLAOU, Greek-born mathematician, born in 1943. He works at STANFORD University, Stanford, CA.

[3] Jenõ Pál (Eugene Paul) WIGNER, Hungarian-born physicist, 1902–1995. He received the Nobel Prize in Physics in 1963. He worked at Princeton University, Princeton, NJ.

[4] Thierry PAUL, French mathematician. He works at École Normale Supérieure, Paris, France.

LERAY. I could not find a simple way of using that idea, until I understood that a simpler problem was to mix materials with similar properties (what I later called small-amplitude homogenization).

The first hint occurred in 1984, when I discovered that in the problem modeled on the stationary Stokes equation with a force field in $\mathbf{u} \times curl(\mathbf{v}_0 + \lambda \mathbf{v}_n)$ that I have described in Lecture 37, the correcting matrix M had a factor λ^2, and that I could compute the correction from the behavior at infinity of quadratic quantities in the Fourier transform.

The second hint occurred in 1986, when I heard a talk by Stephen COWIN[5] on bone evolution, and it seemed to me that the tensors that he was computing using methods from stereometry should be replaced by second or fourth derivatives of functions defined in a similar way to the ones that David BERGMAN had been using in an isotropic setting (again small-amplitude homogenization was behind this). The third hint occurred in the fall of 1986, when I tried to check what LANDAU[6] and LIFSCHITZ[7] had written about the conductivity of mixtures, which I had been aware of twelve years before (and at that time I had dismissed their computations as nonsense). I realized that their formula was just one of the bounds of Zvi HASHIN and SHTRIKMAN, for which I had given the first mathematical proof in 1980 by using a method based on correctors together with the compensated compactness theorem 38.1, which is now often called the translation method (with rarely a mention that I had introduced it). Actually, there is something like H-measures hidden in the formal argument of Zvi HASHIN and SHTRIKMAN to show that the bounds must hold, while their argument that these bounds are attained required little change, and using that part of their argument was clear to me, but as they proposed to apply it to mixtures where the variations in conductivities are small, I checked their result with Gilles FRANCFORT and François MURAT (again using the compensated compactness theorem 38.1), and their formula appeared to be accurate. It was quite a miracle if one considers the gaps in their derivation, and I understood then what a framework for small-amplitude homogenization should be, and from the previous hints I could now guess easily how I was going to prove a mathematical version of their result, using these H-measures which I immediately knew how to use for that purpose, before I had a clear idea of how I was going to define them correctly.

For a scalar sequence u_n converging weakly to 0 in $L^2_{loc}(\Omega)$, and $\varphi \in C_c(\Omega)$, the Fourier transform of $\varphi \, u_n$ is bounded in $C_0(R^N)$ and converges pointwise to 0, and therefore $\mathcal{F}(\varphi \, u_n)$ converges to 0 in $L^2_{loc}(R^N)$ strong by the Lebesgue dominated convergence theorem; from the hints, I knew that I needed to describe how the information contained in $|\mathcal{F}(\varphi \, u_n)|^2$ was going to infinity,

---

[5] Stephen C. COWIN, American mathematician. He works at City College, New York, NY.

[6] Lev Davidovich LANDAU, Russian physicist, 1908–1968. He received the Nobel Prize in Physics in 1962. He worked in Moscow, Russia.

[7] Evgenii Mikhailovich LIFSCHITZ, Russian physicist, 1915–1985. He worked in Moscow, Russia.

and therefore for a continuous function ψ defined on the sphere S^{N-1}, I looked at the limit

$$L(\varphi, \psi) = \lim_{n \to \infty} \int_{\xi \in R^N} |\mathcal{F}(\varphi u_n)(\xi)|^2 \psi\left(\frac{\xi}{|\xi|}\right) d\xi. \qquad (42.1)$$

Of course, using separability arguments and a diagonal procedure, one can extract a subsequence such that the preceding limit exists for every $\varphi \in C_c(\Omega)$ and every $\psi \in C(S^{N-1})$, and for φ given there exists a nonnegative Radon measure μ_φ on S^{N-1} such that the limit is $\langle \mu_\varphi, \psi \rangle$, but although the dependence with respect to φ is not straightforward, I was sure from what I knew about the local character of homogenization and the hints, that the limit was given by a nonnegative Radon measure μ on $\Omega \times S^{N-1}$,

$$L(\varphi, \psi) = \langle \mu, |\varphi|^2 \otimes \psi \rangle, \text{ written formally as } \int_{\Omega \times S^{N-1}} |\varphi(\mathbf{x})|^2 \psi(\xi) \, d\mu(\mathbf{x}, \xi).$$
$$(42.2)$$

Of course, the hints had also told me that these measures in $(\mathbf{x}, \xi)$ could handle partial differential equations, and in order to compare with the compensated compactness theorem 38.1, there was an obvious generalization for the case of a sequence U^n converging weakly to 0 in $L^2_{loc}(\Omega; R^p)$; for two indices j and k, I found it more natural to take two different test functions in $\mathbf{x}$, $\varphi_1, \varphi_2 \in C_c(\Omega)$, and to consider the limit

$$L_{jk}(\varphi_1, \varphi_2, \psi) = \lim_{n \to \infty} \int_{\xi \in R^N} \mathcal{F}(\varphi_1 U_j^n)(\xi) \overline{\mathcal{F}(\varphi_2 U_k^n)(\xi)} \psi\left(\frac{\xi}{|\xi|}\right) d\xi, \qquad (42.3)$$

which I expected to be given by a Radon measure μ_{jk} on $\Omega \times S^{N-1}$,

$$L_{jk}(\varphi_1, \varphi_2, \psi) = \langle \mu_{jk}, \varphi_1 \overline{\varphi}_2 \otimes \psi \rangle, \text{ written formally as} \atop \int_{\Omega \times S^{N-1}} \varphi_1(\mathbf{x}) \overline{\varphi_2(\mathbf{x})} \psi(\xi) \, d\mu_{jk}(\mathbf{x}, \xi). \qquad (42.4)$$

As L_{jk} is linear in φ_1, antilinear in φ_2, and linear in ψ, it was reasonable to think of a Radon measure ν_{jk} on $\Omega \times \Omega \times S^{N-1}$ with $L_{jk}(\varphi_1, \varphi_2, \psi) = \int_{\Omega \times \Omega \times S^{N-1}} \varphi_1(\mathbf{x}) \overline{\varphi_2(\mathbf{y})} \psi(\xi) \, d\nu_{jk}(\mathbf{x}, \mathbf{y}, \xi)$, and therefore it was important to show that the support of ν_{jk} was included in the diagonal $\{(\mathbf{x}, \mathbf{y}, \xi) \mid \mathbf{x} = \mathbf{y}\}$. However, a general linear continuous mapping from $C_c(\Omega)$ (with the sup norm) into $\mathcal{M}(S^{N-1})$, the space of Radon measures on S^{N-1}, is given by an operator with a distribution kernel, according to the *kernel theorem* of Laurent SCHWARTZ, and I expected this kernel to be a measure by a positivity argument: if $\varphi_2 = \varphi_1$ and $\psi \geq 0$, then $L_{jk}(\varphi_1, \varphi_2, \psi) \geq 0$. Jacques-Louis LIONS had told me that he had obtained with Lars GÅRDING a simple proof of Laurent SCHWARTZ's kernel theorem, so the crucial step was to show that $L_{jk}(\varphi_1, \varphi_2, \psi)$ only depends upon $\varphi_1 \overline{\varphi_2}$ and ψ. This was obtained by a *commutation lemma*, and some kind of pseudo-differential calculus, but because the classical theory of pseudo-differential operators of Joseph KOHN and Louis NIRENBERG requires smooth coefficients, and I wanted to be careful with the hypothesis of smoothness of coefficients, I developed the theory which I needed.

I assume that all functions are extended by 0 outside Ω, so that one works on R^N. With any $b \in L^\infty(R^N)$ one associates the operator M_b of multiplication by b, i.e.

$$(M_b v)(\mathbf{x}) = b(\mathbf{x})v(\mathbf{x}) \text{ a.e. } \mathbf{x} \in R^N; \tag{42.5}$$

M_b is linear continuous from $L^2(R^N)$ into itself, and its norm is the $L^\infty(R^N)$ norm of b. With any $a \in L^\infty(R^N)$ one associates the operator P_a defined by $\mathcal{F}P_a = M_a\mathcal{F}$, i.e.

$$\mathcal{F}P_a v(\boldsymbol{\xi}) = a(\boldsymbol{\xi})\mathcal{F}v(\boldsymbol{\xi}) \text{ a.e. } \boldsymbol{\xi} \in R^N; \tag{42.6}$$

P_a is linear continuous from $L^2(R^N)$ into itself, and its norm is the $L^\infty(R^N)$ norm of a. In the quantity which I had considered, I only used functions ψ defined on the sphere S^{N-1} and extended to $R^N \setminus 0$ as homogeneous functions of order 0, and I then wanted the limit of $\int_{\boldsymbol{\xi} \in R^N} \mathcal{F}(P_\psi M_{\varphi_1} U_j^n)\overline{\mathcal{F}M_{\varphi_2} U_k^n}\,d\boldsymbol{\xi}$, which by Plancherel's formula is $\int_{\boldsymbol{\xi} \in R^N} P_\psi M_{\varphi_1} U_j^n \overline{M_{\varphi_2} U_k^n}\,d\boldsymbol{\xi}$; as the limit of $\int_{\boldsymbol{\xi} \in R^N} M_{\varphi_1} P_\psi U_j^n \overline{M_{\varphi_2} U_k^n}\,d\boldsymbol{\xi}$ obviously depends only upon $\varphi_1 \overline{\varphi_2}$ and ψ, it remained to show that $P_\psi M_{\varphi_1} U_j^n - M_{\varphi_1} P_\psi U_j^n$ converges strongly to 0 in $L^2(R^N)$, and as U_j^n converges weakly to 0 in $L^2(R^N)$, this would be a consequence of the commutator $P_\psi M_{\varphi_1} - M_{\varphi_1} P_\psi$ being a compact operator from $L^2(R^N)$ into itself. It was not too difficult to prove a *first commutation lemma*

$$\begin{array}{l} \text{if } a \in C(S^{N-1}), b \in C_0(R^N) \text{ then the commutator} \\ P_a M_b - M_b P_a \text{ is a compact operator on } L^2(R^N), \end{array} \tag{42.7}$$

using the fact that *Hilbert–Schmidt*[8] *operators* (which have a kernel in $L^2(R^N \times R^N)$) are compact, and that *uniform limits of compact operators are compact*; as a consequence of a *commutation lemma* of COIFMAN, ROCHBERG[9] and WEISS, this result is actually true for $b \in VMO(R^N)$, and one can therefore extend my theory to use functions in $L^\infty \cap VMO$. The "pseudo-differential" operators of order 0 which I use have symbols of the form

$$\begin{array}{l} s(\mathbf{x}, \boldsymbol{\xi}) = \sum_k a_k(\boldsymbol{\xi})b_k(\mathbf{x}) \text{ with} \\ a_k \in C(S^{N-1}), b_k \in C_0(R^N) \text{ for all } k \text{ and } \sum_k \|a_k\|\,\|b_k\| < \infty, \end{array} \tag{42.8}$$

where the norms are sup norms; I define the *standard operator* S of symbol s by

$$S = \sum_k P_{a_k} M_{b_k}, \tag{42.9}$$

---

[8] Erhard SCHMIDT, German mathematician, 1876–1959. He worked in Berlin, Germany.

[9] Richard H. ROCHBERG, American mathematician, born in 1941. He works at WASHINGTON University, Saint Louis, MO.

which corresponds to

$$\mathcal{F}Sv(\boldsymbol{\xi}) = \int_{\mathbf{x} \in R^N} s\left(\mathbf{x}, \frac{\boldsymbol{\xi}}{|\boldsymbol{\xi}|}\right) u(x)e^{-2i\pi(\mathbf{x}.\boldsymbol{\xi})} \, dx \text{ a.e. } \boldsymbol{\xi} \in R^N \tag{42.10}$$
$$\text{for } v \in L^2(R^N) \cap L^1(R^N),$$

and I say that a linear continuous operator L from $L^2(R^N)$ into itself has symbol s if $L - S$ is a compact operator on $L^2(R^N)$; this is the case for the operator

$$L_0 = \sum_k M_{b_k} P_{a_k}, \tag{42.11}$$

which corresponds to

$$L_0v(\mathbf{x}) = \int_{\boldsymbol{\xi} \in R^N} s\left(\mathbf{x}, \frac{\boldsymbol{\xi}}{|\boldsymbol{\xi}|}\right) \mathcal{F}u(\boldsymbol{\xi})e^{+2i\pi(\mathbf{x}.\boldsymbol{\xi})} \, d\boldsymbol{\xi} \text{ a.e. } \mathbf{x} \in R^N \tag{42.12}$$
$$\text{for } v \in L^2(R^N) \cap \mathcal{F}L^1(R^N),$$

which specialists of linear partial differential equations prefer. In my framework it is more natural to apply first an operator of multiplication in $\mathbf{x}$ in order to have a function defined on all R^N, so that one can apply the Fourier transform, but it is for the *second commutation lemma* that the choice of S is more crucial.

The H-measure μ associated with the chosen subsequence of $\mathbf{U^n}$ is a $p \times p$ matrix whose entries are (complex) Radon measures on $\Omega \times S^{N-1}$, and μ is *Hermitian*[10] nonnegative. By taking $\psi = 1$, one sees that the integral of μ_{jk} in $\boldsymbol{\xi}$ gives the limit of $U_j^n\overline{U_k^n}$, i.e. if $U_j^n\overline{U_k^n} \rightharpoonup \pi_{jk}$ in the sense of measures, then for every $\varphi \in C_c(\Omega)$ one has $\langle \pi_{jk}, \varphi \rangle = \langle \mu_{jk}, \varphi \otimes 1 \rangle$. With this calculus modulo compact operators at hand, one can improve the compensated compactness theorem 38.1 by the *localization principle*: if the functions A_{jk} are continuous and if U^n satisfies

$$\sum_{j=1}^N \sum_{k=1}^p \frac{\partial(A_{jk}U_k^n)}{\partial x_j} \to 0 \text{ in } H_{loc}^{-1}(\Omega) \text{ strong}, \tag{42.13}$$

then one has

$$\sum_{j=1}^N \sum_{k=1}^p \xi_j A_{jk}\mu_{k\ell} = 0 \text{ for } \ell = 1, \dots, p; \tag{42.14}$$

the converse is actually true: if $\sum_{j=1}^N \sum_{k=1}^p \xi_j A_{jk}\mu_{k\ell} = 0$ for all ℓ, then $\sum_{j=1}^N \sum_{k=1}^p \frac{\partial(A_{jk}U_k^n)}{\partial x_j} \to 0$ in $H_{loc}^{-1}(\Omega)$ strong (if R_j is the *Riesz operator* which has symbol $i\frac{\xi_j}{|\xi|}$, then the condition is equivalent to $\sum_{j=1}^N \sum_{k=1}^p R_j A_{jk}U_k^n \to 0$ in $L_{loc}^2(\Omega)$ strong). One sees then that what H-measures do is to compute the limits of sequences $v_n\overline{w_n}$, where v_n and w_n are obtained from U^n by applying "pseudo-differential" operators of the class introduced; of course they do

---

[10] Charles HERMITE, French mathematician, 1822–1901. He worked in Paris, France.

not make Young measures obsolete as H-measures cannot see the limits of nonquadratic quantities, and therefore the compensated compactness theorem 38.1 has been *improved* (as one can consider equations with continuous coefficients if they are written in conservation form), but not so much the compensated compactness method, which is only *strengthened* by the addition of the theory of H-measures.

The question of *small-amplitude homogenization* consists, for example, in looking at elliptic problems of the form

$$div(A^n grad\, u_n) = f$$
$$A^n = A^\infty + \gamma\, B^n \text{ and } B^n \rightharpoonup 0 \text{ weak}\star \qquad (42.15)$$

in which case the effective coefficient A^{eff} is analytic in γ (as was first noticed by Sergio SPAGNOLO for the symmetric case), and

$$A^{eff} = A^\infty + \gamma^2 C_2 + \gamma^3 C_3 + \dots. \qquad (42.16)$$

The H-measure of B^n permits us to compute the coefficient of γ^2 in the expansion; the reason is that if one takes Dirichlet conditions for a slightly larger open set, for example, for $\lambda \in R^N$ one can choose f (depending on γ) so that

$$grad\, u_n = \lambda + \gamma\, grad\, v_n + \gamma^2 grad\, w_n + O(\gamma^3) \text{ in } \Omega, \qquad (42.17)$$

then

$$div(A^\infty grad\, v_n + B^n \lambda) = 0$$
$$div(A^\infty grad\, w_n + B^n grad\, v_n) = 0, \qquad (42.18)$$

so that $grad\, v_n, grad\, w_n$ converge weakly to 0 and $C_2 \lambda$ is the weak limit of $B^n grad\, v_n$. If we were on R^N, the mapping $B^n \mapsto grad\, u_n$ would be given by a "pseudo-differential" operator, and the limit of $B^n grad\, v_n$ could be computed using the H-measure associated with B^n, but another way to prove the result is to use the localization principle.

A similar method enters the computation of the correction M in the problem modeled on the stationary Stokes equation with a force in $\mathbf{u} \times curl(\mathbf{v}_0 + \lambda \mathbf{v}_n)$, that I discussed in Lecture 37.

Although the small-amplitude homogenization is important in many instances, a crucial step is to realize that H-measures can describe the *transport of oscillations/concentration effects* (which are the usual words involved when one looks at the difference between strong and weak convergence). I first considered a first-order differential operator

$$\sum_{j=1}^{N} b_j \frac{\partial u_n}{\partial x_j} = f_n \text{ in } R^N, \qquad (42.19)$$

with $u_n \rightharpoonup 0$ weakly in $L^2(R^N)$ and $f_n \to 0$ strongly in $H_{loc}^{-1}(R^N)$, and $b_j \in C_0^1(R^N)$ for $j = 1, \dots, N$; if the sequence is associated with the H-measure μ, then the localization principle asserts that

$$P(\mathbf{x}, \boldsymbol{\xi})\mu = 0 \text{ with } P(\mathbf{x}, \boldsymbol{\xi}) = \sum_{j=1}^{N} b_j(x)\xi_j \qquad (42.20)$$

(notice that if the b_j are complex there may well be no points in the zero set of P). I assume that the coefficients b_j are real, so that multiplying the equation by $\overline{u_n}$ and taking the real part, one has

$$\sum_{j=1}^{N} b_j \frac{\partial |u_n|^2}{\partial x_j} = 2\Re(f_n \overline{u_n}) \text{ in } R^N. \qquad (42.21)$$

One then assumes that $f_n \rightharpoonup 0$ weakly in $L^2(R^N)$, and for a real $a \in C^1(S^{N-1})$, one applies the operator P_a to the equation and one obtains

$$P_a f_n = \sum_{j=1}^{N} P_a \partial_j (M_{b_j} u_n) - \sum_{j=1}^{N} P_a(\partial_j M_{b_j})u_n = \\ \sum_{j=1}^{N} \partial_j ((P_a M_{b_j} - M_{b_j} P_a)u_n) + \sum_{j=1}^{N} M_{b_j} \partial_j P_a u_n + \qquad (42.22)\\ \sum_{j=1}^{N} ((\partial_j M_{b_j})P_a - P_a(\partial_j M_{b_j}))u_n.$$

One then needs a *second commutation lemma*, which requires a little more smoothness either on all the b_j or on a (and one uses a commutation result of Alberto CALDERÓN for avoiding smoothness hypotheses on b_j):

$P_a M_{b_j} - M_{b_j} P_a$ maps $L^2(R^N)$ into $H^1(R^N)$
$\frac{\partial}{\partial x_j}(P_a M_{b_j} - M_{b_j} P_a)$ has symbol $\xi_j \sum_{k=1}^{N} \frac{\partial a}{\partial \xi_k} \frac{\partial b_j}{\partial x_k} = \xi_j\{a, b_j\} = \{a, \xi_j b_j\}$
$$\qquad (42.23)$$
where the *Poisson bracket* $\{g, h\}$ of two functions on $R^N \times S^{N-1}$ is

$$\{g, h\} = \sum_{k=1}^{N} \frac{\partial g}{\partial \xi_k} \frac{\partial h}{\partial x_k} - \frac{\partial g}{\partial x_k} \frac{\partial h}{\partial \xi_k}. \qquad (42.24)$$

Therefore one has

$$\sum_{j=1}^{N} b_j \frac{\partial (P_a u_n)}{\partial x_j} + K u_n = P_a f_n \text{ in } R^N, \text{ and the symbol of } K \text{ is } \{a, P\},$$
$$\qquad (42.25)$$

and one deduces that

$$\sum_{j=1}^{N} b_j \frac{\partial ((P_a u_n)\overline{u_n})}{\partial x_j} + (K u_n)\overline{u_n} = 2\Re(P_a f_n)\overline{u_n} \text{ in } R^N. \qquad (42.26)$$

One then assumes that

$$U^n = (u_n, f_n) \text{ corresponds to a H-measure } \nu, \text{ so that } \nu_{11} = \mu, \qquad (42.27)$$

and one applies the last equation to a test function $\varphi \in C_c^1(R^N)$, and one gets $-\langle a\,\mu, \sum_{j=1}^{N} \partial_j(b_j \varphi)\rangle + \langle\{a, P\}\mu, \varphi\rangle = 2\langle\Re(a\,\nu_{12}), \varphi\rangle$, so that if one defines $\Phi(\mathbf{x}, \boldsymbol{\xi}) = a(\boldsymbol{\xi})\varphi(\mathbf{x})$, one has

$$\langle \mu, \{\Phi, P\} - div \, b \, \Phi \rangle = 2\langle \Re \nu_{12}, \Phi \rangle, \qquad (42.28)$$

which extends by linearity and density to all $\Phi \in C_c^1(R^N \times S^{N-1})$, and this is a first-order differential equation (i.e. a transport equation) for μ, written in weak formulation.

The method applies to linear differential systems endowed with a *sesquilinear balance relation* for their complex solutions (even if u_n is real, $P_a u_n$ takes complex values in general); in principle it applies also to semi-linear equations, but what the source term ν_{12} is in these cases is not clear.

I failed to find the way to use H-measures for proving theorems of *compactness by averaging*, but it is precisely for this purpose that Patrick GÉRARD introduced *independently* the same objects (actually he introduced them for functions taking values in Hilbert spaces); he called these objects *microlocal defect measures*, and if I agree with the term microlocal, I do not like the term defect: the transport theorem for the wave equation shows that H-measures give the way to describe what a *beam of light* is, for example (and it is *polarized light* if one uses Maxwell's equation) and the important physical quantities carried along it, nothing that looks like a defect.

H-measures use *no characteristic lengths*, and they are useful for phenomena where the frequency is not so important as long as it is high. The transport equation for the wave equation says that in the limit of high frequency one obtains *geometrical optics*, but *refraction effects* or *grazing rays* in the *geometric theory of diffraction* of Joe KELLER are frequency dependent, and one needs other objects.

My idea for taking care of one characteristic length was to add a new coordinate and consider H-measures in R^{N+1}, but Patrick GÉRARD had another idea, quite related, and he introduced the *semi-classical measures* on $\Omega \times R^N$ by looking for a sequence ε_n tending to 0 at the quantities

$$\lim_{n \to \infty} \int_{\xi \in R^N} |\mathcal{F}(\varphi \, u_n)(\xi)|^2 \psi(\varepsilon_n \xi) \, d\xi. \qquad (42.29)$$

There are various technical improvements of this idea, and *known deficiencies of this or other variants*, but it is not well understood yet how to handle situations with *many length scales*; obviously this is of importance for fluids in general, and for oceanography in particular.

I have preferred to sketch the existing mathematical tools before looking at more precise mathematical models in oceanography, but time was a little short for doing that. I hope that nevertheless these lecture notes will stimulate many to investigate more on all the questions which I have only sketched.

[Taught on Friday April 30, 1999.]

Biographical Information

[In a reference a–*b*, a is the lecture number, 0 referring to the Introduction (in the Preface), and b the footnote number in that lecture.]

Abbreviations and Mathematical Notation

Abbreviations for states: For those not familiar with geography, I have used UK = United Kingdom, and I have mentioned a few states from the United States of America: CA = California, CT = Connecticut, IL = Illinois, IN = Indiana, MA = Massachusetts, MD = Maryland, MN = Minnesota, MO = Missouri, NJ = New Jersey, NY = New York, OH = Ohio, PA = Pennsylvania, TX = Texas, UT = Utah, VA = Virginia, WA = Washington, WI = Wisconsin.

- a.e.: almost everywhere
- $B(x, r)$: open ball centered at x and radius $r > 0$, i.e. $\{y \in E \mid ||x-y||_E < r\}$ (in a normed space E)
- BMO(R^N): space of functions of bounded mean oscillation on R^N, i.e. seminorm $||u||_{BMO} < \infty$, with $||u||_{BMO} = \sup_{cubes\ Q} \frac{\int_Q |u-u_Q|\, dx}{|Q|} < \infty$ ($u_Q = \frac{\int_Q u\, dx}{|Q|}$, $|Q| = meas(Q)$)
- $[b]$: jump of the quantity b through a curve of discontinuity
- $[B, C]$: commutator of two operators $B, C \in \mathcal{L}(E, E)$, i.e. $B\,C - C\,B$
- $\{b, c\}$: Poisson bracket of two functions $b(\mathbf{x}, \boldsymbol{\xi}), c(\mathbf{x}, \boldsymbol{\xi}) \in C^1(R^N \times R^N)$, i.e. $\sum_{j=1}^{N} \frac{\partial b}{\partial \boldsymbol{\xi}_j} \frac{\partial c}{\partial \mathbf{x}_j} - \frac{\partial b}{\partial \mathbf{x}_j} \frac{\partial c}{\partial \boldsymbol{\xi}_j}$
- $BV(R^N)$: space of functions of bounded variation in R^N, whose partial derivatives (in the sense of distributions) belong to $\mathcal{M}(R^N)$ and have finite total mass
- $BV_{loc}(R^N)$: space of functions whose partial derivatives (in the sense of distributions) belong to $\mathcal{M}(R^N)$
- $C(\Omega)$: space of scalar continuous functions in Ω
- $C(\Omega; R^m)$: space of continuous functions from Ω into R^m
- $C(\overline{\Omega})$: space of continuous and bounded functions on $\overline{\Omega}$
- $C_c(\Omega)$: space of continuous functions with compact support in Ω
- $C_c^k(\Omega)$: space of functions of class C^k with compact support in Ω

- $C^k(\Omega)$: space of continuous functions with continuous derivatives up to order k in $\Omega \subset R^N$
- $C^k(\overline{\Omega})$: restrictions to $\overline{\Omega}$ of functions in $C^k(R^N)$
- $C^{0,\alpha}(\Omega)$: space of Hölder continuous functions of order $\alpha \in (0,1)$ (Lipschitz continuous functions if $\alpha = 1$), i.e. bounded functions for which there exist M such that $|u(x) - u(y)| \leq M\,|x - y|^\alpha$ for all $x, y \in \Omega$
- $C^{k,\alpha}(\Omega)$: space of functions of $C^k(\Omega)$ whose derivatives of order k belong to $C^{0,\alpha}(\Omega)$
- $conv(A)$: convex hull of $A \subset R^N$
- $\times$: apart from general Cartesian products, denotes the cross product in R^3, i.e. $(a \times b)_i = \sum_{jk} \varepsilon_{ijk} a_j b_k$
- curl: rotational operator $(curl(u))_i = \sum_{jk} \varepsilon_{ijk} \frac{\partial u_j}{\partial x_k}$
- D^α: $\frac{\partial^{\alpha_1}}{\partial x_1^{\alpha_1}} \cdots \frac{\partial^{\alpha_N}}{\partial x_N^{\alpha_N}}$ (for a multi-index α with α_j nonnegative integers, $j = 1, \ldots, N$)
- $\mathcal{D}'(\Omega)$: space of distributions T in Ω, dual of $\mathcal{D}(\Omega) = C_c^\infty(\Omega)$ (equipped with its natural topology), i.e. for every compact $K \subset \Omega$ there exists $C(K)$ and an integer $m(K) \geq 0$ with $|\langle T, \varphi \rangle| \leq C(K)\sup_{|\alpha|\leq m(K)} \|D^\alpha \varphi\|_\infty$ for all $\varphi \in C_c^\infty(\Omega)$ with support in K
- $deg(F; \Omega, \mathbf{a})$: Brouwer topological degree for a continuous function F from a bounded open set $\Omega \subset R^N$ into R^N, defined for $\mathbf{a} \in R^N \setminus F(\partial\Omega)$, i.e. algebraic count of the number of solutions of $F(\mathbf{z}) = \mathbf{a}$, equal to $\sum_{z_j | F(\mathbf{z_j}) = \mathbf{a}} sign(det(\nabla F)(\mathbf{z_j}))$ if $\nabla F(\mathbf{z_j})$ is invertible at each of the solutions of $F(\mathbf{z_j}) = \mathbf{a}$
- det: determinant
- div: divergence operator $div(u) = \sum_i \frac{\partial u_i}{\partial x_i}$
- $\mathcal{F}$: Fourier transform, $\mathcal{F}f(\boldsymbol{\xi}) = \int_{R^N} f(\mathbf{x}) e^{-2i\pi(\mathbf{x}.\boldsymbol{\xi})}\,d\mathbf{x}$
- $\overline{\mathcal{F}}$: inverse Fourier transform, $\overline{\mathcal{F}}f(\boldsymbol{\xi}) = \int_{R^N} f(\mathbf{x}) e^{+2i\pi(\mathbf{x}.\boldsymbol{\xi})}\,d\mathbf{x}$,
- $grad(u)$: gradient operator, $grad(u) = \left(\frac{\partial u}{\partial x_1}, \ldots \frac{\partial u}{\partial x_N} \right)$
- $H^s(R^N)$: Sobolev space of functions such that $(1 + |\boldsymbol{\xi}|^2)^{s/u} \mathcal{F}u \in L^2(R^N)$ ($L^2(R^N)$ for $s = 0$, $W^{s,2}(R^N)$ for s a positive integer)
- $H^s(\Omega)$: space of restrictions to Ω of functions from $H^s(R^N)$ (for $s \geq 0$)
- $H_0^s(\Omega)$: for $s \geq 0$, closure of $C_c^\infty(\Omega)$ in $H^s(\Omega)$
- $H^{-s}(\Omega)$: for $s \geq 0$, dual of $H_0^s(\Omega)$
- $H(div; \Omega)$: space of functions $u \in L^2(\Omega; R^N)$ with $div(u) \in L^2(\Omega)$ (for an open set $\Omega \subset R^N$)
- $H(curl; \Omega)$: space of functions $u \in L^2(\Omega; R^3)$ with $curl(u) \in L^2(\Omega; R^3)$ (for an open set $\Omega \subset R^3$)
- $\mathcal{H}^1(R^N)$: Hardy space of functions $f \in L^1(R^N)$ with $R_j f \in L^1(R^N)$, $j = 1, \ldots, N$
- $ker(A)$: kernel of a linear operator $A \in \mathcal{L}(E, F)$, i.e. $\{e \in E \mid A\,e = 0\}$
- $\mathcal{L}$: Laplace transform $\mathcal{L}f(p) = \int_0^\infty f(t) e^{-pt}\,dt$
- $\mathcal{L}(E, F)$: space of linear continuous operators M from the normed space E into the normed space F, i.e. with $\|M\|_{\mathcal{L}(E,F)} = \sup_{e \neq 0} \frac{\|M\,e\|_F}{\|e\|_E} < \infty$

- $L^p(A)$; $L^\infty(A)$: Lebesgue space of (equivalence classes of a.e. equal) measurable functions u with $||u||_p = \left(\int_A |u(\mathbf{x})|^p\, d\mathbf{x}\right)^{1/p} < \infty$ if $1 \le p < \infty$; with $||u||_\infty = \inf\{M \mid |u(\mathbf{x})| \le M \text{ a.e. in } A\} < \infty$ (spaces considered for the induced $(N-1)$-dimensional Hausdorff measure if $A = \partial\Omega$ for a smooth open set $\Omega \subset R^N$)
- $L^p_{loc}(\Omega)$: (equivalence classes of) measurable functions whose restriction to every compact $K \subset \Omega$ belongs to $L^p(K)$ (for $1 \le p \le \infty$)
- $L^p(0,T;E)$: (weakly or strongly) measurable functions u from $(0,T)$ into a separable Banach space E, such that $t \mapsto ||u(t)||_E$ belongs to $L^p(0,T)$ (for $1 \le p \le \infty$)
- loc: for any space Z of functions in Ω, Z_{loc} is the space of functions u such that $\varphi u \in Z$ for all $\varphi \in C^\infty_c(\Omega)$
- M_b: for $b \in L^\infty(R^N)$, operator $\in \mathcal{L}\big(L^2(R^N), L^2(R^N)\big)$ of multiplication by b, i.e. $(M_b u)(\mathbf{x}) = b(\mathbf{x})u(\mathbf{x})$ a.e. $\mathbf{x} \in R^N$
- Mf: maximal function of f, i.e. $Mf(x) = \sup_{r>0} \dfrac{\int_{B(\mathbf{x},r)} |f(\mathbf{y})|\, d\mathbf{y}}{|B(\mathbf{x},r)|}$
- $\mathcal{M}(\Omega)$: space of Radon measures μ in Ω, dual of $C_c(\Omega)$ (equipped with its natural topology), i.e. for every compact $K \subset \Omega$ there exists $C(K)$ with $|\langle \mu, \varphi \rangle| \le C(K)||\varphi||_\infty$ for all $\varphi \in C_c(\Omega)$ with support in K
- meas(A): Lebesgue measure of A
- ∇u: for an open set $\Omega \subset R^N$ and $u \in W^{1,1}(\Omega; R^m)$ it is the $m \times N$ matrix with entries $\frac{\partial u_i}{\partial x_j}$, $i = 1,\ldots, m$, $j = 1,\ldots, N$ (∇ is called nabla)
- $|\cdot|$: norm in H
- $||\cdot||$: norm in V
- $||\cdot||_*$: dual norm in V'
- p': conjugate exponent of $p \in [1,\infty]$, i.e. $\frac{1}{p} + \frac{1}{p'} = 1$
- p^*: Sobolev exponent of $p \in [1, N)$, i.e. $\frac{1}{p^*} = \frac{1}{p} - \frac{1}{N}$ for $\Omega \subset R^N$ and $N \ge 2$
- P_a: for $a \in L^\infty(R^N)$, ("pseudo-differential") operator $\in \mathcal{L}\big(L^2(R^N), L^2(R^N)\big)$ equal to $\mathcal{F}^{-1}M_a\mathcal{F}$, i.e. $\mathcal{F}(P_a u)(\boldsymbol{\xi}) = a(\boldsymbol{\xi})\mathcal{F}u(\boldsymbol{\xi})$ a.e. $\boldsymbol{\xi} \in R^N$
- R_j: Riesz operator, $\mathcal{F}(R_j f)(\boldsymbol{\xi}) = \frac{i\xi_j}{|\boldsymbol{\xi}|}\mathcal{F}f(\boldsymbol{\xi})$
- $R(A)$: range of a linear operator $A \in \mathcal{L}(E,F)$, i.e. $\{f \in F \mid f = A e \text{ for some } e \in E\}$
- $\mathcal{S}(R^N)$: Schwartz space of functions $u \in C^\infty(R^N)$ with $x^{\boldsymbol{\alpha}} D^{\boldsymbol{\beta}} u$ bounded for all multi-indices $\boldsymbol{\alpha}, \boldsymbol{\beta}$ with α_j, β_j nonnegative integers for $j = 1,\ldots, N$
- $\mathcal{S}'(R^N)$: temperate distributions, dual of $\mathcal{S}(R^N)$, i.e. $T \in \mathcal{D}'(R^N)$ and there exists C and an integer $m \ge 0$ with $|\langle T, \psi \rangle| \le C \sup_{|\boldsymbol{\alpha}|,|\boldsymbol{\beta}|\le m} ||x^{\boldsymbol{\alpha}} D^{\boldsymbol{\beta}}\psi||_\infty$ for all $\psi \in \mathcal{S}(R^N)$
- $\star$: convolution product $(f \star g)(\mathbf{x}) = \int_{\mathbf{y}\in R^N} f(\mathbf{x} - \mathbf{y})g(\mathbf{y})\, d\mathbf{y}$
- VMO: space of functions of vanishing mean oscillation on R^N, closure of $C_c(R^N)$ in BMO(R^N)
- $W^{m,p}(\Omega)$: Sobolev space of functions in $L^p(\Omega)$ whose derivatives (in the sense of distributions) of length $\le m$ belong to $L^p(\Omega)$
- $W^{m,p}(\Omega; R^m)$: Sobolev space of functions from Ω into R^m whose components belong to $W^{m,p}(\Omega)$

- x^{α}: $x_1^{\alpha_1} \ldots x_N^{\alpha_N}$ for a multi-index α with α_j nonnegative integers for $j = 1, \ldots, N$
- $X(\Omega)$: space of distributions $f \in H^{-1}(\Omega)$ with $\frac{\partial f}{\partial x_j} \in H^{-1}(\Omega)$ for $j = 1, \ldots, N$

- α; $|\alpha|$: multi-index $(\alpha_1, \ldots, \alpha_N)$; its length is $|\alpha_1| + \ldots + |\alpha_N|$
- Δ: Laplacian $\sum_{j=1}^{N} \frac{\partial^2}{\partial x_j^2}$
- δ_{ij}: Kronecker symbol, equal to 1 if $i = j$ and equal to 0 if $i \neq j$ (for $i, j = 1, \ldots, N$)
- ε_{ijk}: for $i, j, k \in \{1, 2, 3\}$, completely antisymmetric tensor, equal to 0 if two indices are equal, and equal to the signature of the permutation $123 \mapsto ijk$ if indices are distinct (i.e. $\varepsilon_{123} = \varepsilon_{231} = \varepsilon_{312} = +1$ and $\varepsilon_{132} = \varepsilon_{321} = \varepsilon_{213} = -1$)
- ϱ_{ε}: smoothing sequence, with $\varrho_{\varepsilon}(\mathbf{x}) = \frac{1}{\varepsilon^N} \varrho_1\left(\frac{\mathbf{x}}{\varepsilon}\right)$ with $\varepsilon > 0$ and $\varrho_1 \in C_c^{\infty}(R^N)$ with $\int_{\mathbf{x} \in R^N} \varrho_1(\mathbf{x}) \, d\mathbf{x} = 1$
- Λ_1: Zygmund space, $|u(\mathbf{x}+\mathbf{h}) + u(\mathbf{x}-\mathbf{h}) - 2u(\mathbf{x})| \leq M \, |\mathbf{h}|$ for all $\mathbf{x}, \mathbf{h} \in R^N$

References

[1]   BIRKHOFF G., A source book in classical analysis. Harvard University Press, Cambridge, MA (1973).

[2]   CARLEMAN T., Problèmes mathématiques dans la théorie cinétique des gaz. Publications Scientifiques de l'Institut Mittag-Leffler, Almqvist & Wiksells, Uppsala (1957).

[3]   DAFERMOS C. M., Hyperbolic conservation laws in continuum physics, Grundlehren der Mathematischen Wissenschaften, 325. Springer-Verlag, Berlin (2000).

[4]   DAUTRAY R. & LIONS J.-L., Mathematical analysis and numerical methods for science and technology, Vol. 1. Physical origins and classical methods. Springer-Verlag, Berlin-New York (1990).

[5]   DAUTRAY R. & LIONS J.-L., Mathematical analysis and numerical methods for science and technology, Vol. 2. Functional and variational methods. Springer-Verlag, Berlin-New York (1988).

[6]   DAUTRAY R. & LIONS J.-L., Mathematical analysis and numerical methods for science and technology, Vol. 3. Spectral theory and applications. Springer-Verlag, Berlin (1990).

[7]   DAUTRAY R. & LIONS J.-L., Mathematical analysis and numerical methods for science and technology, Vol. 4. Integral equations and numerical methods. Springer-Verlag, Berlin (1990).

[8]   DAUTRAY R. & LIONS J.-L., Mathematical analysis and numerical methods for science and technology, Vol. 5. Evolution problems. I. Springer-Verlag, Berlin (1992).

[9]   DAUTRAY R. & LIONS J.-L., Mathematical analysis and numerical methods for science and technology, Vol. 6. Evolution problems. II. Springer-Verlag, Berlin (1993).

[10]   FEYNMAN R. P., LEIGHTON R. B. & SANDS M., The Feynman lectures on physics, Vol. 1: Mainly mechanics, radiation, and heat, Vol. 2: The electromagnetic field, Vol. 3: Quantum mechanics. Addison-Wesley, Reading, MA (1963), (1964), (1965).

[11]   FRISCH U., Turbulence. The legacy of A. N. Kolmogorov. Cambridge University Press, Cambridge (1995).

[12]   GALDI G. P., An introduction to the mathematical theory of the Navier-Stokes equations, Vol. 1. Linearized steady problems. Springer Tracts in Natural Philosophy, 38. Springer-Verlag, New York (1994).

[13]   GALDI G. P., An introduction to the mathematical theory of the Navier-Stokes equations, Vol. 2. Nonlinear steady problems. Springer Tracts in Natural Philosophy 39, Springer-Verlag, New York (1994).

[14]   GATIGNOL R., Théorie cinétique des gaz à répartition discrète de vitesses. Lecture Notes in Physics 36, Springer, Berlin (1975).

[15]   GILL A. E., Atmosphere-ocean dynamics. Academic Press, New York (1982).

[16]   LADYZHENSKAYA O. A., The mathematical theory of viscous incompressible flow. Mathematics and its applications, Vol. 2 Gordon and Breach, Science Publishers, New York-London-Paris (1969).

[17]   LEMARIÉ-RIEUSSET P.-G., Recent developments in the Navier–Stokes problem. Chapman & Hall/CRC Research Notes in Mathematics, 431. Chapman & Hall/CRC, Boca Raton, FL (2002).

[18]   LEWANDOWSKI R., Analyse mathématique en océanographie: Essai sur la modélisation et l'analyse mathématique de quelques modèles de turbulence utilisés en océanographie. Collection Recherches en Mathématiques Appliquées, Masson, Paris (1997).

[19]   LIONS J.-L., Quelques méthodes de résolution des problèmes aux limites non linéaires. Dunod; Gauthier-Villars, Paris (1969).

[20]   MOHAMMADI B. & PIRONNEAU O., Analysis of the k-epsilon turbulence model. RAM: Research in Applied Mathematics, Masson, Paris; John Wiley & Sons, Ltd., Chichester (1994).

[21]   SCHWARTZ L., Un mathématicien aux prises avec le siècle. Éditions Odile Jacob, Paris (1997). A mathematician grappling with his century. Birkhäuser Verlag, Basel (2001).

[22]   STEIN E., Singular integrals and differentiability properties of functions. Princeton University Press, Princeton NJ (1970).

[23]   TARTAR L., Topics in nonlinear analysis. Publications Mathématiques d'Orsay 78, 13, Université de Paris-Sud, Département de Mathématiques, Orsay (1978).

Index

Editor in Chief: Franco Brezzi

Editorial Policy

1. The UMI Lecture Notes aim to report new developments in all areas of mathematics and their applications - quickly, informally and at a high level. Mathematical texts analysing new developments in modelling and numerical simulation are also welcome.

2. Manuscripts should be submitted (preferably in duplicate) to
   Redazione Lecture Notes U.M.I.
   Dipartimento di Matematica
   Piazza Porta S. Donato 5
   I – 40126 Bologna
   and possibly to one of the editors of the Board informing, in this case, the Redazione about the submission. In general, manuscripts will be sent out to external referees for evaluation. If a decision cannot yet be reached on the basis of the first 2 reports, further referees may be contacted. The author will be informed of this. A final decision to publish can be made only on the basis of the complete manuscript, however a refereeing process leading to a preliminary decision can be based on a pre-final or incomplete manuscript. The strict minimum amount of material that will be considered should include a detailed outline describing the planned contents of each chapter, a bibliography and several sample chapters.

3. Manuscripts should in general be submitted in English. Final manuscripts should contain at least 100 pages of mathematical text and should always include
   - a table of contents;
   - an informative introduction, with adequate motivation and perhaps some historical remarks: it should be accessible to a reader not intimately familiar with the topic treated;
   - a subject index: as a rule this is genuinely helpful for the reader.

4. For evaluation purposes, manuscripts may be submitted in print or electronic form (print form is still preferred by most referees), in the latter case preferably as pdf- or zipped ps- files. Authors are asked, if their manuscript is accepted for publication, to use the LaTeX2e style files available from Springer's web-server at
   ftp://ftp.springer.de/pub/tex/latex/mathegl/mono.zip

5. Authors receive a total of 50 free copies of their volume, but no royalties. They are entitled to a discount of 33.3 % on the price of Springer books purchased for their personal use, if ordering directly from Springer.

6. Commitment to publish is made by letter of intent rather than by signing a formal contract. Springer-Verlag secures the copyright for each volume. Authors are free to reuse material contained in their LNM volumes in later publications: A brief written (or e-mail) request for formal permission is sufficient.